Recent Progress in Plasma Science and Technology

Recent Progress in Plasma Science and Technology

Editor: Louis Murphy

MURPHY & MOORE
www.murphy-moorepublishing.com

Published by Murphy & Moore Publishing
1 Rockefeller Plaza,
New York City, NY 10020, USA
www.murphy-moorepublishing.com

Recent Progress in Plasma Science and Technology
Edited by Louis Murphy

International Standard Book Number: 978-1-63987-483-5 (Hardback)

Cataloging-in-Publication Data

Recent progress in plasma science and technology / edited by Louis Murphy.
p. cm.
Includes bibliographical references and index.
ISBN 978-1-63987-483-5
1. Plasma (Ionized gases). 2. Ionized gases. 3. Physics. I. Murphy, Louis.
QC718 .R43 2022
530.44--dc23

Contents

Preface

Plasma is one of the four fundamental states of matter, along with the solid, liquid and gaseous states. It consists of a gas of ions, which are atoms or molecules with a net electrical charge. The presence of free charged particles makes plasma responsive to electromagnetic fields. This property of plasma is used in many modern technological devices, such as plasma televisions and plasma etching. Partially ionized plasmas, which contain a certain amount of neutral particles, is found in neon signs and lightning. Plasma is the third most abundant form of ordinary matter in the universe, behind dark matter and dark energy. It is found in stars; including the Sun. Plasma science is an upcoming field of science that has undergone rapid development over the past few decades. The various studies that are constantly contributing towards advancing technologies and evolution of this field are examined in detail. Scientists and students actively engaged in this field will find this book full of crucial and unexplored concepts.

This book has been the outcome of endless efforts put in by authors and researchers on various issues and topics within the field. The book is a comprehensive collection of significant researches that are addressed in a variety of chapters. It will surely enhance the knowledge of the field among readers across the globe.

It gives us an immense pleasure to thank our researchers and authors for their efforts to submit their piece of writing before the deadlines. Finally in the end, I would like to thank my family and colleagues who have been a great source of inspiration and support.

Editor

1

Physicochemical Analysis of Argon Plasma-Treated Cell Culture Medium

Claudia Bergemann, Cornelia Hoppe, Maryna Karmazyna, Maxi Höntsch, Martin Eggert, Torsten Gerling and Barbara Nebe

Abstract

The effects of cold plasma under atmospheric pressure are being explored for medical applications. It was found that plasma effects on cells correspond to a plasma–medium interaction; thus, plasma-treated cell culture medium alone is able to influence the cell behavior. Here, we discovered that the liquid-mediated effect of atmospheric-pressure argon plasma on mouse liver epithelial cells persists up to 21 days of storage; i.e., the liquid preserves the characteristics once induced by the argon plasma. Earlier investigations of our group revealed that temperature and pH, hydrogen peroxide production and oxygen content can be excluded as initiators of the detrimental biological changes. As we found here, the increased osmolality in the media caused by plasma treatment can also be excluded as a reason for the observed cell effects. Conversely, we found changes in the components of cell culture medium by fast protein liquid chromatography (FPLC) and decreased cell viability in plasma-treated media independent of the presence of fetal calf serum (FCS) during plasma treatment. The persistent biological effect on plasma-treated liquids observed here could open up new medical applications. Stable plasma-treated liquids could find application for dermatological, dental, or orthopedic therapy.

Keywords: Non-thermal atmospheric-pressure plasma jet, Plasma–liquid interaction, Osmolality, Size-exclusion chromatography, Cell viability

1. Introduction

Plasma medicine is an emerging field of interdisciplinary research combining physics, biology, and clinical medicine [1,2]. In the first applications, gas plasma was used in the form of hot plasma for cauterization [1,3]. Currently, the effects of cold plasma under atmospheric pressure are being explored for medical applications. Several experiments showed the achievement of blood coagulation and bacterial decontamination, with little effect on the surrounding tissue [4–9], as well as the promotion of wound healing and tissue regeneration [10–13]. This proliferative reaction was correlated with the secretion of growth factors induced by plasma treatment [14]. Conversely, adverse effects such as anti-proliferative and anti-tumorigenic effects were reported [15–18]. Also, the detachment of cells from the substrate, an effect that accompanies the change to a rounded cell morphology, and apoptosis were reported for animal cell lines like CHO-K1 (Chinese hamster ovary epithelial cells), 3T3 (mouse fibroblasts), BAEC (bovine aorta endothelial cells), H5V (mouse endothelial cells), and RASMC (rat aorta smooth muscle cells) [1,19,20]. This is in line with the observation that the expression of cell adhesion molecules is changed upon plasma treatment [21]. Detrimental cell effects like DNA damage were shown by Kalghatgi et al. [22]. Plasma treatment of epithelial cells led to cell detachment and additional DNA damage, which was proven by the phosphorylation of the histone protein H2AX [22].

It was discovered that these effects can be mediated by the use of plasma-treated liquids alone [23]. Plasma treatment of cell culture medium resulted in DNA damage to the subsequently incubated cells [22,24]. Similar to the direct treatment, liver epithelial cells lost their substrate attachment when incubated with argon plasma-treated medium alone [25]. In addition, cell–cell contact proteins, e.g., the tight junction protein zonula occludens (ZO-1), as well as the cell membrane morphology were shown to be impaired [25,26]. Based on these observations, it could be concluded that plasma effects are not only solely based on the direct plasma–cell interaction but also on the interaction between plasma and medium, and these plasma-induced changes in the complex Dulbecco's modified Eagle's medium (DMEM) were long lasting: up to 7 days and more [26].

Further analysis revealed that neither lipid oxidation [24] nor the generation of ozone [27] was responsible for these effects. Also, changes in temperature, pH-value, or the concentration of hydrogen peroxide could be excluded as a reason [25,26]. In addition, it was shown that decreasing the oxygen concentration of the medium as a result of plasma treatment was not the reason for the respective cell defects [26]. Several publications speculated that the generation of reactive oxygen species (ROS) through plasma treatment would be an explanation for DNA damage [20,28,29]. However, the cell-damaging effect of plasma-treated medium could be detected after 1 h as well as up to 7 days after treatment [22,25,26], indicating that the component(s) remained stable for a long time. This result excludes the short-lived ROS as an inducing agent.

The present work aimed to examine the contribution of different medium components, e.g., fetal calf serum and gentamicin, to the detrimental cell effects resulting from plasma treatment. Furthermore, physicochemical parameters other than temperature, pH, oxygen concentration, or the presence of hydrogen peroxide, which had already been excluded as a reason [25, 26], were investigated. Taking into consideration that the plasma effect is mediated only by liquids [22,23,25,26,29], the cell culture medium was treated with argon plasma and cells were incubated afterward with the medium. The plasma treatment was performed using an atmospheric-pressure argon plasma jet.

2. Material and methods

2.1. Argon plasma source

The atmospheric-pressure plasma jet kINPen®09 (neoplas tools GmbH, Greifswald) was used for all experiments. This plasma source consists of a quartz capillary (inner diameter of 1.6 mm) with a needle electrode (diameter of 1 mm). A high-frequency voltage of 1.1 MHz/2–6 kV is applied at this electrode. Argon gas (purity 99.996%) is used as a feed gas with a gas flow of 1.9 slm. A gas discharge is ignited at the tip of the high-voltage needle exciting the argon gas. In this way, low-temperature plasma is generated and blown out of the capillary. The so-called plasma jet outside the nozzle has a length of 12–14 mm and is about 1 mm wide. The temperature at the visible tip of the plasma jet does not exceed 50°C. Details for the characterization of the plasma source are given by Weltmann *et al.* [12,26].

2.2. Argon plasma treatment of cell culture media

One-hundred microliters of the defined cell culture medium (see below) was supplied per well in a 96-well plate (Greiner Bio-One). The kINPen®09 was mounted vertically and the quartz capillary was positioned at the top edge of each well in the 96-well plate. The distance between the tip of the quartz capillary and the 96-well plate does not exceed 1 mm. In this way, the argon plasma had an immediate contact with the cell culture medium (Figure 1).

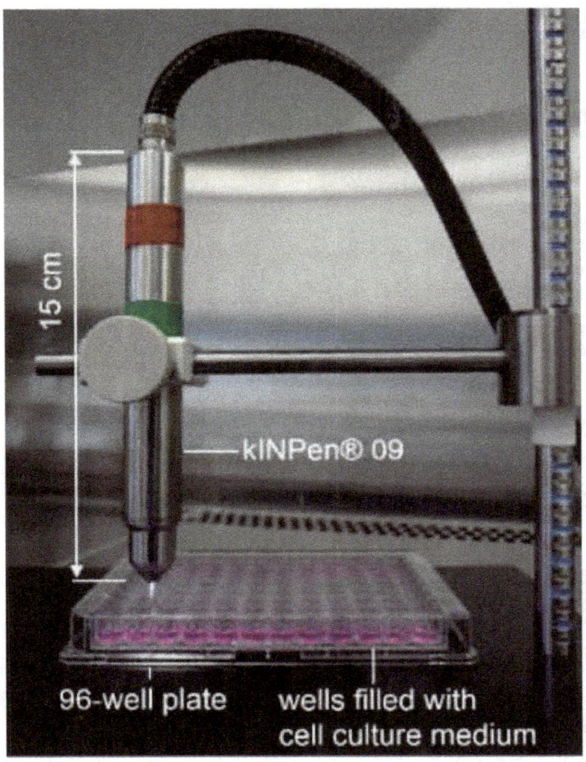

Figure 1. Experimental setup for argon plasma treatment of cell culture media. The kINPen®09 was mounted vertically. The distance between the tip of the quartz capillary and the 96-well plate does not exceed 1 mm.

Dulbecco's modified Eagle's medium (DMEM; Invitrogen) was used for all experiments. For cell culture experiments, DMEM is usually supplemented with 10% fetal calf serum (PAA Gold, PAA Laboratories) and 1% gentamicin (Ratiopharm). To explore the effect of the supplements on plasma treatment results, pure DMEM without supplements or with one supplement was treated as well. We refer to the different media as follows: pure DMEM = DMEM; DMEM with 10% FCS and 1% gentamicin = DMEM+FCS+Genta; DMEM with 10% FCS = DMEM+FCS; DMEM with 1% gentamicin = DMEM+Genta. The different media underwent argon plasma treatment for 60 or 120 s as indicated for the specific experiment. Argon gas treatment using the kINPen®09 without igniting the plasma was used as a control. Untreated medium was included as a further control. After plasma treatment, the different samples and controls were pooled in separate tubes (1.5 ml, Eppendorf) and analyzed for their physicochemical parameters. For subsequent cell experiments, the medium was stored at 37°C and 5% CO_2 for at least 24 h to ensure that the cell effects observed were neither caused by the changed oxygen concentrations nor by the persistence of hydrogen peroxide; these were shown to be balanced over 24 h after plasma treatment [26]. For investigations regarding the persistent effect of plasma-treated medium on cells, the medium was stored at 37°C and 5% CO_2 for 7 or 21 days. For all experiments, with the exception of the size-exclusion chromatography, the supplements omitted during plasma treatment were added afterward to ensure analogous conditions.

2.3. Physicochemical analysis of argon plasma-treated cell culture media

2.3.1. Osmolality

To acquire the loss of solvent during argon plasma treatment, the osmotic strength was utilized by the measurement of the freezing point depression. The respective medium (50 μl) was applied to an Osmomat 030 (Gonotec) and analyzed in triplicate. The determination of osmolality was performed in three independent experiments for DMEM+FCS+Genta, which had been argon plasma-treated for 60 or 120 s, and for the same medium after argon flow for 60, 120, or 180 s versus the untreated control.

2.3.2. Size-exclusion chromatography

Size-exclusion chromatography was performed by gel filtration on a fast performance liquid chromatography (FPLC) system equipped with a UV detector (ÄKTA purifier, GE Healthcare). After argon plasma treatment, samples were pooled and the protein concentration was measured. The samples were diluted with distilled water to a concentration of 4 mg/ml protein, sterile-filtrated (0.2 μm), applied to a size-exclusion column (Superdex 200, 10/300 GL, GE Healthcare) and separated with a constant flow of 0.5 ml/min for 60 min, using a phosphate buffer (34.3 mM Na_2HPO_4, 14.5 mM NaH_2PO_4, 200 mM NaCl, pH 7.14). The samples were monitored by the UV detector (280 nm) and collected in fractions of 500 μl. Chromatograms were conducted for DMEM, DMEM+FCS, and DMEM+Genta, which were argon plasma-treated for 120 s, for the untreated control and for the same medium after argon flow for 120 s.

2.4. Epithelial cell experiments

2.4.1. Cell culture of mHepR1

The epithelial cell line mHepR1 (murine hepatocytes) [30] was used throughout the experiments. These immortalized mHepR1 cells represent a clone of the HepSV40 line derived from transgenic mice [31] and exhibit characteristic markers for epithelial cells like the tight-junction-associated protein ZO-1. The epithelial cells were cultured in DMEM supplemented with 10% FCS and 1% gentamicin at 37°C and 5% CO_2. Near confluence, cells were detached with 0.25% trypsin/0.38% EDTA (Invitrogen) for 10 min. After stopping trypsinization by the addition of cell culture medium, an aliquot of 100 μL was put into 10 mL of CASY® ton buffer solution (Roche Innovatis) and the cell number was measured in the counter CASY® Model DT (Schärfe System).

2.4.2. Immunofluorescence staining of ZO-1

Into a 24-well plate (Greiner), which was provided with a round coverslip (diameter of 12 mm, MENZEL) and 1 ml DMEM+FCS+Genta, 5×10^4 cells/well were seeded. The mHepR1 cells were incubated for 2 days at 37°C and 5% CO_2 to achieve a confluent cell layer for ZO-1 observation. Then the culture medium was replaced by plasma-treated or control DMEM+FCS+Genta, prepared as described in paragraph 2.2. and stored for 21 days at 37°C and 5% CO_2. The plate was incubated for another 24 h at 37°C and 5% CO_2 followed by immunofluorescence staining of the cells for ZO-1 protein. The cells were permeabilized with ice-cold acetone (−20°C, 200 ml, Lab-Scan) for 5 min and, after washing three times with phosphate buffer solution (PBS; PAA Laboratories), the cells were incubated with rabbit anti-ZO-1 antibody (diluted 1:100, Invitrogen) for 30 min at room temperature. After washing again with PBS, the cells were incubated with fluorescein-conjugated Alexa Fluor 488 secondary antibody goat anti-rabbit (1:100, Invitrogen) for 30 min at room temperature in the dark to avoid fading of the fluorescent dye. Finally, the cells were embedded in mounting medium (FluroShield, Sigma Aldrich) with a coverslip. The ZO-1 protein was analyzed by the inverted confocal laser scanning microscope LSM 780 (Carl Zeiss, Oberkochen, Germany), equipped with a 63× oil-immersion differential interference contrast (DIC) objective.

2.4.3. Cell viability

To monitor the viability of the cells in plasma-treated medium, the CellTiter 96® AQueous Non-Radioactive Cell Proliferation Assay (Promega) was used. For each medium sample and the control (prepared as described in paragraph 2.2.), three wells containing 100 μl of medium were prepared in a 96-well plate and 2×10^4 mHepR1 cells were added to each well. After an incubation time of 24 h (37°C and 5% CO_2), 20 μl of the MTS solution (Promega) was added to each well and incubated for 2 h (37°C and 5% CO_2) with manual shaking for every 30 min. Afterward, 60 μl of each well was transferred into a fresh 96-well plate and the absorption was determined at 492 nm (reference 620 nm) using a microplate reader (Anthos). The appropriate medium without cells served as a blank. The viability of cells was calculated corresponding to cells in the untreated control medium (100%). The determination of cell viability was performed in three independent experiments for DMEM+FCS+Genta, which had been argon plasma-treated for 60 or 120 s, for the untreated control and for the same medium after argon flow for 60, 120, or 180 s. Three additional experiments were performed for DMEM and DMEM +FCS, which were argon plasma-treated for 120 s, for the untreated control and for the same media after argon flow for 120 s. The supplements omitted during plasma treatment were added before cell culture to ensure analogous conditions.

2.4.4. Scanning Electron Microscopy (SEM)

Into a 24-well plate (Greiner), which was provided with a round coverslip and 1 ml DMEM +FCS+Genta, 5×10^4 cells/well were seeded. The mHepR1 cells were incubated for 2 days at 37°C and 5% CO_2 to achieve a confluent cell layer. Then the culture medium was replaced by plasma-treated DMEM+FCS+Genta (60 s) or control medium, prepared as described in paragraph 2.2. and stored for 7 days at 37°C and 5% CO_2. After 24 h of incubation, the mHepR1 monolayer was washed three times with PBS and subsequently fixed in glutardialdehyde (GDA, Sigma Aldrich, 0.5% in PBS) and stored for at least 24 h at 4°C. Afterward, the mHepR1 monolayer was critical point-dried. For this purpose, the GDA PBS solution was first gradually removed by drainage using acetone with increasing concentrations of 30, 50, 70, 90, and 100% for 10 s, 5, 10, and 15 min and twice for 10 min, respectively. Subsequently, the acetone was substituted by critical point drying (K 850 EMITECH). Then samples were sputter-coated with gold particles for 100°s (layer thickness approximately 20 nm), achieving a conductive dissipation for SEM. The cell surface structure was analyzed with a DSM 960 A (Carl Zeiss), operated at 10 kV at a magnification of 5,000× and a tilt angle of 60°.

3. Results and discussion

The atmospheric-pressure plasma jet kINPen09 [12] was applied throughout the experiments. The same plasma jet was used in our recent work [25], where we found that plasma effects on cells correspond to a plasma–medium interaction, and thus plasma-treated DMEM+FCS +Genta alone is able to open cell–cell contacts in a confluent cell monolayer of mHepR1 cells. Immunostaining of the zonula occludens tight junction protein (ZO-1) showed large openings between cells, which led to the complete degradation of the tight junction proteins. Normally, untreated cells represent clear, continuous ZO-1 bands between adjoining cells. In contrast, large intercellular openings were observed using plasma-treated DMEM+FCS+Genta. This effect was verified also for the medium that was stored up to 7 days after plasma treatment showing a persistent effect of plasma-treated medium on cells [26].

As we discovered here, this long-lasting effect is persistent for up to 21 days after argon plasma treatment. The tight junction protein ZO-1 of murine hepatocytes mHepR1 is shown in Figure 2. After incubation of cells with DMEM+FCS+Genta, which was plasma-treated for 120 s and stored for 21 days, the expression of ZO-1 in the cell contact zone was either reduced in different areas or disappeared due to the retraction in a centripetal direction. In consequence, the cell size increased due to the loss of cell–cell contacts in the monolayer. In normal controls, tight cell–cell contacts could be found between neighboring cells.

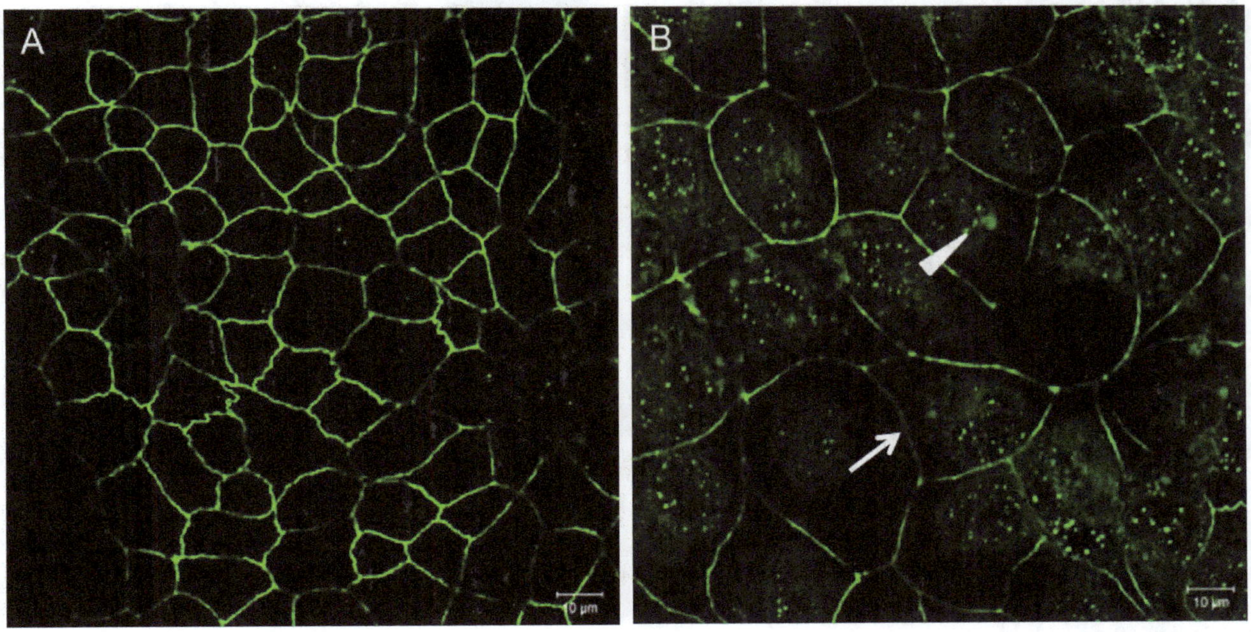

Figure 2. ZO-1 of mHepR1 cells after 21 days. (A) In control cells, the tight cell–cell contacts between neighboring cells are clearly visible. (B) After incubation of cells with plasma-treated DMEM+FCS (120 s) – which was stored for 21 days – either the expression of ZO-1 in the cell contact zone was reduced (arrow) or ZO-1 disappeared due to the retraction in a centripetal direction (arrowhead). Confocal microscopy (LSM 780, Carl Zeiss, bars 10 μm).

Accompanied with the long-lasting effect of plasma-treated medium on the distribution of ZO-1 in mHepR1 cells, we found changes in the cell surface morphology by SEM. The mHepR1 cells exhibited shortened microvilli with a lower density resulting from the plasma-treated cell culture medium [26]. These findings on microvilli shortening were revealed after a 1-day as well as a 7-day storage time of plasma-treated medium (60 s; complete DMEM). SEM images of mHepR1 cells (Figure 3), incubated for 24 h with the plasma-treated (60 s), 7-day stored DMEM+FCS+Genta illustrate this effect. Untreated and even argon-treated mHepR1 cells present elongated microvilli covering the cell surface at a high density. In contrast, the plasma-treated mHepR1 cells showed greatly shortened microvilli and the density over the entire cell surface seems to be reduced.

The formation of cell surface structures, e.g., microvilli, is essential for characteristic functions of specialized cells in tissues. Microvilli increase the cell surface and play an important role in metabolic processes: they regulate cellular functions by external signals as well as Ca^{2+} signaling [32]. Effects similar to those we observed in our work were detected by Pfister and Burstein after the treatment with ophthalmic drugs [33]. They observed the loss of surface epithelial microvilli as well as rupture of the tight junctions on corneal epithelial cells.

As a consequence, the question arises of which parameter that can persist as long as 21 days is changed in cell culture medium due to plasma treatment. As recently shown, plasma effects due to thermal damage of differential proteins in cell culture medium could be excluded for

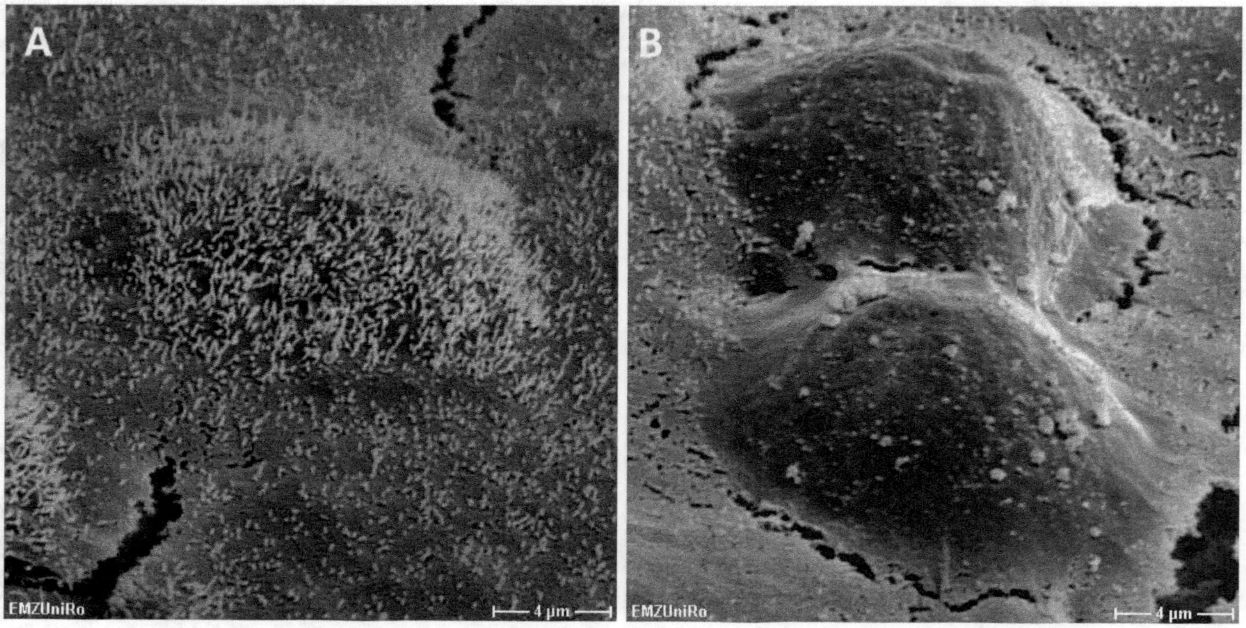

Figure 3. Microvilli of mHepR1 epithelial cells (A) in DMEM+FCS+Genta as control and (B) in the same medium, which was argon plasma-treated for 60 s and stored for 7 days (culture time 24 h). Scanning electron microscopy (SEM, DSM 910, magnification 5,000×, tilt angle 60°, and bar 4 µm).

treatment times below 180 s for the atmospheric-pressure plasma jet kINPen09 [26]. The temperature in the cell culture medium did not exceed 25°C. Plasma–liquid interactions were studied by Oehmigen *et al.* [34] and von Woedtke *et al.* [35] relating to the effective disinfection of liquids. It was shown that the inactivation of bacteria is strongly dependent on the acidification of aqueous liquids (pH decrease). In our work, we could show that the pH of the complete cell culture medium DMEM remained constant. The buffering capacity of sodium hydrogen carbonate in the medium was sufficient to maintain the pH even after a plasma treatment. Furthermore, the redox amphoteric hydrogen peroxide was found in liquids after plasma treatment in various studies [10,34]. As hydrogen peroxide also occurs in the cell metabolism, it can be degraded by various repair and protection mechanisms in the cell. Antioxidants inside the cell are able to detoxify "natural" amounts of ROS. In the presence of an overabundance of ROS, these mechanisms fail. Thus, the concentration of hydrogen peroxide and oxygen in the medium after plasma treatment was investigated earlier. Our investigations revealed that hydrogen peroxide and oxygen concentrations were balanced over the time period of up to 1 day, but the cell behavior was affected by argon plasma-treated medium stored for 1 or 7 days [26] or as described here for 21 days.

These detrimental effects on mHepR1 cells due to argon plasma-treated cell culture medium lead us to investigate other physicochemical parameters, which could change during the plasma treatment of a medium. One parameter to which cells react sensitively is the osmolality [36]. Due to evaporation processes induced by the argon gas flow, the medium volume in the well could be reduced during plasma treatment. Accordingly, the osmotic strength in the medium would increase. To examine this, we utilized the osmolality by measuring the freezing point depression for argon plasma-treated DMEM+FCS+Genta. The results in Figure 4 show an increasing osmolality for the medium after argon plasma treatment for 60 or 120 s. Interestingly, argon gas flow alone did not influence osmolality as high as argon plasma did. Thus, osmolality in the medium after 60 s of plasma treatment is higher (444.11 mOsmol/kg) than in the argon gas-treated control (384.67 mOsmol/kg). To reach the same osmolality level as after 60 s of plasma treatment, the argon gas flow needs to impact on the medium for 180 s.

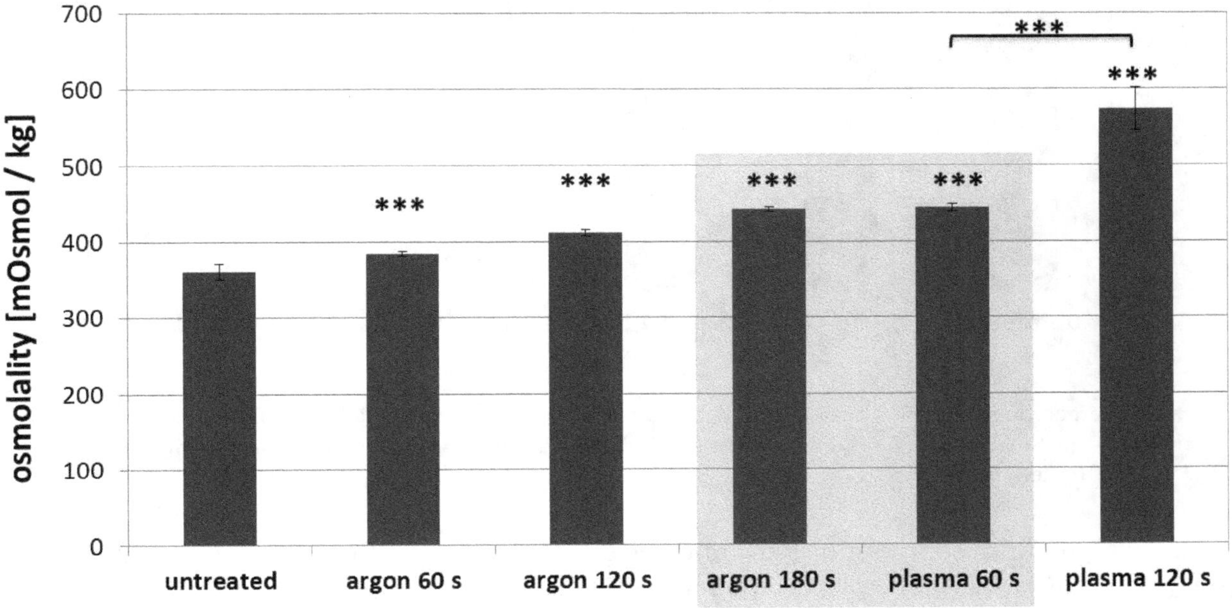

Figure 4. The osmolality of treated DMEM+FCS+Genta increased significantly with longer exposition times. Note that argon plasma treatment (plasma) over 60 s and argon gas treatment (argon) over 180 s result in similar values. (mean ± SD, *** $p < 0.001$, *t*-test, $n = 3$).

To determine the effect of increasing osmolality on the cell behavior, we studied the viability of mHepR1 cells cultured for 24 h in DMEM+FCS+Genta, which was treated similarly to the samples used for osmolality measurements. As can be seen in Figure 5, the viability of the cells was massively reduced due to the 60 s plasma-treated medium. Interestingly, cells in 180 s argon gas-treated medium kept a viability of around 80%, although the osmolality values of 60 s argon plasma versus 180 s argon gas were in the same range.

Thus, increasing osmolality during argon plasma treatment is not the reason for the detrimental cell effects we observed, neither the changing of the pH nor the oxygen concentration nor the persistence of hydrogen peroxide after plasma treatments, which were precluded earlier.

Besides physical parameters, which could be changed due to plasma treatment, alterations in the chemical composition of the cell culture medium were investigated. In a recent study, Kalghatgi *et al.* [24] demonstrated an induction of DNA damage in mammalian breast epithelial cells by plasma-treated cell culture medium. This effect was not reduced if the medium had been stored up to 1 h prior to addition to the cells. The authors hypothesized that this remaining biological effect of plasma-treated cell culture medium is caused by stable organic peroxides

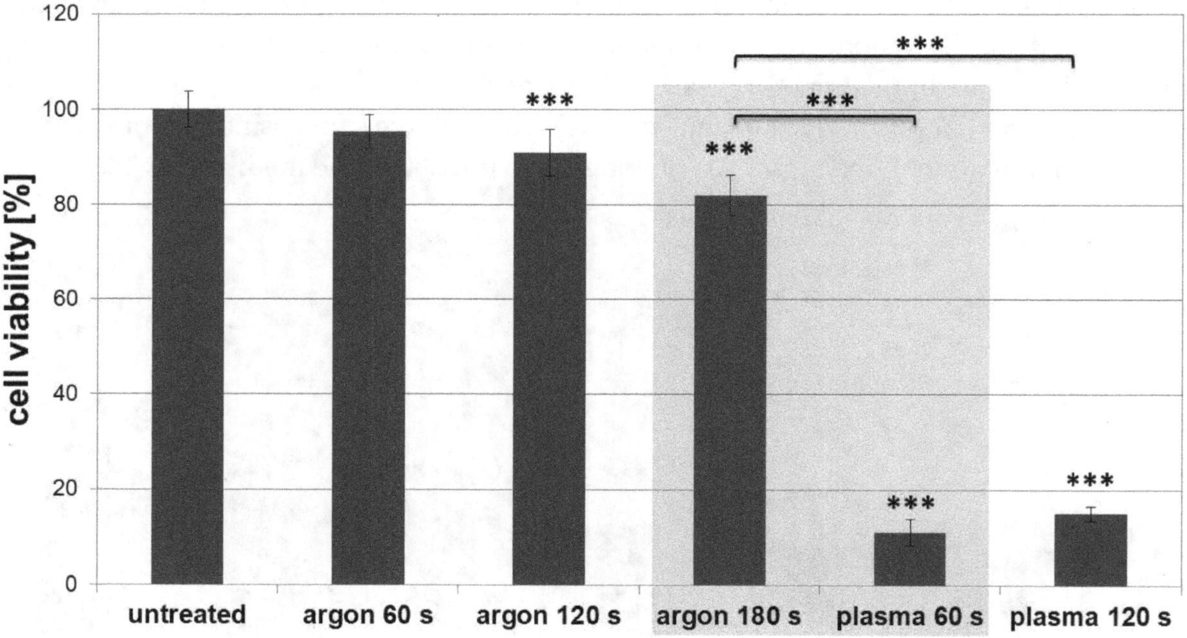

Figure 5. Difference in the viability of cells cultured in argon plasma- and argon gas-treated DMEM+FCS+Genta. Note that argon gas treatment over 180 s did not decrease cell viability as much as the argon plasma treatment did. (mean ± SD, *** $p < 0.001$, t-test, $n = 3$).

made up of medium compounds like amino acids [24]. The formation of stable peroxides from amino acids and proteins by reactive oxygen species like an OH radical is a well-known process [37]. In recent years, an active discussion has begun about the very complex ROS chemistry in plasma-treated liquids and its biological effects, including reactive species like OH radical, hyperoxide anion, and also the relatively stable hydrogen peroxide [1,38,39].

As we showed recently by size-exclusion chromatography on an FPLC analysis system, medium components are modified by argon plasma treatment [26]. We detected an additional peak upon plasma treatment of DMEM+FCS+Genta compared with the argon gas-treated control. In these experiments, we separated the components after plasma treatment by gel filtration. Interestingly, the peak height increased depending on the treatment time of the medium. In the present experiments, we examined the contribution of different medium components, FCS and gentamicin, to this additional peak and to the effects on cells. For this purpose, we investigated plasma treatment for the medium with (DMEM+FCS) and without supplements (DMEM). The different medium samples were analyzed after argon gas flow for 120 s and after argon plasma for 120 s by gel filtration with an FPLC system and compared with untreated controls (Figure 6).

All chromatograms of the analyzed media show peaks at a retention volume of about 22 ml (21.67–22.10 ml) and 26 ml (25.41–25.77 ml). The characteristic peak for albumin at about 15 ml (14.38–14.70 ml) was found in DMEM supplemented with FCS. An additional peak was found for the samples, which were treated for 120 s with argon plasma (see arrow) in basic DMEM, in DMEM+FCS (see Figure 6) and in DMEM+Genta (data not shown). No signal around the retention volume of 20 ml could be detected in untreated media and media that

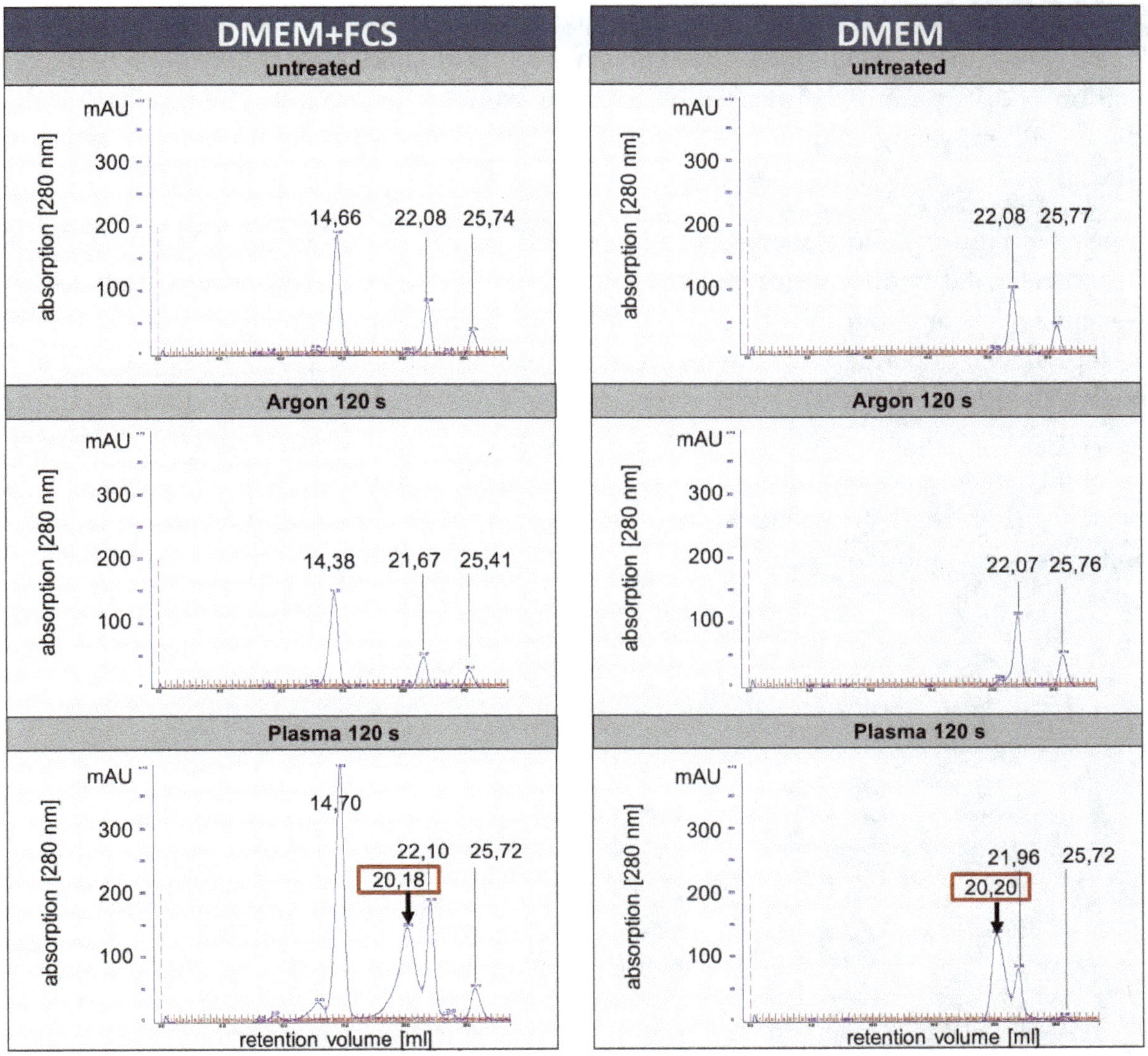

Figure 6. An additional peak at a retention volume of about 20 ml could be detected in the plasma-treated samples using size-exclusion chromatography. Note that cell culture medium DMEM without FCS also showed this signal, indicating that the source of this signal was not the interaction of argon plasma with FCS.

were only exposed to argon gas flow. This result gives rise to the idea that the visible changes in the chromatogram were caused by a change in the basic media (DMEM) independent of the supplements (FCS and gentamicin). Argon plasma seems to initiate the formation of components with higher molecular weight in the basic DMEM that appear at a lower retention time in the chromatogram.

Investigators hypothesized that ROS induces the formation of organic peroxides, which could be responsible for the damaging effects on cells in culture [21,22,37]. This conclusion is mainly based on the comparison of the incubation of cells with plasma-treated PBS, FCS, bovine serum albumin (BSA) or single amino acids [22]. DNA damage could not be shown after incubation of cells with plasma-treated PBS. However, incubation with plasma-treated BSA (in PBS) or FCS (in PBS) resulted in DNA damage of subsequently added cells. Even plasma treatment of PBS containing single amino acids resulted in adverse cell effects. Further analysis revealed a correlation between the peroxidation efficiency of single amino acids and their potential to induce DNA damage in cells [22]. These findings supported the thesis of amino acid peroxide formation and its potential contribution to DNA damage generation. The damage could be mediated either by single amino acids or by FCS, both being usual components of cell culture media.

As shown in Figure 5 and also observed in an earlier work of our group, DMEM+FCS+Genta impaired cell vitality after argon plasma treatment [25]. Based on the results deduced from size-exclusion chromatography (see Figure 6), it was deemed important to analyze the viability of mHepR1 cells dependent on the medium supplements during plasma treatment. For this purpose, DMEM and DMEM+FCS were argon plasma-treated for 60 and 120 s. Controls without treatment or only argon gas flow were also investigated. It was found that the mHepR1 cell viability decreased significantly in plasma-treated media after 24-h incubation for both approaches (Figure 7). In particular, in the presence of FCS during plasma treatment, the inhibitory effect is more obvious.

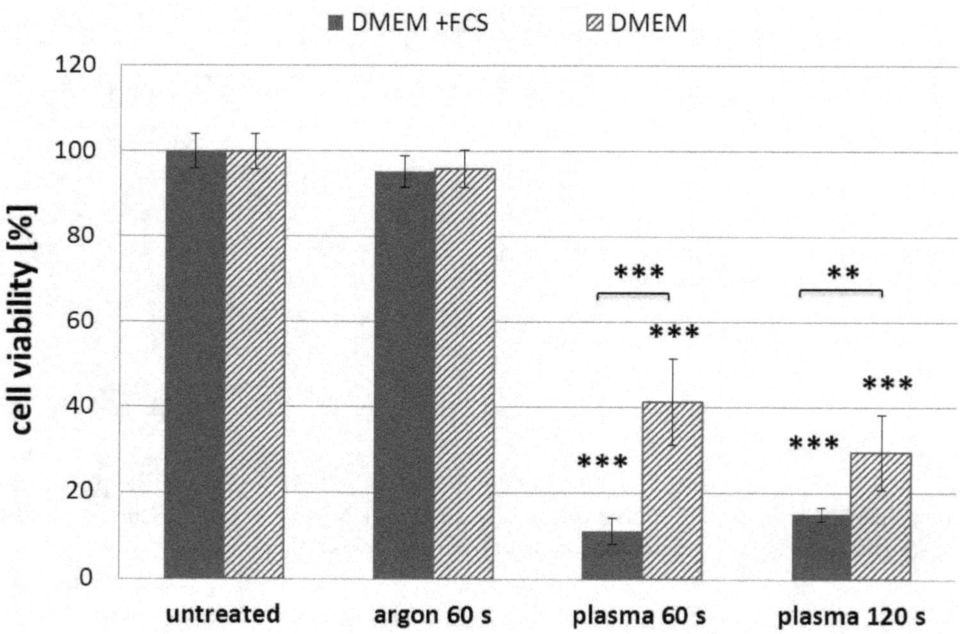

Figure 7. Cell viability after incubation of mHepR1 cells in plasma-treated DMEM and DMEM+FCS. (mean ± SD, ** $p < 0.005$, *** $p < 0.001$, t-test, $n = 3$).

Therefore, it can be assumed that components in the basic DMEM without supplements were changed due to the impact of argon plasma and influenced cell viability.

Based on the hypothesis of Kalghatgi *et al.*, for the generation of stable organic peroxides from amino acids during plasma treatment [22] and our cell viability results in mHepR1 cells in plasma-treated DMEM versus DMEM+FCS, it can be concluded that peroxides formed from single amino acids by plasma treatment could be the reason for the detrimental cell effects we found. However, peroxidated amino acids are semi-reactive compounds and it seems possible that they generate compounds with higher molecular weight during storage, e.g., by polymerization. The identification of the newly generated substances in the cell culture medium awaits further analysis.

There is increasing evidence that plasma treatment promotes the healing process of tissue and accelerates wound healing [13,40–42]. The skin is a complex architectural multilayer cell system and processes concerning wound healing can be examined on cell monolayers as an *in vitro* model system. However, it is important to keep in sight that a cell monolayer *in vitro* is much more sensitive to agents because there are no cells in the "second row" to protect or replace cells in the apical row [43]. Here, we described effects of argon plasma-treated liquids (e.g., cell culture medium) as an indirect approach for plasma application in medicine. Transferred to *in vivo* systems, the opening of cell–cell contacts we observed by plasma-treated liquids could have a positive effect on the penetration of conventional therapeutics (antibiotics and disinfectants) on skin. Often, the application of conventional liquid antiseptics is not sufficient and sustainable as the borders and the surrounding of chronic wounds frequently consist of sclerotic skin, impeding an effectual penetration of these products [44]. With regard to disinfection, direct plasma treatment of living intact and wounded skin was found to be safe for doses even higher than required for inactivation of bacteria [45,46]. Impaired cell adhesion and reduced cell viability due to plasma-treated liquids are important starting points for further investigations concerning cancer therapy. The work presented here, focused on the basic mechanisms in the interaction of plasma-treated cell culture medium with cells and on the components in the treated medium, which are responsible for the persistent biological effect.

The vision could be the establishment of local plasma centers to prepare relatively stable plasma liquids for dermatological (chronic wounds and tumors), dental (peri-implantitis), or orthopedic (joint rinsing) applications to support or replace the conventional therapy.

4. Summary and conclusions

This study focused on the physicochemical analysis of argon plasma-treated cell culture medium DMEM with the additives FCS and gentamicin. In addition, the efficacy of plasma-treated complete cell culture medium DMEM upon storage and its impact on the cell physiology of epithelial mHepR1 cells were ascertained. We discovered that the liquid-mediated effect of atmospheric-pressure argon plasma on mouse liver epithelial cell–cell contacts and cell membrane microvilli persists up to even 21 days or 7 days of storage, respectively. Earlier investigations of our group revealed that temperature and pH (both were constant) as well as hydrogen peroxide production and oxygen content (both decreased within 1 day) can be excluded as initiators of the detrimental biological changes. As we found here, increased osmolality in the media caused by plasma treatment can also be excluded as a reason for the observed cell effects. On the other hand, we found an additional peak in size-exclusion chromatography analysis in basic DMEM after plasma treatment and significantly decreased cell viability in plasma-treated media independent of the presence of FCS during plasma treatment. High molecular compounds generated during plasma treatment of DMEM without

FCS give an impulse for further investigations on the formation, stability, and reaction of amino acid peroxides in this medium. The persistent biological effect on plasma-treated liquids observed here could open up new medical applications. Stable plasma-treated liquids could find applications for dermatological, dental, or orthopedic therapy.

Acknowledgements

The authors acknowledge the financial support of the University Medical Center Rostock, Germany (FORUN 889061, PlasmaBiomedicine) for C.H., M.K., and C.B.. M.H. is grateful for the doctoral stipendium of the state Mecklenburg-Vorpommern (Germany) and of the University of Rostock, Interdisciplinary Faculty, Department of Life, Light and Matter. The authors appreciate the BMBF Germany Pilot Program Campus PlasmaMed (13N11183). They would also like to express their thanks to the Electron Microscopic Center (EMZ) of the University Medical Center Rostock (M. Frank, G. Fulda) for qualified technical assistance.

Author details

Claudia Bergemann[1], Cornelia Hoppe[1,2], Maryna Karmazyna[1,3], Maxi Höntsch[1,4], Martin Eggert[5], Torsten Gerling[6] and Barbara Nebe[1*]

*Address all correspondence to: barbara.nebe@med.uni-rostock.de

1 University Medical Center Rostock, Department of Cell Biology, Schillingallee, Rostock, Germany

2 Technical University Dresden, DFG-Center for Regenerative Therapies, Dresden, Germany

3 Seracell Stem Cell Technology GmbH, Schillingallee, Rostock, Germany

4 Sarstedt AG and Co, Quality Management/EWZ, Nümbrecht, Germany

5 University Medical Center Rostock, Department of Internal Medicine, Center for Extracorporeal Organ Support, Schillingallee, Rostock, Germany

6 Leibniz-Institute for Plasma Science and Technology e.V., Greifswald, Germany

References

[1] von Woedtke T, Reuter S, Masur K, Weltmann KD. Plasmas for medicine. Phys Rep. 2013;530:291–320.

[2] von Woedtke T, Metelmann HR, Weltmann KD. Clinical plasma medicine: State and perspectives of in vivo application of cold atmospheric plasma. Contrib Plasma Phys. 2014;54:104–17.

[3] Fridman G, Friedman G, Gutsol A, Shekhter AB, Vasilets VN, Fridman A. Applied plasma medicine. Plasma Process Polym. 2008;5:503–33.

[4] Fridman G, Peddinghaus M, Ayan H, Fridman A, Balasubramanian M, Gutsol A, et al. Blood coagulation and living tissue sterilization by floating-electrode dielectric barrier discharge in air. Plasma Chem Plasma P. 2006;26:425–42.

[5] Kalghatgi SU, Fridman G, Cooper M, Nagaraj G, Peddinghaus M, Balasubramanian M, et al. Mechanism of blood coagulation by nonthermal atmospheric pressure dielectric barrier discharge plasma. IEEE T Plasma Sci. 2007;35:1559–66.

[6] Daeschlein G, von Woedtke T, Kindel E, Brandenburg R, Weltmann KD, Junger M. Antibacterial activity of an atmospheric pressure plasma jet against relevant wound pathogens in vitro on a simulated wound environment. Plasma Process Polym. 2010;7:224–30.

[7] Daeschlein G, Scholz S, Ahmed R, von Woedtke T, Haase H, Niggemeier M, et al. Skin decontamination by low-temperature atmospheric pressure plasma jet and dielectric barrier discharge plasma. J Hosp Infect. 2012;81:177–83.

[8] Daeschlein G, Scholz S, Ahmed R, Majumdar A, von Woedtke T, Haase H, et al. Cold plasma is well-tolerated and does not disturb skin barrier or reduce skin moisture. J Dtsch Dermatol Ges. 2012;10:509–15.

[9] Daeschlein G, Scholz S, Arnold A, von Podewils S, Haase H, Emmert S, et al. In vitro susceptibility of important skin and wound pathogens against low temperature atmospheric pressure plasma jet (APPJ) and dielectric barrier discharge plasma (DBD). Plasma Process Polym. 2012;9:380–89.

[10] Nosenko T, Shimizu T, Morfill GE. Designing plasmas for chronic wound disinfection. New J Phys. 2009;11:115013.

[11] Nastuta AV, Topala I, Grigoras C, Pohoata V, Popa G. Stimulation of wound healing by helium atmospheric pressure plasma treatment. J Phys D Appl Phys. 2011;44:105204-0.

[12] Weltmann KD, Kindel E, Brandenburg R, Meyer C, Bussiahn R, Wilke C, et al. Atmospheric pressure plasma jet for medical therapy: Plasma parameters and risk estimation. Contrib Plasma Phys. 2009;49:631–40.

[13] Heinlin J, Isbary G, Stolz W, Morfill G, Landthaler M, Shimizu T, et al. Plasma applications in medicine with a special focus on dermatology. J Eur Acad Dermatol. 2011;25:1–11.

[14] Kalghatgi S, Friedman G, Fridman A, Clyne AM. Endothelial cell proliferation is en-

hanced by low dose non-thermal plasma through fibroblast growth factor-2 release. Ann Biomed Eng. 2010;38:748–57.

[15] Vandamme M, Robert E, Lerondel S, Sarron V, Ries D, Dozias S, et al. ROS implication in a new antitumor strategy based on non-thermal plasma. Int J Cancer. 2012;130:2185–94.

[16] Plewa JM, Yousfi M, Frongia C, Eichwald O, Ducommun B, Merbahi N, et al. Low-temperature plasma-induced antiproliferative effects on multi-cellular tumor spheroids. New J Phys. 2014;16:043027.

[17] Kalghatgi S, Kelly C, Cerchar E, Azizkhan-Clifford J. Selectivity of non-thermal atmospheric-pressure microsecond-pulsed dielectric barrier discharge plasma induced apoptosis in tumor cells over healthy cells. Plasma Med. 2011;1:249–63.

[18] Kaushik NK, Kaushik N, Attri P, Kumar N, Kim CH, Verma AK, et al. Biomedical importance of indoles. Molecules. 2013;18:6620–62.

[19] Stoffels E, Sakiyama Y, Graves DB. Cold atmospheric plasma: Charged species and their interactions with cells and tissues. IEEE T Plasma Sci. 2008;36:1441–57.

[20] Kieft IE, Broers JLV, Caubet-Hilloutou V, Slaaf DW, Ramaekers FCS, Stoffels E. Electric discharge plasmas influence attachment of cultured CHO k1 cells. Bioelectromagnetics. 2004;25:362–68.

[21] Haertel B, Wende K, von Woedtke T, Weltmann KD, Lindequist U. Non-thermal atmospheric-pressure plasma can influence cell adhesion molecules on HaCaT-keratinocytes. Exp Dermatol. 2011;20:282–84.

[22] Kalghatgi S, Kelly CM, Cerchar E, Torabi B, Alekseev O, Fridman A, et al. Effects of non-thermal plasma on mammalian cells. Plos One. 2011;6:16270.

[23] Haertel B, Hahnel M, Blackert S, Wende K, von Woedtke T, Lindequist U. Surface molecules on HaCaT keratinocytes after interaction with non-thermal atmospheric pressure plasma. Cell Biol Int. 2012;36:1217–22.

[24] Kalghatgi S, Azizkhan-Clifford J. DNA damage in mammalian cells by atmospheric pressure microsecond-pulsed dielectric barrier discharge plasma is not mediated via lipid peroxidation. Plasma Med. 2011;1:167–77.

[25] Hoentsch M, von Woedtke T, Weltmann KD, Nebe JB. Time-dependent effects of low-temperature atmospheric-pressure argon plasma on epithelial cell attachment, viability and tight junction formation in vitro. J Phys D Appl Phys. 2012;45:025206.

[26] Hoentsch M, Bussiahn R, Rebl H, Bergemann C, Eggert M, Frank M, et al. Persistent effectivity of gas plasma-treated, long time-stored liquid on epithelial cell adhesion capacity and membrane morphology. Plos One. 2014;9:e104559.

[27] Kalghatgi S, Fridman A, Azizkhan-Clifford J, Friedman G. DNA damage in mamma-

lian cells by non-thermal atmospheric pressure microsecond pulsed dielectric barrier discharge plasma is not mediated by ozone. Plasma Process Polym. 2012;9:726–32.

[28] Stoffels E, Kieft IE, Sladek REJ, van den Bedem LJM, van der Laan EP, Steinbuch M. Plasma needle for in vivo medical treatment: Recent developments and perspectives. Plasma Sources Sci Technol. 2006;15:S169–S80.

[29] Dobrynin D, Fridman G, Friedman G, Fridman A. Physical and biological mechanisms of direct plasma interaction with living tissue. New J Phys. 2009;11:115020.

[30] Nebe B, Rychly J, Knopp A, Bohn W. Mechanical induction of beta-1-integrin-mediated calcium signaling in a hepatocyte cell-line. Exp Cell Res. 1995;218:479–84.

[31] Paul D, Hohne M, Pinkert C, Piasecki A, Ummelmann E, Brinster RL. Immortalized differentiated hepatocyte lines derived from transgenic mice harboring Sv40 T-antigen genes. Exp Cell Res. 1988;175:354–62.

[32] Lange K. Role of microvillar cell surfaces in the regulation of glucose uptake and organization of energy metabolism. Am J Physiol-Cell Physiol. 2002;282:C1–C26.

[33] Pfister RR, Burstein N. Effects of ophthalmic drugs, vehicles, and preservatives on corneal epithelium – scanning electron-microscope study. Invest Ophth Visual. 1976;15:246–59.

[34] Oehmigen K, Winter J, Hahnel M, Wilke C, Brandenburg R, Weltmann KD, et al. Estimation of possible mechanisms of *Escherichia coli* inactivation by plasma treated sodium chloride solution. Plasma Process Polym. 2011;8:904–13.

[35] von Woedtke T, Oehmigen K, Brandenburg R, Hoder T, Wilke C, Hähnel M, et al. Plasma-liquid interactions: Chemistry and antimicrobial effects. In: Machala Z, Hensel K, Akishev Y, editors. Plasma for bio-decontamination, medicine and food security: Springer, Netherlands; 2012. pp. 67–78.

[36] Waymouth C. Osmolality of mammalian blood and of media for culture of mammalian cells. In Vitro Cell Dev Biol Plant. 1970;6:109–27.

[37] Gebicki S, Gebicki JM. Formation of peroxides in amino-acids and proteins exposed to oxygen free-radicals. Biochem J. 1993;289:743–49.

[38] van Gils CAJ, Hofmann S, Boekema BKHL, Brandenburg R, Bruggeman PJ. Mechanisms of bacterial inactivation in the liquid phase induced by a remote RF cold atmospheric pressure plasma jet. J Phys D Appl Phys. 2013;46:175203.

[39] Machala Z, Tarabova B, Hensel K, Spetlikova E, Sikurova L, Lukes P. Formation of ROS and RNS in water electro-sprayed through transient spark discharge in air and their bactericidal effects. Plasma Process Polym. 2013;10:649–59.

[40] Barekzi N, Laroussi M. Dose-dependent killing of leukemia cells by low-temperature plasma. J Phys D Appl Phys. 2012;45:422002.

[41] Ermolaeva SA, Varfolomeev AF, Chernukha MY, Yurov DS, Vasiliev MM, Kaminskaya AA, et al. Bactericidal effects of non-thermal argon plasma in vitro, in biofilms and in the animal model of infected wounds. J Med Microbiol. 2011;60:75–83.

[42] Emmert S, Brehmer F, Hänßle H, Helmke A, Mertens N, Ahmed R, et al. Atmospheric pressure plasma in dermatology: Ulcus treatment and much more. Clin Plasma Med. 2013;1:24–29.

[43] Hoene A, Prinz C, Walschus U, Lucke S, Patrzyk M, Wilhelm L, et al. In vivo evaluation of copper release and acute local tissue reactions after implantation of copper-coated titanium implants in rats. Biomed Mater. 2013;8:035009.

[44] Lademann O, Kramer A, Richter H, Patzelt A, Meinke MC, Czaika V, et al. Skin disinfection by plasma-tissue interaction: Comparison of the effectivity of tissue-tolerable plasma and a standard antiseptic. Skin Pharmacol Physiol. 2011;24:284–88.

[45] Dobrynin D, Wu A, Kalghatgi S, Park S, Shainsky N, Wasko K, et al. Live pig skin tissue and wound toxicity of cold plasma treatment. Plasma Med. 2011;1:93–108.

[46] Fluhr JW, Sassning S, Lademann O, Darvin ME, Schanzer S, Kramer A, et al. In vivo skin treatment with tissue-tolerable plasma influences skin physiology and antioxidant profile in human stratum corneum. Exp Dermatol. 2012;21:130–34.

2

Diagnostics of Magnetron Sputtering Discharges by Resonant Absorption Spectroscopy

Nikolay Britun, Stephanos Konstantinidis and Rony Snyders

Abstract

The determination of the absolute number density of species in gaseous discharge is one of the most important plasma diagnostics tasks. This information is especially demanded in the case of low-temperature sputtering discharges since the time- and space-resolved behavior of the sputtered particles in the ground state determines the plasma kinetics and plasma chemistry in this case. Historically, magnetron sputtering is often implied when talking about sputtering discharges due to the popularity and the numerous advantages this technique provides for coating applications. The determination of the absolute density of various atomic and molecular species in magnetron sputtering discharges along with its time and space evolution may be important from several points of view, since it may help to estimate the total flux of particles to a virtual surface in the plasma reactor, to compare the throughputs of two different sputtering systems, to use the absolute particle concentrations as an input data for discharge modeling, etc. This chapter is intended to provide an overview on the advantages and main principles of resonant absorption spectroscopy technique as a reliable tool for in situ diagnostics of the particle density, as well as on the recent progress in characterization of magnetron sputtering discharges using this technique, when the role of reference source is played by another low-temperature discharge. Both continuous and pulsed magnetron sputtering discharges are overviewed. Along with the introduction covering the main principles of magnetron sputtering, the description of the basics of resonant absorption technique, and the selected results related to the particle density determination in direct current and high-power pulsed magnetron sputtering discharges are given, covering both space- and time-resolved density evolutions.

Keywords: Magnetron sputtering, resonant optical absorption spectroscopy, ROAS, atomic absorption, absolute number density

1. Introduction: Magnetron sputtering discharges

Among the families of plasma-related methods involved in thin films growth, those belonging to physical vapor deposition (PVD) are among the most commonly utilized. The PVD-related techniques cover wide range of the deposition methods, including evaporation, vacuum arc deposition, laser ablation, and sputtering. These methods are different from another class referred to as chemical vapor deposition (CVD) in the sense that the source of material is solid or liquid as opposed to a gaseous one in the case of CVD [1]. Thermal evaporation has been the most used PVD process for many years because of the easy handling and relatively high deposition rate comparing to the first known sputtering process, namely, the diode sputtering [2]. The latter process has been known since its first description in 1852 by W. R. Grove, who had performed sputtering using a massive inductive coil [3]. In diode sputtering, ions of the sputter gas, commonly argon, hit a negatively biased cathode (also known as target) with energy up to several hundred electron volts (eV). This energy is high enough to induce ejection of the superficial atoms, which is followed by their condensation on the chamber surfaces, including a potential substrate [4]. In diode sputtering, however, due to the relatively high process pressure needed to ensure the discharge stability (nearly 0.1 Torr), the mean free path of the ejected atoms is in the mm range, which is much lower than the typical substrate-target distance (cm range) resulting in relatively poor quality of the deposited coating.

The limitations of diode sputtering have been overcome in 196× with the works of E. Kay and W. Gill [5,6], who have proposed to utilize magnetic field to create efficient electron trapping above the cathode (target) in order to accelerate the ionization and sputtering itself. The electron trapping region is created due to the fact that the electrons propagate mainly along the magnetic field lines as a result of well-known gyration effect (with gyration radius being in the μm range). The implementation of magnetic field for sputtering, naturally producing the name "magnetron," has been shortly followed by the introduction of planar magnetron sources in 1974 [2,7], for which the presence of permanent magnets beneath the cathode was a distinctive feature, as schematically illustrated in Figure 1. As a result of the electron trapping, these innovations led to essential increase in the ionization degree of the sputter (bulk) gas in the cathode vicinity. As a result, the pressure necessary to maintain the discharge current in this case could be reduced by about one order of magnitude, i.e., down to the mTorr range [5]. Another critical point related to efficiency of magnetron sputtering is the topology of magnetic field. Indeed, the arrangement of magnets beneath the cathode affects significantly the degree of electron confinement above the target. In the so-called balanced magnetron sources, the dense plasma region above the target is roughly comparable to its radius (assuming a planar circular target) [8]. If the substrate is located outside this region, the bombardment of growing film by the plasma ions is essentially reduced (the resulting ion current is <1 mA/cm^2), limiting the benefit of the sputtering process. This effect is different in the so-called unbalanced sources, where the topology of the magnetic field near cathode is different, letting some magnetic field lines reach the substrate. In this case, plasma is not confined completely, and the ion current densities of about 2–10 mA/cm^2 can be reached, which is typically one order of magnitude higher than in the case of balanced magnetrons [9].

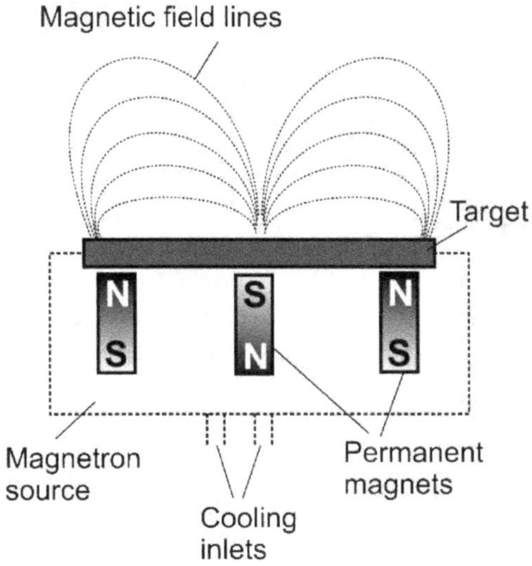

Figure 1. Side view of a typical magnetron source, including planar circular magnetron target, permanent magnets, and cooling system. The magnetic field lines are shown schematically.

In addition to magnetic field configuration, the voltage waveform and its repetition rate also play significant role in magnetron sputtering. The development of the plasma sources utilizing this effect has been mainly driven by the efficient deposition of the insulating compounds using reactive sputtering (see below). Thus, in addition to the well-known direct current magnetron sputtering (DCMS) devices, the radio frequency (RF) magnetron sources mainly working at 13.56 MHz have been introduced. Following the same trend, in the early 1990s, the idea of using pulsed-DCMS (also known as P-DCMS, or pulsed-DC) has been proposed (see, e.g., [10,11]). In most cases, the pulsed-DC power supplies operate successively alternating the negative (sputtering phase) and positive (charge dissipation phase) voltage cycles. The power supply in this case might be unipolar or bipolar, depending on polarity of the utilized pulses. Nowadays, most of magnetron sputtering processes for the synthesis of insulating compound coatings use pulsed magnetron sputtering approach [9].

Further development of the pulsed magnetron sputtering technology has triggered the new family of magnetron sputtering processes, so-called ionized physical vapor deposition (IPVD) techniques [12]. In a typical IPVD discharge, significant fraction of the sputtered atoms is ionized, reaching up to 100% in certain cases. The main idea behind the IPVD techniques is to generate denser plasmas than those appear in conventional magnetron sputtering in order to ionize the sputtered atoms more efficiently (up to the level of number densities ~10^{13}–10^{14} cm^{-3} vs. 10^{8}–10^{11} cm^{-3} in the DCMS case). Among the different approaches targeted to increase the ionization degree, such as using an inductive coil [13], or hollow cathodes [14], the high-power impulse magnetron sputtering technique, or HiPIMS (also known as high-power pulsed magnetron sputtering—HPPMS), is the most notorious IPVD example. HiPIMS discharge uses the pulse duration ranged from few μs to few hundreds μs, while the pulse repetition frequency typically varies from ~10 Hz to ~10 kHz. Under these conditions, the peak current density may reach values of up to several A/cm^{-2} compared to a few mA/cm^{-2} in DCMS, but only during a short time, typically 1–2% of the repetition period [15]. As a result, a dense plasma is generated during the plasma on time enabling not only efficient target sputtering but also high ionization degree of metallic vapor [16], at the same time keeping the average applied power comparable to that of DCMS [15–17]. The schematic comparison of the DCMS, P-DCMS, and HiPIMS techniques in terms of the applied power is shown in Figure 2. Note that the level of time-averaged power applied to the sputtered cathode in each case is roughly the same, which is first of all due to the fact that the permanent magnets beneath the target should not be overheated and reach the Curie temperature (which is as low as about 580 K for widely used neodymium-based magnets). The actual applied power level is typically equal to several hundred Watt and often defined by the magnetron source cooling system (i.e., water temperature) used beneath the target. The interest to the pulsed magnetron discharges and, namely, the HiPIMS discharges from both scientific and application points of view has been continuously increasing since the introduction these techniques [18,19]. The main advantages of the HiPIMS technique, along with the achievements in characterization of these discharges, are described in the numerous works [15, 17,20–23].

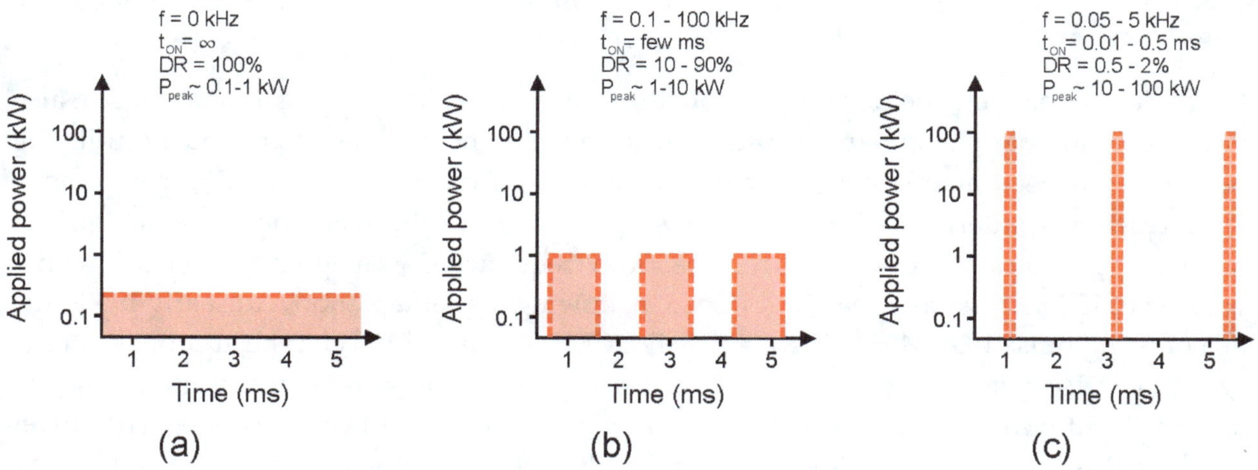

Figure 2. Comparison between the DCMS (a), P-DCMS (b), and HiPIMS (c) magnetron sputtering techniques in terms of the temporal distribution of the applied power. Typical range of the relevant discharge parameters and the values along both horizontal and vertical axes are given for reference only. The quantities f, t_{ON}, DR, and P_{peak} stand for repetition frequency, plasma on time, duty ratio, and peak power, respectively.

Why is the knowledge of the absolute density in plasma and particularly in magnetron sputtering discharges important? There are few main reasons making this parameter critical. First, *particle fluxes* in the discharge can be estimated knowing the density and velocity of each sort of the sputtered species, which is critical for any film growth process. Second, the discharge physics and chemistry can be significantly clarified having the absolute concentration of the relevant species in the discharge estimated properly. Third, the absolute concentration of the discharge species or their fluxes can be used as input information for further kinetic/chemical modeling of the discharge processes. Among the known plasma characterization approaches, resonant optical absorption spectroscopy (ROAS), also known as atomic absorption spectroscopy (AAS) [24,25], represents one of the straightforward ways for the determination of the absolute density of atomic and/or molecular species in gaseous discharges. Utilizing this technique, the density of the absorbers can be determined using an external (reference) light source by measuring the attenuation in its intensity after passing a volume with the absorbing species (see, e.g., Mitchell and Zemansky [24], p. 92). Along with the laser-based plasma diagnostic techniques, such as laser-induced fluorescence (LIF), ROAS technique has been playing a significant role in studying the plasma discharges during the last decades, involving both coherent (such as diode lasers) [26–28], and noncoherent (e.g., hollow cathode lamps—HCL) [29–31] radiation sources. The range of applications of this technique has been continuously increasing due particularly to the improvements of the tunable diode laser systems, namely, their stability, power level, tuning range, etc. [27,32].

In the domain of magnetron sputtering discharges, the determination of number density of the discharge particles by ROAS has been undertaken in the variety of systems, including time-resolved [33] and space-resolved [34] characterization. In particular, among the recent achievements, the works of various research groups devoted to absolute density of sputtered atoms in the DCMS discharges [31,34,35], including those amplified by an RF coil [30], as well as the works of devoted to flux measurements in DCMS [36], should be mentioned. What is related to the utilization of ROAS in the HiPIMS domain, the works dealing with the determination of time-resolved absorption line profile [33], which have been realized using tunable diode laser absorption spectroscopy (TD-LAS), should be emphasized. In addition, a detailed time evolution of the Ar metastable $1s_5$ states (Ar^{met}) has been investigated in a long-pulsed (200 μs) HiPIMS discharge by Vitelaru et al. [26], where the Ar^{met} density growth, followed by its rarefication after about 50 μs, and a gradual refill afterward have been clearly demonstrated. These results are generally in a good agreement with the recent studies in HiPIMS involving LIF [37] and ROAS [38] techniques, which are partially overviewed here. The last two works ([37,38]) are devoted to the short-pulsed HiPIMS discharges studying the propagation of the ground state and metastable sputtered particles, being mainly targeted to the systematic time-resolved characterization of HIPIMS discharges in terms of absolute density of species, 2-D imaging of the main atomic states corresponding to the studied discharge species (LIF+ROAS), as well as to the study of several optically available energy states of the discharge species and their sublevels.

Unifying the recent achievements in the domain of the absolute density measurements for two "extreme" magnetron sputtering cases, namely, DCMS and HiPIMS, this chapter is mainly focused on the particularities of the mentioned sputtering discharges in terms of the absolute density of species where resonant absorption is used as a main diagnostics technique. The resonant absorption method overviewed in this chapter is considered as a reliable tool for the determination of the absolute density of species in the discharge volume and for understanding the sputtering processes at the atomic level.

2. Particularities of resonant absorption

The main principles of the resonant absorption are described in details in the work of Mitchell and Zemansky [24]. This method is based on resonant absorption of radiation emitted by a reference source and absorbed (by atoms or molecules) in an optically thin gaseous discharge. At a definite spectral line the absolute density of states corresponding to a lower level of a chosen spectral transition can be determined by measuring so-called line absorption, i.e., an integral under spectral absorption line of interest, if the effective absorption lengths as well as the line width of both plasma and source spectral lines are known. This approach can be applied for the case when the spectral lines in the discharge are Doppler-limited [31], as well as can be generalized for more general cases [39]. The biggest advantage of ROAS technique is its nonintrusiveness. In this regard, this technique is different from the other methods used for density determination, which are based on the introduction of additional gases to the discharge (such as titration [40]), which may potentially change the electron energy distribution and the discharge kinetics. In addition, the fact ROAS uses external source of radiation and does not depend on the discharge emission itself is critical for characterization of the pulsed discharges, such as HiPIMS, especially during the afterglow time when optical emission spectroscopy (OES), cannot be applied.

As mentioned above, ROAS can be implemented in two different ways, namely, using a gaseous discharge emitting the spectral lines of interest, as well as using a monochromatic laser (e.g., a continuous tunable laser diode). Despite the fact that the laser-based diagnostics has several advantages [21,33], the discharge-based ROAS should be implemented. Among the main reasons for this are the following: (i) discharge-based ROAS setup is normally less expensive than the one based on a tunable laser; (ii) discharge-based ROAS depends on the line width ratio between the plasma and the reference source, which can be easily taken into account [24]; (iii) several spectral transitions can be studied in parallel in this case, as far as the emitters in the reference source coincide with the absorbers of interest, which is not possible with a monochromatic laser; and (iv) in case of a large spectral separation (few nm) between the lines of interest, laser-based absorption setup requires separate laser diodes for each transition, which involves additional calibration procedures. Due to the mentioned advantages, for example, the absolute density at the energy sublevels corresponding to a certain electronic state can be probed in a straightforward way, by simply following the changes in the emission intensity of the corresponding peaks from the reference source, as performed in the numerous studies [30,31,34,35,39]. Let us consider the main relations necessary to understand the ROAS method.

2.1. The basics of resonant absorption method

This chapter deals with the classical ROAS method described by Mitchell and Zemansky [24]. In the case of temperature-limited emission (absorption) lines in the reference source (and in the studied discharge), the application of the resonant absorption method is rather straightforward, and the details of its implementation in magnetron discharges can be found elsewhere [35,38]. In this section, only the essential relations important for understanding the absolute density determination using ROAS method are given. The number density of the absorbing species in the lower (often ground) state can be determined using the following relation:

$$N_{\text{abs}} = 1.2 \cdot 10^{12} k_0 \frac{\delta\sigma^p}{f_{ji}}, \tag{1}$$

where N_{abs} (cm^{-3}) is the absolute number density of the absorbing species in the lower state, k_0 (cm^{-1}) is the absorption coefficient corresponding to the center of the absorption line, $\delta\sigma^{p}$ (cm^{-1}) is the full width at half maximum (FWHM) of the plasma emission line, f_{ji} is the absorption oscillator strength, and j (i) stands for lower (upper) level of a chosen transition, respectively. f_{ji} can be found using the relation [41]:

$$f_{ji} = 1.5 \cdot 10^{-14} \frac{g_i}{g_j} A_{ij} \lambda_{ij}^2, \tag{2}$$

where g is the statistical weight of the corresponding energy level, A_{ij} is the emission probability corresponding to $i \rightarrow j$ transition, and λ_{ij} is the transition wavelength.

The absorption coefficient k_0 in Eq. (1) can be determined based on measurements of line absorption A_L, representing the integral under the absorption line profile. In the case of Doppler broadening of the source emission and plasma absorption lines, A_L can be expressed as follows [34,39]:

$$A_L = \frac{2}{\sqrt{\pi}} \int_0^{\infty} \mathrm{Exp}\left(-x^2\right)\left(1 - \mathrm{Exp}\left(-k_0 L \cdot \mathrm{Exp}\left(-\alpha^2 x^2\right)\right)\right) dx, \tag{3}$$

where L is the effective absorption length (~30 cm in our case), and α is the reference source-to-plasma line broadenings ratio. In the Doppler-limited case $\alpha^2 = T_S / T_P$, where T_S and T_P are the absolute gas temperatures in the reference source and in the plasma, respectively. A_L is normally determined experimentally by measuring the intensities of the corresponding emission lines:

$$A_L = 1 - \frac{\text{transmitted radiation}}{\text{incident radiation}} = 1 - \frac{I_{PS} - I_P}{I_S}, \tag{4}$$

where I_S is the light intensity from the reference source, I_P is the light intensity from the plasma, and I_{PS} is the (partially absorbed) light from the reference source passing through absorption region, along with the light from plasma, if any. The physical meaning of these quantities is illustrated in Figure 3.

After the experimental determination of line absorption A_L, according to Eq. (4), the absorption coefficient k_0 can be found as a result of solution of Eq. (3). This leads to a direct determination (according to Eq. (1)) of the absolute density of species corresponding to the lower atomic (molecular) state of a chosen spectral transition, if the effective absorption length L, as well as the oscillator strength f_{ji} are known. The critical role in the determination of the absorption

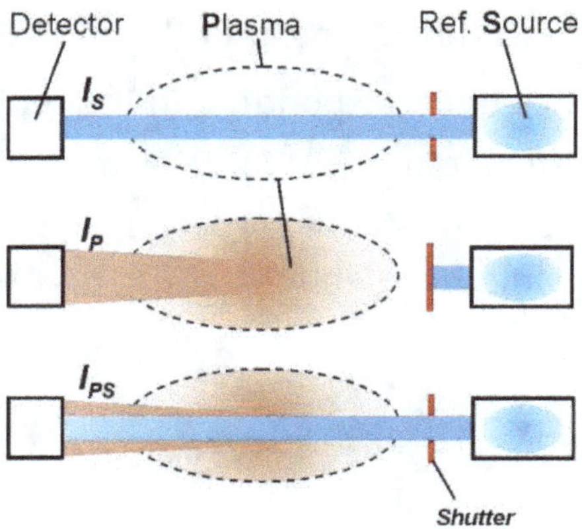

Figure 3. Schematic illustration of the absorption technique and the physical meaning of the quantities I_P, I_S, and I_{PS}.

coefficient is played by the coefficient α, representing the line broadening ratio of the spectral lines, as explained below.

2.2. Role of spectral line broadening

Several important remarks on the line broadening of plasma and reference source spectral lines should be made for proper interpretation of ROAS data. As mentioned above, this chapter deals with discharge-based ROAS when an HCL is primarily used as reference source. Since the gas pressure in HCL normally exceeds few Torr, and the mean free path of the gas particles is in the μm-range, the total particle thermalization can be assumed. The temperature in HCLs under the reviewed conditions, additionally measured by a high-resolution planar Fabry–Perot interferometer [42], is in the range of about 700–1100 K (see Table 1). Despite much lower pressure (~20–30 mTorr) in the magnetron discharges considered here (Table 1), similar considerations are also valid for the *bulk gas* atoms (Ar) in the studied DCMS and HiPIMS cases, which can be considered quasi-thermalized. The Ar temperature can be often assumed to be in the range of 300–500 K in this case, depending on the discharge type and its conditions [21,30,35,43].

Parameter	Value	Comment
	DCMS system parameters	
Target	99.99% Ti	With water cooling
Target diameter/thickness	5 cm/0.5 cm	10 cm/1 cm for Figure 9
Working gas	99.999% Ar	
Working pressure	20 mTorr	30 mTorr for Figure 9
Base pressure	~10^{-6} Torr or less	
Time-averaged power	~200 W	500 W for Figure 9
Monocromator used	Princeton Instrument SpectraPro-500i	Jobin-Yvon HR-460 for Figure 9

Parameter	Value	Comment
Monocromator slit width	10 μm	
Spectral resolution	0.05 nm	
Optical detector used	Princeton Instrument PI-MAX ICCD	Hamamatsu R928 photomultiplier tube
	HiPIMS system parameters	
Target	99.99% Ti	With water cooling
Target diameter/thickness	10 cm/1 cm	
Working gas	99.999% Ar	
Working pressure	20 mTorr	
Base pressure	$<10^{-6}$ Torr	
pulse duration/frequency	20 μs/1 kHz	Unless stated otherwise
Supplied energy per pulse(at 20 μs pulse)	≈0.26 J	Corresponding to ~260 W of time-averaged power
Monocromator used	Andor Shamrock 750	
Monocromator slit width	20–100 μm	
Spectral resolution	0.04–0.1 nm typically	
Optical detector used	Andor iStar DH740-18F ICCD	With $\sim 10^3$ pulses averaged
	Resonant absorption parameters	
Beam diameter	≈1 cm	
Distance above the target, z	≈5 cm	$z = 0..10$ cm for Figure 8, $z = 8$ cm for Figure 9.
Effective absorption length, L	30 cm	Variable for Figure 8, L=25 cm for Figure 9.
Reference source type	Ar-Ba and Ne-Ti hollow cathode lamps	
Reference source pulse duration	40–100 μs	Continuous for Figure 8, 150 μs for Figure 9.
Reference source temperature	≈1100 K (in pulsed mode [38])	950 K for Figure 8 [34], 630 K for Figure 9 [54].
DCMS plasma temperature	580 K (Figure 8), 300 K (Figure 9)	
HiPIMS plasma temperature (Ar)	≈500 K	Based on OES measurements [43]
HiPIMS plasma line width (Ti)	≈5.5 GHz	Based on LIF measurements [44]

Table 1. Main parameters of the DCMS and HiPIMS discharges described in this work and characterized by resonant optical absorption spectroscopy.

The broadening of spectral lines corresponding to the *sputtered* particles may differ from those of the bulk gas (Ar), however. In the DCMS case, despite the fact the sputtered particles are far from thermalization, their velocity distribution function (VDF) can be often approximated by a Gaussian profile. This VDF does not change in time, by definition of DCMS discharge [37], and often can be associated with the particle temperature in the range of 300–700 K for the conditions described here (see Table 1). Due to the directional motion of the sputtered particles in magnetron plasma, the spectral line in the direction of measurements (i.e., along the ROAS line-of-sight) might be significantly broadened, as compared to the one of the reference source. This situation is schematically illustrated in Figure 4. In this case, the resonant absorption will happen only for a part of the total line width, and the ratio between the mentioned broadenings has to be introduced ($\delta\sigma^S/\delta\sigma^P$) to take into account the broadening difference, as described in details by Mitchell and Zemansky [24]. The similar approach should also be utilized in case non-Doppler line broadening dominates in the discharge, as studied recently by Li et al. [39].

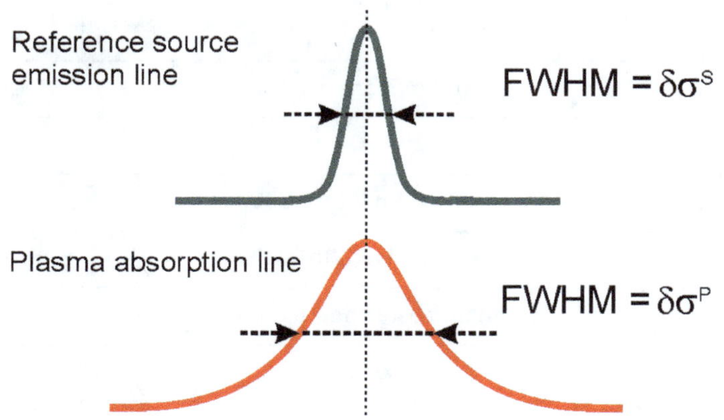

Figure 4. Illustration of the spectral line broadening (FWHM) of the reference source (emission) and plasma (absorption) lines along with their broadening notations.

The situation is yet different for HiPIMS discharges, which are far from thermalization during the plasma on time, especially in the target vicinity [37,43]. The distribution of the velocity component parallel to the target ($v_{||}$) in this case is often different from the Maxwell–Boltzmann one (even though it still can be roughly fitted by a Gaussian profile [33]), and the gas temperature in the classic sense cannot be applied. Because of this, the VDF line width corresponding to a particular velocity component and defining the line width of the emission (absorption) line should be determined separately. The dynamics of the line broadening in HiPIMS has been studied recently by LIF [44,45]. Based on these works, under the conditions described here (20 mTorr, $z = 5$ cm), the VDF width changes only by about 25%, which has rather minor impact on the final density of absorbers determined by Eq. (1). Due to this fact, the plasma absorption line width is assumed constant during the whole HiPIMS discharge period for simplicity, as described in [38,44] (see Table 1). These assumptions are additionally justified by the fact that the absorption coefficient k_0, determined based on Eq. (3) is weakly dependent on the plasma line width $\delta\sigma^P$.

2.3. Experimental setup for resonant absorption

The typical experimental arrangement for characterization of magnetron sputtering discharge by resonant spectroscopy is shown in Figure 5. The main parameters related to the discharges overviewed in this work are summarized in Table 1. Both in DCMS and HiPIMS cases, a cylindrical vacuum chamber with either horizontally or vertically placed balanced magnetron sputtering source holding a planar circular magnetron target has been used. The Ti magnetron targets, 5 or 10 cm in diameter, attached to a magnet-supporting water-cooled copper base, have been utilized. The commercial power supplies have been used to sustain the DCMS discharges, as described elsewhere [30,34]. In the HiPIMS case, the discharge current and voltage along with the pulse parameters have been controlled by the Lab.-made power supply, described by Ganciu et al. [46]. The HiPIMS discharge typically had 20 μs of the pulse duration and 1 kHz of the repetition frequency (thus having 980 μs of the plasma off time). The typical current–voltage waveforms can be found elsewhere [43,44,47]. The base pressure in the magnetron reactor has been kept <10^{-6} Torr, whereas the bulk gas (99.999% Ar) pressure was normally fixed at 20 mTorr during all the ROAS measurements, unless stated otherwise. In order to avoid deposition on the quartz window surface, the special collimators (Ni tubes) with high length-to-radius ratio have been installed in front of each viewport inside the reactor.

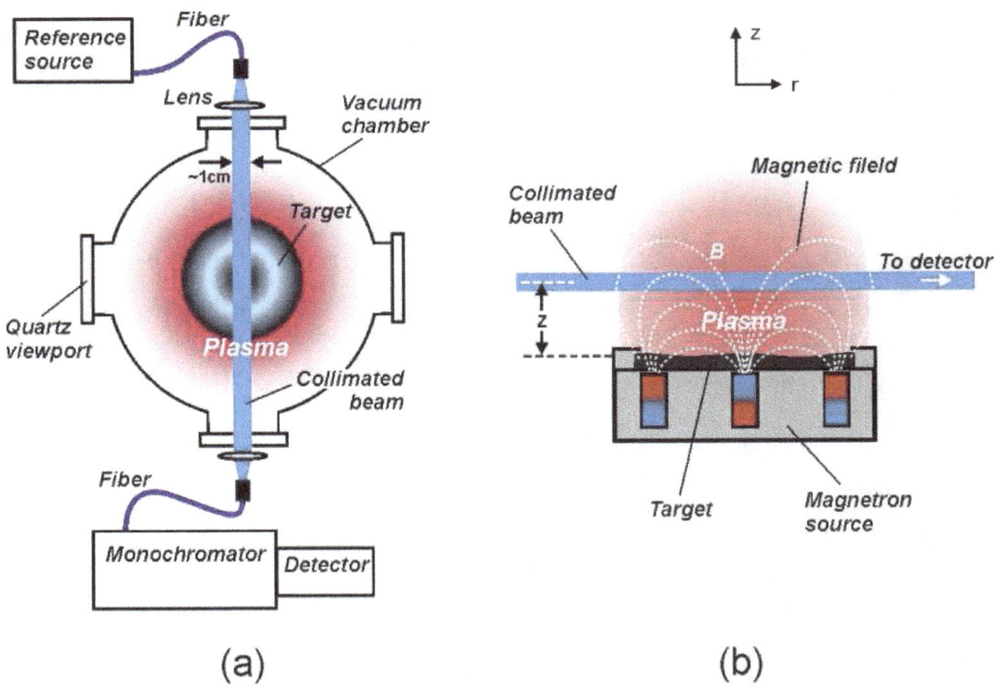

Figure 5. Top (a) and side (b) view of the experimental setup utilized for absolute density measurements in a magnetron sputtering discharge by ROAS. Role of the reference source is played by a hollow cathode lamp. Reproduced with permission from Britun et al. [38].

The Perkin Elmer hollow cathode lamps (one with Ti cathode filled by Ne, and another one with Ba cathode filled by Ar) have been used as the reference sources. The HCLs have been running either in the continuous or pulsed regime (mainly in the HiPIMS case) in order to increase the emission intensity and make it comparable with the HiPIMS plasma emission, aiming at reducing the measurements error during the plasma on time [43,48]. Apart from this, the pulse regime of HCL source favors the ionization of the cathode material inside the lamp, which is critical for analysis of the Ti^+ states [30]. During the measurements, the HCLs have a pulse duration typically ranging from 40 to 100 μs. About 10^3 pulses from the reference source have been averaged by the detector during the measurements of the absorption coefficient. The HCL pulse sequence was triggered by the HiPIMS power supply using the external transistor-transistor logic (TTL) trigger and was time shifted relatively to the HiPIMS plasma pulse using an additional analog TGP-110 pulse generator.

In some cases, when the density of absorbers in the studied discharge is too low, an improved detection scheme, including a triple optical fiber and allowing simultaneous acquisition of the I_S, I_P, and I_{PS} signals, may significantly increase the reliability of the obtained data. Such a scheme, illustrated in Figure 6, allows to avoid the instabilities of the plasma discharge and/or of the reference source during the acquisition time (which may last for several minutes) because the signals necessary to calculate the line absorption are taken simultaneously in this case, as described by Britun et al. [48]. The diagnostics setup in this case contains a reference emission source, a multichannel optical fiber, a group of lenses for beam collimation before passing the vacuum chamber, and a monochromator with an optical detector, according to Figure 5(a).

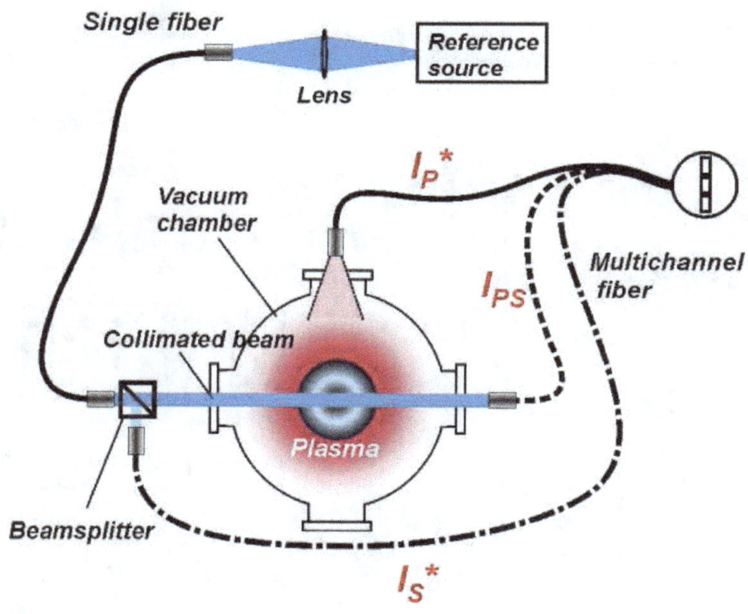

Figure 6. Scheme of the experimental setup for absorption measurements with increased reliability utilizing simultaneous acquisition of the plasma and source optical signals (I_P, I_S, and I_{PS}) necessary for calculation of the line absorption A_L.

A monochromator equipped with an intensified charge-coupled device (ICCD) camera and connected to the discharge reactor by an optical fiber has been used for spectral acquisition in most cases. The spectral resolution of the monochromator was typically ≈0.05–0.1 nm. The typical emission spectra acquired in Ar-Ti HiPIMS discharge at the end of plasma on time are presented in Figure 7. From the general emission spectrum (Figure 7(a)), the relevant spectral regions containing Ti and Ti^+ emission lines are visible (Figure 7(b–d)). The representative emission lines used in this work for resonant absorption are marked by arrows (see also Table 2). In most cases, the studied emission lines are well-resolved from the neighboring peaks. Due to this fast, rather low spectral resolution is normally sufficient for ROAS measurements. The total procedure for the number density determination consisted of the emission peaks intensities measurements (I_S, I_P, and I_{PS} values at each condition), A_L value determination by Eq. (4), determination of the absorption coefficient k_0 as a result of solving Eq. (3), followed by calculation of the absolute density of absorbers according to Eq. (1).

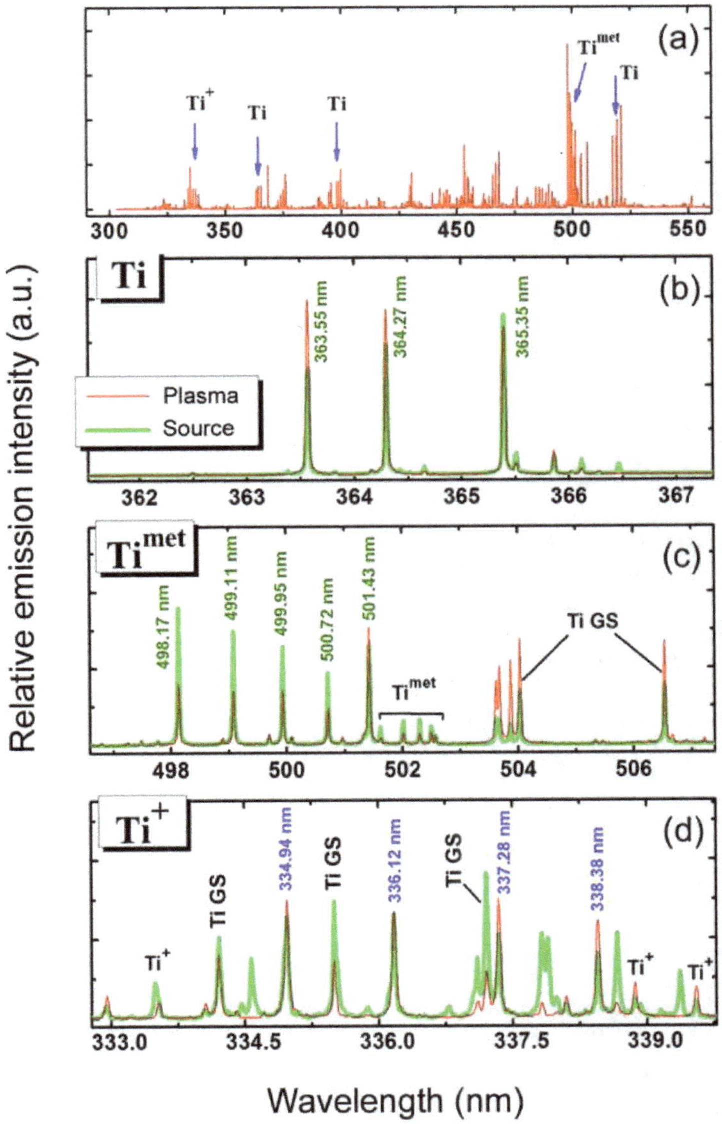

Figure 7. General emission spectrum (a) obtained in Ar-Ti HiPIMS discharge showing the spectral regions of interest used for the determination of the absolute density of different atomic species. The spectral regions corresponding to Ti ground state atoms (b), Ti metastables (c), as well as Ti ions (d) are shown separately. The HiPIMS and HCL spectra (b–d) are normalized arbitrary. Reproduced with permission from Britun et al. [38].

Spectral parameter Studied atom	Nature of the probed state	Energy of the probed state (eV)	Emission wavelength λ (nm)	Spectral transition (lower–upper level, *j*–*i*)	Statistical weight of the lower level, g_j	Oscillator strength, f_{ji} [a]
Ti	Ground	0.000	363.55	$3d^2 4s^2\ a^3F_2 - 3d^2(^3F)4s4p(^1P^\circ)\ y^3G^\circ_3$	5	0.20
		0.021	364.27	$3d^2 4s^2\ a^3F_3 - 3d^2(^3F)4s4p(^1P^\circ)\ y^3G^\circ_4$	7	0.16
		0.048	365.35	$3d^2 4s^2\ a^3F_4 - 3d^2(^3F)4s4p(^1P^\circ)\ y^3G^\circ_5$	9	0.14
Ti^{met}	Metastable	0.813	501.43	$3d^3(^4F)4s\ a^5F_1 - 3d^3(^4F)4p\ y^5G^\circ_2$	3	0.38
		0.818	500.72	$3d^3(^4F)4s\ a^5F_2 - 3d^3(^4F)4p\ y^5G^\circ_3$	5	0.26
		0.826	499.95	$3d^3(^4F)4s\ a^5F_3 - 3d^3(^4F)4p\ y^5G^\circ_4$	7	0.22
		0.836	499.11	$3d^3(^4F)4s\ a^5F_4 - 3d^3(^4F)4p\ y^5G^\circ_5$	9	0.17
		0.848	498.17	$3d^3(^4F)4s\ a^5F_5 - 3d^3(^4F)4p\ y^5G^\circ_6$	11	0.22
Ti^+	Ground	0.000	338.38	$3d^2(^3F)4s\ a^4F_{3/2} - 3d^2(^3F)4p\ z^4G^\circ_{5/2}$	4	0.28
		0.012	337.28	$3d^2(^3F)4s\ a^4F_{5/2} - 3d^2(^3F)4p\ z^4G^\circ 7/2$	6	0.25
		0.028	336.12	$3d^2(^3F)4s\ a^4F7/2 - 3d^2(^3F)4p\ z^4G^\circ 9/2$	8	0.26
		0.049	334.94	$3d^2(^3F)4s\ a^4F9/2 - 3d^2(^3F)4p\ z^4G\ ^\circ 11/2$	10	0.27
Ti^{+met}	Metastable	0.574	376.13	$3d^2(^3F)4s\ a^2F_{5/2} - 3d^2(^3F)4p\ z^2F^\circ_{5/2}$	6	0.22
		0.607	375.93	$3d^2(^3F)4s\ a^2F_{7/2} - 3d^2(^3F)4p\ z^2F^\circ_{7/2}$	8	0.21
Ar^{met}	Metastable	11.548	811.53	$3s^2 3p^5 4s(1s_5) - 3p^5 4p(2p_9)$[b]	5	0.51
		11.723	794.82	3s23p54s(1s $_3$)–3p54p(2p $_4$)[b]	1	0.56

[a]The values of the oscillator strength are taken from Crintea et al. [75] for Ar and Wiese and Fuhr [76] otherwise.

[b]Paschen notations are used for Ar energy levels [75].

Table 2. Spectral parameters of the optical transitions for Ti, Ti^{met}, Ti^+, and Ar^{met} atomic states used in this work for ROAS diagnostics. Spectroscopic data are taken from the online NIST spectral database [74] unless stated otherwise. Reproduced with permission from Britun et al. [38].

3. Particle density behavior in DCMS discharges

3.1. Spatial characterization of the DCMS discharge

Even though the main plasma parameters of the DCMS discharges do not alter in time, the plasma density as well as the density of the sputtered species may vary significantly as a function of the spatial position in the discharge. The first parameter in this case defines mainly the excitation level of the bulk and sputtered species as well as their ionization degree, whereas the (ground state) density of the sputtered species is mainly affected by the applied power, sputtering yield, target size, diffusion processes in the discharge volume, as well as by the other processes. The effects of ground state density depletion also may take place, as recently shown for DCMS discharges by LIF [49], but these effects are rather minor, especially comparing with HiPIMS case [37,38], since the ionization degree in the DCMS discharges generally remains at the level of few percent [15].

The optical diagnostics applied to a DCMS discharge clearly visualizes the plasma region corresponding to the confined electrons, which is represented by the bright area near the target surface observable by a naked eye, as shown in Figure 8. This region clearly corresponds to a spatial segment formed by the magnetic lines in the target vicinity, as shown in Figure 8(b), where the simulated magnetron magnetic field lines are drawn. The target diameter in this case is equal to 5 cm, but nevertheless the obtained data can be compared to the typical absolute density measured with larger targets (see Table 1), as soon as the target materials are the same and the current density are known, due to the scalability of the magnetron discharges [15]. In addition to this, as shown by Britun et al. [34], the emission lines in the plasma region decay exponentially along z direction, where an exponential dependence with significantly lower decrement is found at higher z, which corresponds to the diffusion region.

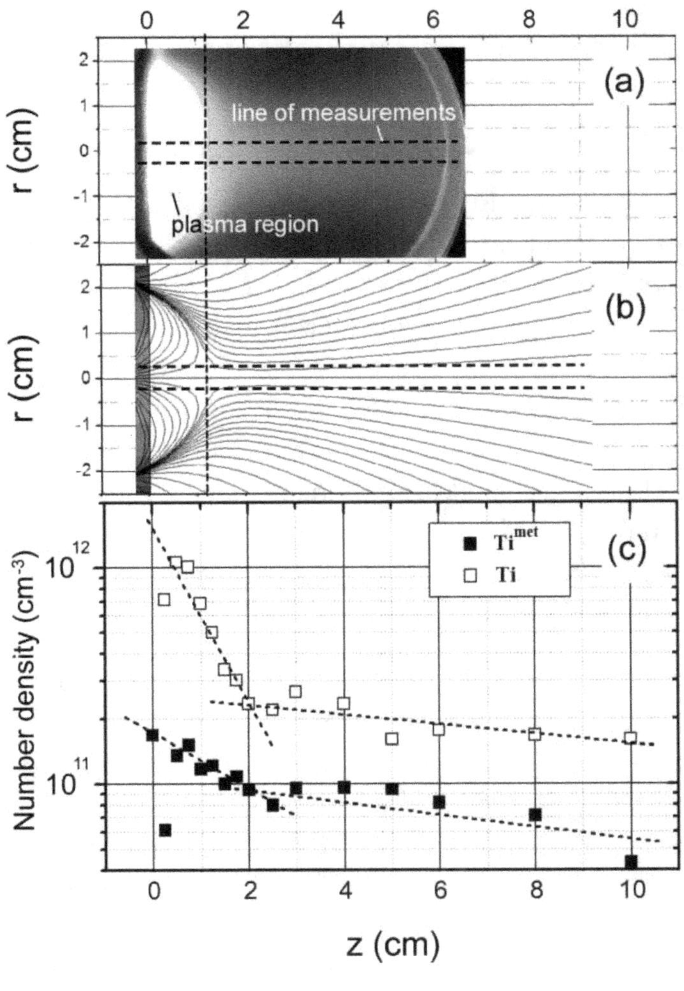

Figure 8. A photograph of the actual DCMS discharge operating at 4 mTorr of Ar pressure (a) along with the simulated magnetic field lines (b), and the distribution of the absolute density of Ti ground state atoms and Ti metastables measured by ROAS as a function of z at 30 mTorr of Ar pressure (c). Reproduced with permission from Britun et al. [34].

In addition, the ground state density of the sputtered Ti as well as Ti metastables, measured by ROAS, follows the trend found for the emission lines by Britun et al. [34], as shown in Figure 8(c). As one can observe, the number density of Ti neutrals reaches nearly 10^{12} cm^{-3} near the target at the examined conditions (see Table 1), dropping about 5 times already at z = 2 cm, which is mainly due to the diffusion and the angular particularities of sputtering [50]. The Ti metastable atoms (the energy level is about 0.8 eV, see Table 2) reveal the same tendency having the total density 2–5 times smaller. The presented absolute values of Ti density are in the good agreement with related measurements performed in DCMS under the similar conditions [31]. We have to note that the found particularities of spatial distribution for the sputtered particles in DCMS should resemble those for HiPIMS discharges, as far as the magnetic field topology and the average power are similar. In HiPIMS discharge, however, the additional physical effects, such as the time-dependent broadening of the emission [22] and absorption spectral lines [44], cross-field ion transport [15,51], ionization-induced density depletion near the target [37], time-dependent current-induced changes of the magnetic field [52], etc., may not allow a direct comparison between the DCMS and the HiPIMS discharges in terms of the absolute density of particles. These differences should be even more essential with increasing of the plasma density, i.e., the magnetron current density, as briefly mentioned below.

3.2. Absolute density in reactive DCMS discharge

Another interesting possibility that ROAS provides is visualization and control of the physical processes during the so-called reactive sputtering. This process is mainly known for synthesis of oxide, nitride, and oxynitride thin films [21,53]. Reactive sputtering normally involves standard magnetron sputtering sources, where a certain percentage of, e.g., molecular oxygen or nitrogen, is added to the bulk gas (typically Ar) in order to finally synthesize oxide or nitride compounds. Physically, during the reactive sputtering, oxidation (nitriding, etc.) of the cathode superficial layer takes place, which is often referred to as "poisoned" regime. This process is characterized by a significant drop of the density of metallic species in the discharge volume [54]. Apart from this, a hysteresis effect (in terms of discharge voltage or metal atom density) depending on the direction of change of the reactive gas content is often considered to be among the main characteristics of reactive sputtering [15,55–57].

The example of measurements of the absolute number density of Ti atoms sputtered in a DCMS discharge at z = 8 cm away from the target using resonant absorption is given in Figure 9. Molecular nitrogen has been added to the bulk gas in this case in order to synthesize nitride films, as described in details in [54]. As we can see, as a result of nitrogen addition, a clear drop in the density of the metal (Ti) atoms can be detected. Due to the target poisoning, the density of sputtered Ti drops by a factor of about 4 when the nitrogen content in the discharge exceeds ≈5%. The relative density of Ti ions in this case is measured using mass-spectrometry, demonstrating the same effect. Based on this example, we can conclude that the (real-time) measurements of the absolute density using absorption spectroscopy, contrary to the well-known plasma emission monitoring method [58], may have additional benefits for reactive sputtering since the emission is always defined by electron excitation and may be weak at the high distances from the target. In addition, the spatial selectivity of ROAS may also be better than that of OES, since the collimated beam used in this case can be highly focalized and thin, especially if a laser is used as a reference source [27].

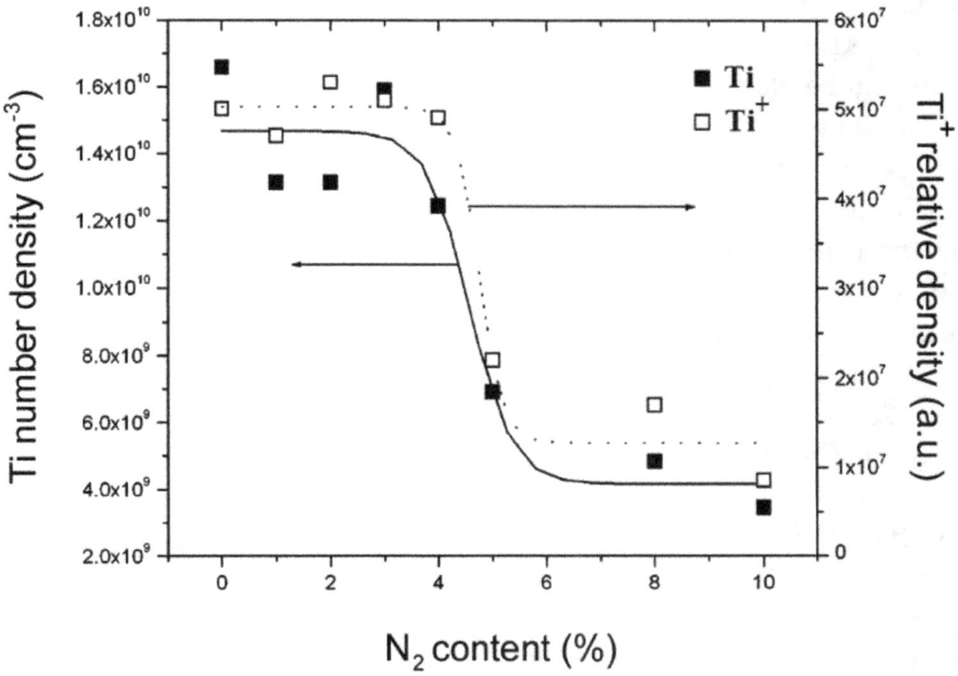

Figure 9. Evolution of the absolute Ti ground state density measured by ROAS (filled squares) and Ti^+ relative density obtained by mass spectrometry (open squares) in a reactive DCMS discharge as a function of the nitrogen content. Ar pressure = 30 mTorr, applied power = 500 W, z = 8 cm. Reproduced with permission from Konstantinidis et al. [54].

4. Particle density dynamics in HiPIMS discharges

HiPIMS discharges, possessing very short plasma on time, as illustrated in Figure 2, demonstrate different physics from that known for the DCMS discharges. The main reason for this is significantly elevated plasma density as well as the ionization degree related to both the bulk gas and the sputtered particles in HiPIMS, as mentioned in the Introduction and analyzed in the numerous works. Despite the fact that DCMS and HiPIMS sputtering processes can be realized using the same magnetron source, the direct comparison between these discharges is rather difficult, especially at the high values of the peak power applied to HiPIMS (or cathode current density), even though the time-averaged level of the applied power might be the same. In this section, for the sake of illustration of the dynamics of HiPIMS discharge in terms of the absolute density of species, the time-resolved evolution of the main discharge species in an Ar-Ti short-pulsed HiPIMS discharge analyzed by ROAS is presented. The experimental details on the HiPIMS system used for diagnostics are listed in Table 1, whereas the corresponding spectroscopic data are available in Table 2, or can be found elsewhere [31,38].

4.1. Time-resolved particle density evolution in HiPIMS

The time-resolved evolution of the absolute densities of Ti and Ti^+ atomic species sputtered in a 20-μs pulse Ar-Ti HiPIMS discharge measured at $z = 5$ cm above the target is shown in Figure 10. The total density in this case represents the sum of populations of the corresponding ground state sublevels (see Table 2). As we can observe from this figure, after a small density decrease (depletion) during the plasma on time, the Ti density is peaking at the time delay $\Delta t \approx 100$ μs and reaches the value of about 6×10^{11} cm^{-3}. The number density is found to be proportional to the pulse duration in this case, as far as the other discharge parameters are the same, which should be due to the change in the pulse energy (0.13 vs. 0.26 J). The pulse duration effect is also valid for the Ti ions in HiPIMS (not shown). The decrease in Ti density at the end of the plasma on time (roughly twice) is likely related to the VDF broadening of the sputtered particles during this time, neglected during the density calculation, as explained in [38]. In this case, an increase in the VDF width during the plasma on time observed by Palmucci et al. [44] results only in ~20% of the total density drop, which is less than the depletion visible in Figure 10(a) for Ti. This implies a presence of the additional factors responsible for density depletion, such as ionization of the sputtered particles, as discussed by Britun et al. [37]. Quantitative estimation of the different contributions is barely possible in this case, primarily due to the strong plasma emission in this time interval, which results in high deviation of the obtained data. The error bar shown in Figure 7(a) represents a standard deviation of the data points corresponding to several independent ROAS measurements, whereas the density data correspond to the values averaged among several measurements, as well as among three Ti ground state energy sublevels listed in Table 1.

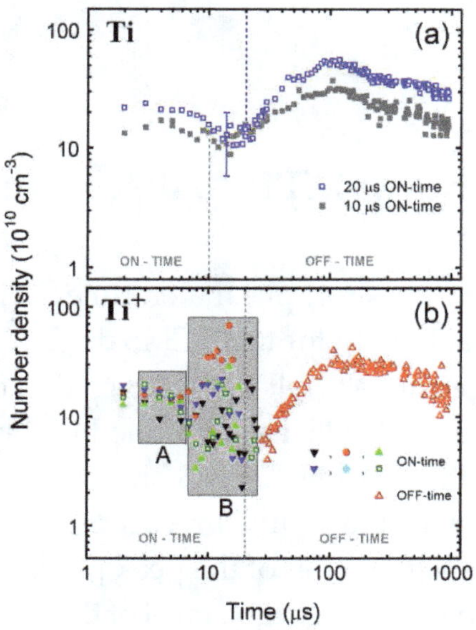

Figure 10. Time-resolved evolution of the Ti (a) and Ti^+ (b) number density in Ar-Ti HiPIMS discharge measured by ROAS. The A and B time intervals correspond to the density depletion and the current self-organization, respectively. Six independent measurements have been undertaken in the case of Ti^+ continuously showing the density instabilities in the interval B. Ar pressure = 20 mTorr, $z = 5$ cm. Reproduced with permission from Britun et al. [38].

The evolution of the Ti ion density shown in Figure 10(b) reveals similar tendencies, especially during the plasma off time, where the density has a maximum at 100–200 μs, reaching ~3 × 10^{11} cm^{-3}. During the plasma on time, two phenomena can be observed: the first one is the beginning of density depletion interval, which is similar to that found for Ti neutrals (called interval A), and the second one which characterized by the strong instabilities of the measured Ti^+ density at the end of the plasma on time, resembling the stochastic density behavior (called interval B). Note that six independent density measurements made during the plasma on time always lead to the different Ti^+ density evolution. Since Ti ionization is likely defined by the electrons presented in the discharge volume during this time ($Ti + e \rightarrow Ti^+ + 2e$), the latter phenomenon is probably related to the discharge current self-organization at the end of the plasma on time, resulting in a nonperiodic rotation of the excitation zones ("spokes") along the target race-track area, as reported by Kozyrev et al. [59] and studied by the other groups [60–62]. Apparently, the uncertainties in the measured Ti^+ density during the interval B persist even after the averaging of the signal over 10^3 pulses by the ICCD detector. This implies an existence of the additional factors responsible for the strong uncertainty of Ti^+ density, such as the high plasma-to-source emission peak ratio at the end of plasma on time. Indeed, a strong domination of the plasma emission intensity over the emission from the reference source dramatically decreases the stability of ROAS measurements. In our case, the typical emission intensity from plasma may surpass the one coming from the reference source by a factor of about 100 for Ti ions, and of about 30 for Ti neutrals, as illustrated in Figure 11. As follows from the recent analysis performed by Britun et al. [48], the relative error of the ROAS measurements in this case may easily exceed 100%, even if the relative error of the emission intensity measurements is only about 5% (typical for ICCD detectors), which may explain the instabilities observed during the HiPIMS on time for ions. Because of the mentioned instabilities, the plasma parameters such as Ti ionization degree, which can be defined for Ti as

$$\eta = \frac{[Ti^+]}{[Ti^+] + [Ti]} \tag{5}$$

(where $[X]$ stand for the absolute number density of specie X), are rather difficult to determine during the interval B. In the plasma off time, however, η reaches the level of about 0.4, which is typical for HiPIMS off time [63], taking into account rather low pulse energy used in this work ($E_P = 0.26$ J).

The summary of the time evolution of several atomic species studied in a 20-μs Ar-Ti HiPIMS discharge at $z = 5$ cm is given in Figure 12. From this figure, the main stages of HiPIMS discharge can be distinguished. First, after the discharge ignition, the excitation of Ar and Ti atoms remaining from the previous plasma pulse is taking place, as a result of propagation of the secondary electrons away from the cathode. The background Ti density is found to be ~2 × 10^{11} cm^{-3}, whereas the density of Ti metastables is much lower (~10^{10} cm^{-3}) for the considered conditions. The propagation of the hot electrons [64,65] results in formation of the Ar and Ti metastable states (Ar^{met}, Ti^{met}) as well as in the intensive Ti ionization during the plasma on

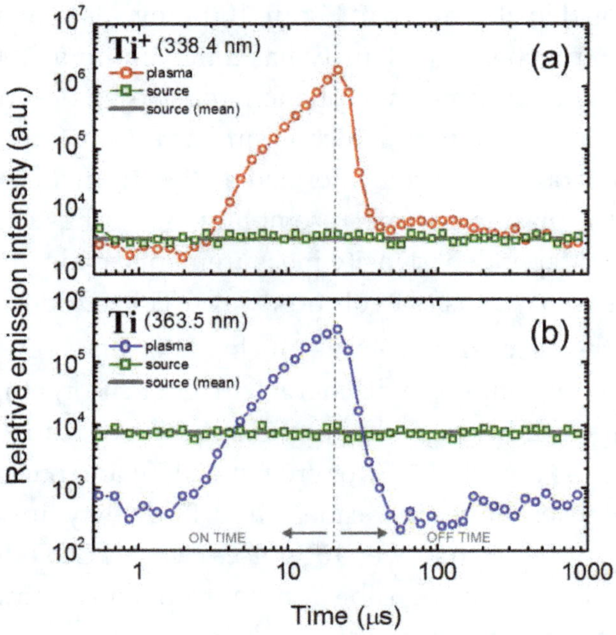

Figure 11. Typical example of the time evolution of Ti^+ (a) and Ti (b) emission line intensity in Ar-Ti HiPIMS discharge comparing to the intensity of the same emission lines in the reference source (HCL) showing a dramatic increase in the emission intensity at the end of the HiPIMS pulse. Ar pressure = 20 mTorr, *z* ~1 cm.

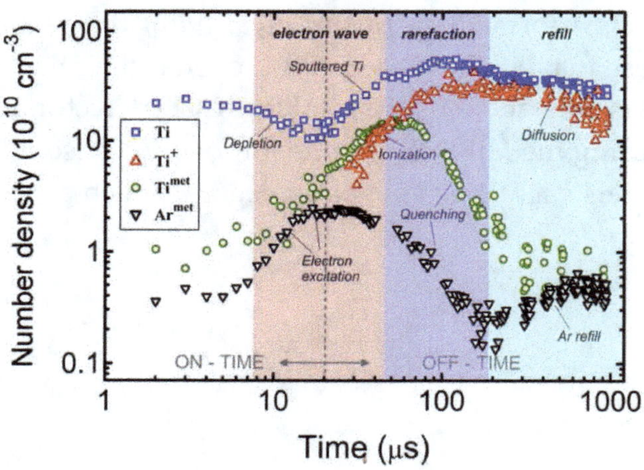

Figure 12. Summary of time evolution of the absolute density corresponding to the main discharge species in Ar-Ti HiPIMS discharge measured by ROAS. Three main time intervals corresponding to the secondary electron wave propagation, rarefaction, and refill are indicated. Ar pressure = 20 mTorr, *z* = 5 cm. Reproduced with permission from Britun et al. [21].

time–beginning of the plasma off time, which is accompanied by the inversion of population of the Ti energy sublevels, as discussed below. At the beginning of the plasma off time, the density of metal ions increases (red triangles in Figure 12), corresponding to their transport toward the volume analyzed by ROAS. Second, the depletion of Ti density at the end of the plasma on time is also visible. This depletion can be induced by the VDF broadening of Ti neutrals at the end of the plasma pulse [44]. Apart from this, a strong ionization of the sputtered neutrals in this region should also contribute to the observed density depletion [37,38]. Third, the strong excitation and ionization above the target are followed by rarefication of the bulk gas, resulted by the sputtered Ti species incoming to the studied volume, which is clearly observable together with a decrease in the Ar^{met} density, and also supported by the recent 2D density mapping results [37]. This process should be accompanied by electron cooling, as a result of collisions with the incoming heavy particles [64]. The ionization of sputtered Ti is taking place during the same time interval (20–100 μs), presumably as a result of the direct electron impact $(Ti + e \rightarrow Ti^{+} + 2e)$. The role of Penning ionization of Ti $(Ti + Ar^{met} \rightarrow Ti^{+} + Ar + e)$ in this case is supposed to be negligible, due to the much higher measured increase in Ti^{+} density ($\Delta[Ti^{+}] \approx 2 \times 10^{11}$ cm^{-3}) comparing to a small decrease in the density of Ar^{met} ($\Delta[Ar^{met}] \approx 2 \times 10^{10}$ cm^{-3}) during this time. Finally, in addition to the gas rarefication above the target, the quenching of both Ti and Ar metastable states by the sputtered Ti should also contribute in the observable decrease of Ti^{met} and Ar^{met} density during the 50- to 200-μs time interval. Afterward (at $\Delta t > 200$ μs), the refill of Ar to this area is observed, which is consistent with the Ar refill time estimated in the related works [26,66]. The Ti and Ti^{+} ground state densities gradually decrease during this time, due mainly to the diffusion of these species. The described processes are summarized in Table 3 where the main kinetic reactions along with the possible reasons for the observed processes are given.

Time	Particle	Processes	Main reaction(s)	Evidence
0–5 μs	Electrons	Generation		
	Ar	Ionization + acceleration toward the cathode	$Ar + e \rightarrow Ar^{+} + 2e$	Related works [15]
	Ti	Have their "background" density level, remaining from previous plasma pulse.		
	Ti^{met}			
	Ti^{+}			
5–20 μs	Electrons	**Electron wave** propagation (secondary hot electrons—e^{sec}) (few km/s)		LIF imaging data [37], Langmuir probe measurements [77]
	Ar^{met}	Generation by e^{sec}	$Ar + e \rightarrow Ar^{met} + e$	LIF imaging data [37]
	Ti	VDF broadening and strong ionization, resulting in density depletion		LIF imaging data [37], measured VDF width [44]
	Ti^{met}	Generation by e^{sec}	$Ti + e \rightarrow Ti^{met} + e$	Presence of Ar^{met} during this interval [37]
	Ti^{+}	Generation by e^{sec} (?)	$Ti + e \rightarrow Ti^{+} + 2e$	Not visible, probably due to the discharge current instabilities [59,60]
20–50 μs	Electrons	**Electron wave** (gradual cooling)		LIF imaging data [37], inversion of sublevel populations [67] (Figures 13 and 14).

<table>
<tr><th>Time</th><th>Particle</th><th>Processes</th><th>Main reaction(s)</th><th>Evidence</th></tr>
<tr><td rowspan="4"></td><td>Ar^{met}</td><td>1. Saturation, rarefication
2. Quenching by incoming Ti neutrals</td><td>2. $Ar^{met} + Ti \rightarrow Ar + Ti$</td><td>1. Decrease in [Ar^{met}], Related studies [15,26]
2. LIF imaging data [37]</td></tr>
<tr><td>Ti</td><td>1. Propagation of sputtered Ti
2. Ti VDF relaxation</td><td></td><td>1. LIF imaging data [37]2. Related studies [44,45]</td></tr>
<tr><td>Ti^{met}</td><td>Follow the growth of [Ti]</td><td>$Ti + e \rightarrow Ti^{met} + e$</td><td>Same growth rate for Ti and Ti^{met} (Figure 12)</td></tr>
<tr><td>Ti^{+}</td><td>Generation (following Ti wave)</td><td>1. $Ti + e \rightarrow Ti^{+} + 2e$
2. $Ar^{met} + Ti \rightarrow Ti^{+} + Ar + e$ (Penning ionization)</td><td>1. Faster growth of [Ti^{+}] and slower growth of [Ti] at the same time
2. Negligible, due to very small drop in [Ar^{met}] (Figure 12)</td></tr>
<tr><td rowspan="5">50–200 μs</td><td>Electrons</td><td>Thermalization</td><td>$X + e^{hot} \rightarrow X + e^{cold}$,
where X is mainly Ti</td><td>Loss of the inversion of sublevel populations (Figure 14), Langmuir probe data [64,65], modeling [78]</td></tr>
<tr><td>Ar^{met}</td><td>Rarefication</td><td></td><td>Figure 12</td></tr>
<tr><td>Ti</td><td>Diffusion</td><td></td><td>Gradual decrease of [Ti] (Figure 12)</td></tr>
<tr><td>Ti^{met}</td><td>Quenching, diffusion</td><td>$Ti^{met} + X \rightarrow Ti + X$,
where X is a heavy particle</td><td>Permanent density drop, as well as [Ti^{met}]/[Ti] ratio drop (Figure 12)</td></tr>
<tr><td>Ti^{+}</td><td>Density saturation (production-loss balance)</td><td></td><td>Figure 12</td></tr>
<tr><td rowspan="5">200–1000 μs</td><td>Electrons</td><td>Thermalized</td><td></td><td>Related works [64,65,78]</td></tr>
<tr><td>Ar^{met}</td><td>Refill</td><td></td><td>Increase in [Ar^{met}], refill time estimations [26,66]</td></tr>
<tr><td>Ti</td><td>Continuing diffusion</td><td></td><td>Gradual decrease of [Ti] (Figure 12)</td></tr>
<tr><td>Ti^{met}</td><td>Reaching background [Ti^{met}] level ~10^{10} cm^{-3}.</td><td></td><td>Figure 12</td></tr>
<tr><td>Ti^{+}</td><td>Slight drop after 600 μs, might be due to:</td><td>1. $X^{+} + Ar + e \rightarrow X + Ar^{met}$
2. $X^{+} + e \rightarrow X$
where $X = Ar,\ Ti$</td><td>1. [Ti^{+}] drop + [Ar^{met}] increase
2. [Ti^{+}] drop (Figure 12)</td></tr>
</table>

Table 3. Summary of the processes related to the absolute density evolution of the discharge species measured by ROAS in a 20-μs Ar-Ti HiPIMS discharge (see Figure 12). Partially reproduced with permission from Britun et al. [38].

4.2. Inversion of the energy sublevel populations

An interesting effect can be observed if the measured density is represented separately for each energy sublevel at the studied electronic state, as shown in Figure 13. As one can observe, a clear inversion among the studied energy states is achieved for five Ti sublevels corresponding to the metastable Ti a^5F_J states (see Table 2). An inversion of the considered energy sublevels (i.e., when the population of the state corresponding to higher energy is found to be higher) takes place in Figure 13(a) during the 20- to 100-μs time interval. The maximum of the mentioned inversion for Ti^{met} case is located around $\Delta t \approx 60$ μs. Similar but much weaker effect for Ar metastables (only two levels are considered) is also visible in Figure 13(b).

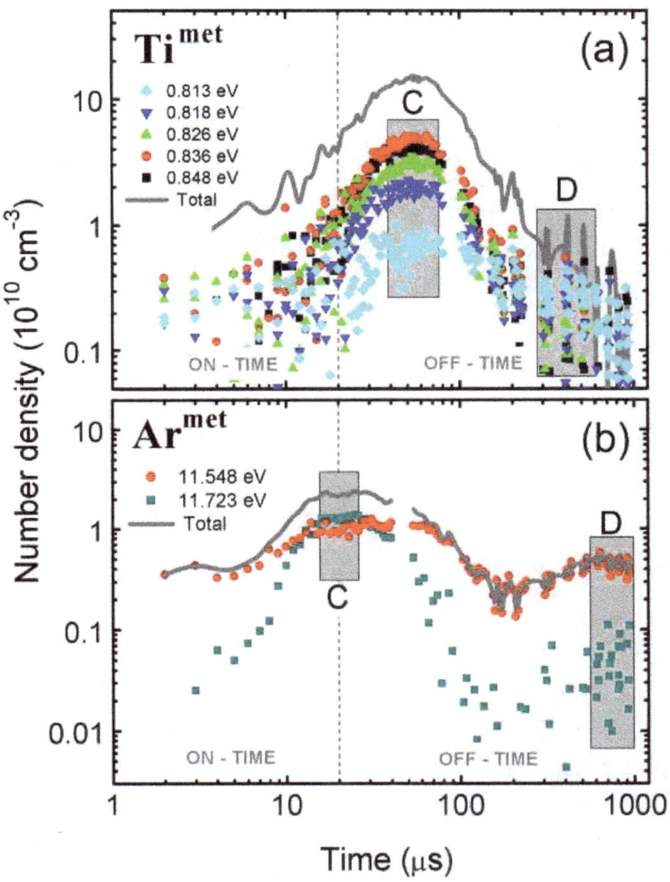

Figure 13. Time-resolved number density evolution for Ti (a) and Ar (b) metastable atoms measured by ROAS in Ar-Ti HiPIMS showing the populations of the corresponding energy sublevels (see Table 2). The total density evolution is given by gray lines for comparison. Ar pressure = 20 mTorr, z = 5 cm. Reproduced with permission from Britun et al. [38].

Despite the fact that the inversion effect is rather new for HiPIMS, it is well known for the other low-temperature discharges, such as Ti hollow cathode plasma [67], and RF-amplified DCMS discharge [30]. In the former case, it is explained by the (partial) equilibrium of the heavy discharge species with the energetic electrons during the short plasma on time interval. Besides this, the inversion of the sublevel populations might be also responsible for the apparent density depletion in the magnetron target vicinity detected in the DCMS discharges

[30,49,68,69]. The sublevel inversion can be explained based on the Boltzmann distribution of the corresponding energy sublevel populations. Assuming the presence of the dense flux of hot electrons during certain time in the discharge, the heavy particles should be intensively excited by the electrons during this time. Under these conditions, for the close energy sublevels, the factor $\Delta E / kT$ in the Boltzmann distribution may be negligible due to rather high effective (excitation) temperature of the involved particles. This should result finally in the proportionality between the level population N_i and its statistical weight g_i, as shown by the following expressions:

$$\frac{N_i}{g_i} = \frac{N_j}{g_j} \cdot \mathrm{Exp}\left(-\frac{E_i - E_j}{kT_{\mathrm{exc}}} \right) \equiv \frac{N_j}{g_j} \cdot \mathrm{Exp}\left(-\frac{\Delta E}{kT_{\mathrm{exc}}} \right) \tag{6}$$

where i (upper) and j (lower) stand for two close levels, $N_{i,j}$ is the level populations, $g_{i,j}$ - is their statistical weights, ΔE is the energy difference between two levels, and T_{exc} is the excitation temperature. Under $\Delta E \ll kT_{\mathrm{exc}}$, we obtain the following:

$$\frac{N_i}{g_i} \approx \frac{N_j}{g_j}\left(1 - \frac{\Delta E}{kT_{\mathrm{exc}}} \right) \approx \frac{N_j}{g_j} \tag{7}$$

which, after normalization to $\frac{N_j}{g_j}$, results in $N_i \sim g_i$. Note that this effect has the same nature as the "saturated LIF" mode described by Amorim et al. [70].

The sublevel inversion is even more clear if a "stabilized" modification of the resonant absorption method is implemented, when a pulsed reference source is synchronized with the time gate of the ICCD detector, as discussed above. The relative error of the ROAS measurements can be significantly reduced in this case, and fine spectral effects in the discharge, such as inversion of the energy sublevels, can be easily visualized. The example of these measurements using three Ti ground state energy sublevels is given in Figure 14. In Figure 14(a), the experimental ROAS results, clearly showing the presence of the Ti ground state sublevel inversion (roughly between 20 and 300 μs), are shown. At the same time, a sketch of the same level populations drawn in accordance to the inversion model described by Eqs. (6) and (7) is given in Figure 14(b) for the sake of clarification of the experimental data. We have to note that before the inversion interval, the studied Ti sublevels do not perfectly follow the Boltzmann distribution, as the middle level is mostly populated in this case. This fact might be due to the complicated HiPIMS discharge kinetics, which is not well studied yet. Let us also note that the inversion interval might be significantly shifted in time depending on position of the ROAS line of sight, and the inversion should happen much earlier if the measurements are performed closer to the target surface.

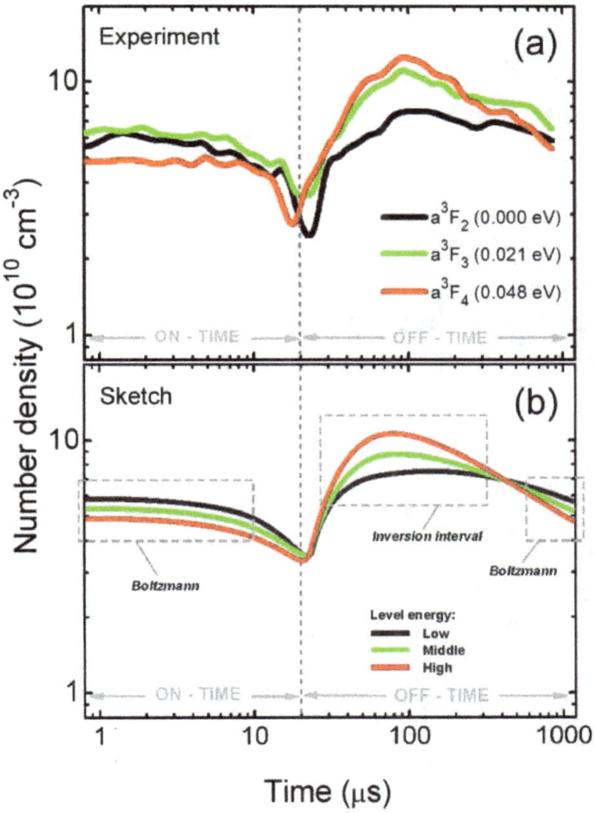

Figure 14. The time-resolved populations of three ground state Ti energy sublevels (a) measured in Ar-Ti HiPIMS discharge by ROAS using the dynamic source triggering. Ar pressure = 20 mTorr, *z* = 5 cm. A sketch (b) illustrating the dynamics of the Ti sublevels population as a function of time in HiPIMS. The beginning of the inversion interval should depend on the observation point (*z*) and should start earlier if the ROAS line-of-sight is located closer to the target.

Fitting the experimental data corresponding to Ti sublevels, the Ti *excitation* temperature can be obtained, which is a measure of excitation of the corresponding energy sublevels by the hot electrons. In our case, the estimations lead to T_{exc} ~0.25 eV for Ti, and to the close value for Ti^{met}. In general, a precise ROAS measurement of the energy sublevel populations for a certain atomic state should allow a reliable determination of the excitation temperature T_{exc} for the heavy discharge particles after their interaction with hot electrons, as well as the determination of its temporal evolution in the discharge. Representing a measure of excitation of the heavy particles by electrons in the discharge volume, T_{exc} should be proportional to the HiPIMS pulse energy, secondary electron emission coefficient, and should be generally higher in the target vicinity [22].

4.3. Density dynamics in reactive HiPIMS

It is also interesting to consider some selected phenomena in the case of reactive HiPIMS discharge. The behavior of the number density of the sputtered species in reactive HiPIMS should follow the same tendency as in the reactive DCMS case (shown in Figure 9), i.e., a density drop by several times in the poisoned regime of sputtering should be expected. Let us consider the behavior of the absolute density of *reactive* species here. The presented example is related to the time-behavior of the absolute density of the atomic oxygen measured by ROAS. This question has been particularly studied by Vitelaru et al. [71] using a tunable diode laser and giving the dynamics of the atomic oxygen in the HiPIMS discharge, however, without measuring its absolute density, which is a very important parameter, as mentioned in Section 1. In the considered case, a discharge-based ROAS with simultaneous signal acquisition (see Figure 6) has been applied for measurements of the absolute density of O atoms in the HiPIMS discharge volume. Due to the spectral limitations, however, the ground state O atoms are not accessible by ROAS (the corresponding transitions are located in the deep UV range [72]), so only the O *metastable* atoms (O^{met}, 3s $^5S^\circ_2$ level with the energy = 9.15 eV) are analyzed in this case. A microwave surfaguide discharge [73] operating at 2.5 GHz and using Ar–7% O_2 gas mixture has been utilized as a reference source in this case.

Analyzing the O^{met} atoms in any discharge, one should bear in mind that these are excited species which keep the dynamics of the plasma electrons as well as the dynamics of the ground state O_2 and O. The time evolution of O^{met} atoms measured by ROAS at $z = 5$ cm in the HiPIMS discharges working with Ar–20% O_2 gas mixture and using Ti and Ag targets is shown in Figure 15. The typical stages of nonreactive sputtering discharge can be observed in this case (compare to Figure 12). The first noticeable processes are the O^{met} density depletion at the beginning of the plasma on time (visible in the Ti target case only) followed by the density increase at the end of the plasma on time. These processes are followed by the rarefication of O^{met} and the gas refill afterward (exists both for Ti and Ag target cases). We would like to emphasize that both in Ti and Ag cases the O^{met} density starts to increase already at the end of the plasma on time, unlike in the nonreactive case. Such an increase correlates well with the time-resolved dynamics of the Ar metastables in the nonreactive case, as shown in Figure 12. This similarity, as well as the proximity of the excitation thresholds for these species (11.55 eV for Ar^{met} vs. 9.15 eV for O^{met} [74]) point out on the role of the secondary electrons in formation of the O^{met} state, which should take place already during the plasma on time (i.e., $O_2 + e \rightarrow O + O + e$, followed by $O + e \rightarrow O^{met} + e$). The correlation with Ar^{met} is additionally confirmed by the gas rarefication and refill processes observed both for Ar and atomic O during the plasma off time. The different density values in the density maxima of O^{met} ($\approx 5 \times 10^8$ cm^{-3} for Ti vs. $\approx 3 \times 10^9$ cm^{-3} for Ag) may be related to the differences in the ion-induced secondary electron emission (ISEE) coefficients in these cases, assuming the electron impact to be a primary mechanism for O^{met} formation. Interestingly, the absolute number density below 10^8 cm^{-3} has been detected in these measurements, approving the high reliability of ROAS with simultaneous signal detection. (see Figure 6).

Since in the case of Ar–20% O_2 gas mixture magnetron target is supposed to be completely oxidized [47], the sputtering of oxide material from (i.e., M_xO_y molecules, where M is a metal atom) followed by formation of O atoms and finally O^{met} should be another possible mechanism of the observed O^{met} density elevation at the end of the plasma on time. This mechanism, however, is supposed to be rather minor since the wave of the sputtered species arrives in the volume of measurements much later, having the maximum at $\Delta t \approx 100$ μs, as shown in Figure 12.

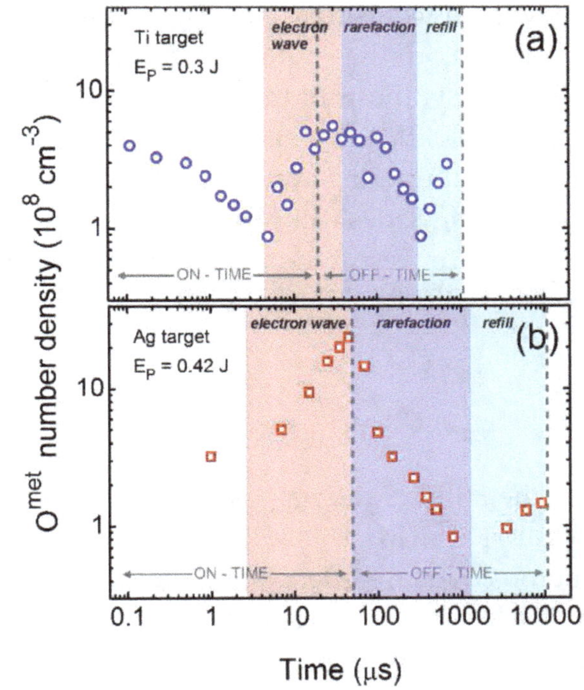

Figure 15. Time-resolved evolution of the absolute density of O metastable atoms (3s $5S^0_2$ state) measured by ROAS in HiPIMS discharge using Ar–20% O_2 gas mixture for the case of Ti (a) and Ag (b) cathode. Ar pressure = 20 mTorr, z = 5 cm. Pulse duration, discharge frequency, and pulse energy (E_P) are, respectively, 20 ms, 1 kHz, 0.3 J (a) and 50 ms, 0.1 kHz, 0.42 J (b). Reproduced with permission from Britun et al. [22].

5. Summary and conclusions

5.1. DCMS discharges

Resonant absorption analysis of the direct current sputtering discharges shows that the absolute density of the sputtered particles is ranged in a relatively narrow interval, being always in the range of 10^{10}–10^{12} cm^{-3} in the Ti case (assuming the average power level to be several hundred Watt). These values may be somewhat different varying the cathode materials with essentially different sputtering yield and/or in the sputtering systems, which use significantly higher power levels. The spatial distribution of the sputtered particles is mainly determined by their diffusion (at high pressure), by the angular directivity of sputtering (low pressure), as well as by the magnetic field topology (critical mainly for the sputtered ions), which lead to nearly exponential decay of the observed density of species as a function of the distance from cathode, so the density might decay at least by one order of magnitude already at few cm away from the target surface (for 5 cm diameter round target).

One of the advantages of using ROAS technique for diagnostics of sputtering processes is its ability to probe the species in a wide range of absolute density, starting from roughly 10^7–10^8 cm^{-3}, where the signal-to-noise ratio of the used detector is a limiting factor, and finishing by the values several orders of magnitude higher, when the increase of the optical thickness of the volume of interest becomes the limiting factor. The bottom sensitivity level may be additionally affected by the type of the absorption scheme used (e.g., single-pass, multipass, with or without simultaneous signal acquisition, etc.). The variety of the ROAS implementation schemes allow real time monitoring of the sputtered species also during *reactive* sputtering, which is very demandable, as the synthesis of the compound (oxides, nitrides, etc.) coatings is one of the main sputtering workflow nowadays. Since the density drop between the "metallic" and "reactive" mode of sputtering normally drop by about one order of magnitude, the absolute density levels in both modes can be easily real time monitored by absorption technique, which might be useful for research-oriented as well as for industry applications. This conclusion is valid both for DCMS (as shown in this work), pulsed-DC, as well as for HiPIMS discharges.

5.2. HiPIMS discharges

The physical phenomena in the HiPIMS discharges are rather different from those observed in DCMS and P-DCMS cases, first of all in terms of the plasma density and much shorter timing (μs scale). As a result, numerous dynamic effects should be taken into account explaining the properties of the HiPIMS discharges. Among these effects, the following time-dependent phenomena should be mentioned: the propagation of the fast electron wave the dynamic gas rarefaction and refill, the ionization and excitation of various discharge particles, quenching of the excited states as a result of collisions, the anomalous electron transport, etc. ROAS technique, being essentially limited by only the time-resolution of the used optical detector, suits very well for systematic characterization of these discharges in terms of the absolute density. In addition, the modern ICCD detectors typically provide ns time resolution, which surpasses the typical HiPIMS time scale by few orders of magnitude.

In terms of the time-resolved evolution of the discharge particles density, there are several processes in HiPIMS discharge, which can be visualized using ROAS technique. The time-resolved summary of these processes is given in Table 3, considering the main discharge species studied in the Ar-Ti nonreactive HiPIMS case. The presence of such dynamic effects as the density depletion, gas rarefaction, gas refill, quenching of the metastable states, ionization, as well as the propagation of the electron wave away from the target surface can be concluded based on the undertaken ROAS analysis. Despite a rather high HiPIMS repetition rate considered in this work (1 kHz), the gradual relaxation for the majority of the listed processes till the end of the plasma off time can be observed. Even though the total range of the measured absolute density of the discharge species reaches nearly three orders of magnitude in the considered case, the detected densities are still well above the detection threshold (which is estimated to be roughly 10^7–10^8 cm^{-3} with the detection scheme used). The sensitivity of the used ROAS method can be additionally enhanced using the lock-in detection technique, dynamic triggering of reference source, and related approaches.

One of these approaches, namely, the simultaneous acquisition of the emission signals using a triple optical fiber, has been utilized for time-resolved detection of the oxygen metastable atoms during the reactive HiPIMS process. The results reveal the presence of the same discharge phases (such as gas rarefaction and refill, as well as the density depletion in some cases) as detected in the nonreactive HiPIMS case. The absolute density of the O metastable atoms detected in this case is still above the mentioned ROAS sensitivity threshold, being in the range of 10^8–10^9 cm^{-3}. The optical diagnostics of the reactive HiPIMS discharges, however, is a relatively new area and the obtained results require additional verification by the other methods, such as LIF, in order to be understood completely.

5.3. General remarks

The resonant absorption technique represents a powerful diagnostic tool suitable for the determination of the absolute density of the relevant species in sputtering discharges, such as DCMS, P-DCMS, HiPIMS, as well as the other types of magnetron discharges. This technique has several modifications allowing its optimization depending on the discharge geometry etc. For example, using a plasma discharge as a reference source, ROAS setup can be rather simple, requiring only the additional optics for reference source beam collimation, and (sometimes) additional synchronization between the source and the discharge of interest. Both direct current and pulsed reference sources can be used for resonant absorption purposes. In addition to this, in case a broad range spectral detector, such as ICCD, is utilized, a parallel detection of several spectral transitions as well as the corresponding absolute densities becomes possible. These advantages open new possibilities to study the basic effects in the discharge volume. Among these effects are the population of the metastable levels of the various discharge species, the population of the atomic energy sublevels of the atoms and their time-resolved dynamics, the study of the excitation temperature across the discharge volume, etc. These possibilities are important for studying magnetron sputtering processes at the fundamental level, whereas rather basic configuration of the ROAS setup may be helpful in industry, e.g., for real-time density monitoring of the relevant species during a certain process.

The sensitivity of the absorption method can also be significantly improved by varying the detection schemes, such as lock-in amplification (improving the sensitivity by 1–2 orders of magnitude typically), multipass methods, as well as by simultaneous signal acquisition, dynamic source triggering, etc. Using the simultaneous signal acquisition for example, the sensitivity below 10^8 cm^{-3} has been achieved for O metastable atoms in a reactive HiPIMS discharge case, as illustrated in this chapter. This value is far below the typical concentrations of the sputtered atoms in the discharge (10^{10}–10^{12} cm^{-3}) achieved under the laboratory conditions, which is enough for most applications. For very weakly populated energy states, like those appearing in molecular discharges, the multipass absorption, including the ring-down spectroscopy method, could be suggested, which are, being out of the scope of this chapter, still based on the same fundamental principles.

Acknowledgements

This work is supported by the Belgian Government through the "Pôle d'Attraction Interuniversitaire" (PAI, P7/34, "Plasma-Surface Interaction," Ψ). N. Britun is a postdoctoral researcher; S. Konstantinidis is a research associate of the Fonds National de la Recherche Scientifique (FNRS), Belgium.

Author details

Nikolay Britun[1*], Stephanos Konstantinidis[1] and Rony Snyders[1,2]

*Address all correspondence to: nikolay.britun@umons.ac.be

1 Chimie des Interactions Plasma–Surface (ChIPS), CIRMAP, Université de Mons, Mons, Belgium

2 'Materia Nova' Research Center, Parc Initialis, Mons, Belgium

References

[1] S. M. Rossnagel, J. Vac. Sci. Technol. A. 21, S74 (2003).

[2] G. Bräuer, B. Szyszka, M. Vergöhl, and R. Bandorf, Vacuum 84, 1354 (2010).

[3] W. R. Grove, Philos. Trans. R. Soc. London 142, 87 (1852).

[4] P. Sigmund, *Sputtering by Particle Bombardment I. Physical Sputtering of Single-Element Solids* (Springer, Berlin, 1981).

[5] W. D. Gill and E. Kay, Rev. Sci. Instrum. 36, 277 (1965).

[6] E. Kay, J. Appl. Phys. 34, 760 (1963).

[7] J. S. Chapin., Res. Dev. 25, 37 (1974).

[8] R. D. Arnell and P. J. Kelly, Surf. Coat. Technol. 112, 170 (1999).

[9] P. J. Kelly and R. D. Arnell, Vacuum 56, 159 (2000).

[10] S. Schiller, K. Goedicke, J. Reschke, V. Kirchhoff, S. Schneider, and F. Milde, Surf. Coat. Technol. 61, 331 (1993).

[11] J. Musil, J. Lestina, J. Vlcek, and T. Tolg, J. Vac. Sci. Technol. A. 19, 420 (2001).

[12] U. Helmersson, M. Lattemann, J. Bohlmark, A. P. Ehiasarian, and J. T. Gudmundsson, Thin Solid Films 513, 1 (2006).

[13] S. M. Rossnagel and J. Hopwood, Appl. Phys. Lett. 63, 3285 (1993).

[14] A. Tonegawa, H. Taguchi, and K. Takayama, Nucl. Instrum. Methods Phys. Res. B. 55, 331 (1991).

[15] J. T. Gudmundsson, N. Brenning, D. Lundin, and U. Helmersson, J. Vac. Sci. Technol. A. 30, 030801 (2012).

[16] K. Sarakinos, J. Alami, and S. Konstantinidis, Surf. Coat. Technol. 204, 1661 (2010).

[17] J. T. Gudmundsson, Vacuum 84, 1360 (2010).

[18] I. K. Fetisov, A. A. Filippov, G. V. Khodachenko, D. V. Mozgrin, and A. A. Pisarev, Vacuum 53, 133 (1999).

[19] V. Kouznetsov, K. Macák, J. M. Schneider, U. Helmersson, and I. Petrov, Surf. Coat. Technol. 122, 290 (1999).

[20] U. Helmersson, M. Lattemann, J. Bohlmark, A. P. Ehiasarian, and J. T. Gudmundsson, Thin Solid Films 513, 1 (2006).

[21] N. Britun, T. Minea, S. Konstantinidis, and R. Snyders, J. Phys. D. Appl. Phys. 47, 224001 (2014).

[22] N. Britun, S. Konstantinidis, and R. Snyders, Plasma Process. Polym. DOI: 10.1002/ppap.201500051 (2015).

[23] K. Sarakinos, J. Alami, and S. Konstantinidis, Surf. Coat. Technol. 204, 1661 (2010).

[24] A. C. Mitchell and M. W. Zemansky, *Resonance Radiation and Excited Atoms* (Plenum Press, Cambridge, 1971).

[25] G. F. Kirkbright and M. Sargent, *Atomic Absorption and Fluorescence Spectroscopy* (Academic Press, London, 1974).

[26] C. Vitelaru, D. Lundin, G. D. Stancu, N. Brenning, J. Bretagne, and T. Minea, Plasma Sources Sci. Technol. 21, 025010 (2012).

[27] A. Zybin, J. Koch, H. D. Wizemann, J. Franzke, and K. Niemax, Spectrochim. Acta Part B At. Spectrosc. 60, 1 (2005).

[28] D. S. Baer and R. K. Hanson, J. Quant. Spectrosc. Radiat. Transf. 47, 455 (1992).

[29] L. De Poucques, J.-C. Imbert, C. Boisse-Laporte, J. Bretagne, M. Ganciu, L. Teulé-Gay, and M. Touzeau, Plasma Sources Sci. Technol. 15, 661 (2006).

[30] S. Konstantinidis, A. Ricard, M. Ganciu, J. P. Dauchot, C. Ranea, and M. Hecq, J. Appl. Phys. 95, 2900 (2004).

[31] M. Gaillard, N. Britun, Y. M. Kim, and J. G. Han, J. Phys. D. Appl. Phys. 40, 809 (2007).

[32] TOPTICA Photonics Webpage: http://www.toptica.com/products/research_grade_diode_lasers.html

[33] C. Vitelaru, L. de Poucques, T. M. Minea, and G. Popa, J. Appl. Phys. 109, 053307 (2011).

[34] N. Britun, M. Gaillard, L. Schwaederlé, Y. M. Kim, and J. G. Han, Plasma Sources Sci. Technol. 15, 790 (2006).

[35] N. Britun, S. Ershov, A.-A. El Mel, S. Konstantinidis, A. Ricard, and R. Snyders, J. Phys. D. Appl. Phys. 46, 175202 (2013).

[36] C. Vitelaru, L. de Poucques, T. M. Minea, and G. Popa, Plasma Sources Sci. Technol. 20, 045020 (2011).

[37] N. Britun, M. Palmucci, S. Konstantinidis, and R. Snyders, J. Appl. Phys. 117, 163302 (2015).

[38] N. Britun, M. Palmucci, S. Konstantinidis, and R. Snyders, J. Appl. Phys. 117, 163303 (2015).

[39] L. Li, A. Nikiforov, N. Britun, R. Snyders, and C. Leys, Spectrochim. Acta Part B At. Spectrosc. 107, 75 (2015).

[40] A. Ricard, V. Monna, and M. Mozetic, Surf. Coat. Technol. 174–175, 905 (2003).

[41] W. Lochte-Holtgreven, editor, *Plasma Diagnostics* (North-Holland publishing company, Amsterdam, 1968).

[42] N. Britun, J. G. Han, and S.-G. Oh, Plasma Sources Sci. Technol. 17, 045013 (2008).

[43] N. Britun, M. Palmucci, S. Konstantinidis, M. Gaillard, and R. Snyders, J. Appl. Phys. 114, 013301 (2013).

[44] M. Palmucci, N. Britun, S. Konstantinidis, and R. Snyders, J. Appl. Phys. 114, 113302 (2013).

[45] N. Britun, M. Palmucci, and R. Snyders, Appl. Phys. Lett. 99, 131504 (2011).

[46] M. Ganciu, M. Hecq, S. Konstantinidis, J.-P. Dauchot, M. Touzeau, L. de Poucques, and J. Bretagne, 2007/0034498 A1 (2007).

[47] M. Palmucci, N. Britun, T. Silva, R. Snyders, and S. Konstantinidis, J. Phys. D. Appl. Phys. 46, 215201 (2013).

[48] N. Britun, M. Michiels, and R. Snyders, Rev. Sci. Instrum. 86, (2015) In press.

[49] N. Britun, M. Gaillard, and J. G. Han, J. Phys. D. Appl. Phys. 41, 185201 (2008).

[50] A. Goehlich, D. Gillmann, and H. F. Döbele, Nucl. Instrum. Methods Phys. Res. B. 164–165, 834 (2000).

[51] D. Lundin, P. Larsson, E. Wallin, M. Lattemann, N. Brenning, and U. Helmersson, Plasma Sources Sci. Technol. 17, 035021 (2008).

[52] J. Bohlmark, U. Helmersson, M. VanZeeland, I. Axnäs, J. Alami, and N. Brenning, Plasma Sources Sci. Technol. 13, 654 (2004).

[53] J. Musil, P. Baroch, J. Vlček, K. H. Nam, and J. G. Han, Thin Solid Films 475, 208 (2005).

[54] S. Konstantinidis, A. Ricard, R. Snyders, H. Vandeparre, J. P. Dauchot, and M. Hecq, Surf. Coat. Technol. 200, 841 (2005).

[55] D. Depla, J. Haemers, G. Buyle, and R. De Gryse, J. Vac. Sci. Technol. A. 24, 934 (2006).

[56] M. Audronis and V. Bellido-Gonzalez, Thin Solid Films 518, 1962 (2010).

[57] D. Depla and S. Mahieu, editors, *Reactive Sputter Deposition* (Springer, 2008).

[58] P. P. Woskov, K. Hadidi, P. Thomas, K. Green, and G. Flores, Waste Manag. 20, 395 (2000).

[59] A. V. Kozyrev, N. S. Sochugov, K. V. Oskomov, A. N. Zakharov, and a. N. Odivanova, Plasma Phys. Rep. 37, 621 (2011).

[60] A. Hecimovic, M. Böke, and J. Winter, J. Phys. D. Appl. Phys. 47, 102003 (2014).

[61] A. P. Ehiasarian, A. Hecimovic, T. De Los Arcos, R. New, V. Schulz-Von Der Gathen, M. Boke, J. Winter, and M. Bke, Appl. Phys. Lett. 100, 114101 (2012).

[62] J. Andersson, P. Ni, and A. Anders, Appl. Phys. Lett. 103, 054104 (2013).

[63] A. Ehiasarian, A. Vetushka, A. Hecimovic, and S. Konstantinidis, J. Appl. Phys. 104, 083305 (2008).

[64] A. Pajdarová, J. Vlček, P. Kudláček, and J. Lukáš, Plasma Sources Sci. Technol. 18, 025008 (2009).

[65] P. Poolcharuansin and J. W. Bradley, Plasma Sources Sci. Technol. 19, 025010 (2010).

[66] D. Lundin, N. Brenning, D. Jädernäs, P. Larsson, E. Wallin, M. Lattemann, M. A. Raadu, and U. Helmersson, Plasma Sources Sci. Technol. 18, 045008 (2009).

[67] D. Ohebsian, N. Sadeghi, C. Trassy, and J. M. Mermet, Opt. Comm. 32, 81 (1980).

[68] N. Nafarizal, N. Takada, K. Shibagaki, K. Nakamura, Y. Sago, and K. Sasaki, Jpn. J. Appl. Phys. 44, L737 (2005).

[69] N. Nafarizal, N. Takada, and K. Sasaki, Jpn. J. Appl. Phys. 48, 126003 (2009).

[70] J. Amorim, G. Baravian, and J. Jolly, J. Phys. D. Appl. Phys. 33, R51 (2000).

[71] C. Vitelaru, D. Lundin, N. Brenning, and T. Minea, Appl. Phys. Lett. 103, 104105 (2013).

[72] R. Payling and P. Larkins, *Optical Emission Lines of the Elements (CD-ROM)* (Wiley & Sons Inc., New York, 2000).

[73] T. Godfroid, J. P. Dauchot, and M. Hecq, Surf. Coat. Technol. 174–175, 1276 (2003).

[74] NIST Atomic Spectra Database Lines Form: http://physics.nist.gov/PhysRefData/ASD/lines_form.html

[75] D. L. Crintea, U. Czarnetzki, S. Iordanova, I. Koleva, and D. Luggenhölscher, J. Phys. D. Appl. Phys. 42, 045208 (2009).

[76] W. L. Wiese and J. R. Fuhr, J. Phys. Chem. Ref. Data 4, 263 (1975).

[77] K. B. Gylfason, J. Alami, U. Helmersson, and J. T. Gudmundsson, J. Phys. D. Appl. Phys. 38, 3417 (2005).

[78] T. Kozák and A. Dagmar Pajdarová, J. Appl. Phys. 110, 103303 (2011).

3

A Technique for Time-Resolved Imaging of Millimeter Waves Based on Visible Continuum Radiation from a Cs-Xe DC Discharge — Fundamentals and Applications

Mikhail S. Gitlin

Abstract

The chapter presents a review of a highly sensitive technique for time-resolved imaging and measurement of the 2D intensity profiles of millimeter waves (MMW) based on the use of visible continuum radiation (VCR) from the positive column (PC) of a medium pressure Cs-Xe DC discharge (VCRD technique). The review focuses on the operating principles, fundamentals, and applications of this new technique. The design of a discharge tube and an experimental setup which were used to create a wide homogeneous PC plasma slab are described. The MMW effects on the plasma slab are studied. The mechanism of microwave-induced variations in the VCR intensity and the causes of violation of the local relation between the visible continuum emissivity and the MMW intensity are discussed. The main characteristics, e.g., spatial and temporal resolution, and sensitivity of the VCRD technique have been evaluated. Experiments on imaging of the field patterns of horn antennas and quasioptical beams demonstrated that the VCRD technique can be used for a good-quality imaging of the MMW beams in the entire MM-wavelength band. The VCRD technique was applied for imaging of output field patterns of the MMW electron tubes and determination of some of their characteristics, as well as for active real-time imaging and nondestructive testing using MM waves.

Keywords: Microwaves, Imaging, Electric discharges, Continuum

1. Introduction

Millimeter waves (MMWs) are electromagnetic (EM) waves with wavelengths ranging from 1 to 10 mm in free space. Millimeter waves are widely used in radar, navigation, telecommunications, remote sensing, plasma heating and diagnostics, material processing, spectroscopy, etc. Imaging and nondestructive testing (NDT) with millimeter waves is also of great interest for scientific, industrial, security, and biomedical applications [1–10]. In many MMW applications, as well as for engineering of MMW devices and systems, it is necessary to measure the spatial profiles of MMW radiation in real time. Electronic, electro-optic (EO), and thermographic techniques are commonly used for measuring and imaging the MMW field profiles. The electronic techniques employ receivers with semiconductor [4–9], superconducting [10], and glow discharge [11–13] MMW detectors. Electro-optic techniques use EO sensors for MM waves [14, 15]. Mechanically scanned receiver systems are widespread. However, a mechanically scanned receiver cannot be applied for a time-resolved imaging of MMW field profiles. Time-resolved images of MMW beams can be obtained with two-dimensional (2D) antenna arrays with MMW detectors [4–8, 12, 13]. The images of MMW beams obtained using 2D antenna arrays are of rather poor quality, because of the element-to-element gain and sensitivity variations, large element-to-element spacing, and small number of sensor elements. Because of the strong scattering and reflection of millimeter waves from antenna arrays, they can hardly be used for near-field imaging. Moreover, 2D antenna arrays are expensive and sophisticated systems. A live electro-optic camera has recently been used for imaging of traveling millimeter waves at video frame rates [15]. However, a live 2D electro-optic imaging system is a very complex system as well. Small aperture of the modern live EO sensor is another disadvantage of this technique. In thermographic techniques of measuring of the MMW beam profiles, thermo-sensitive phosphor or liquid crystal films, thermal paper, and dielectric sheets are used as screens, which are irradiated with millimeter waves. Thermographic techniques, which use thermo-sensitive phosphor [16, 17] or liquid crystal films [18], provide images of MMW intensity profiles in the visible region. A very simple way of obtaining MMW beam images in the visible region is based on the utilization of thermal papers. However, only rough time-integrated images of the field profiles of high-energy MMW beams can be obtained by using thermal paper [19]. In another variant of the thermographic technique, a flat sheet of dielectric material (e.g., paper, polyvinylchloride) is used as an absorbing screen for MMWs [20–23]. A spatial distribution of the microwave-induced variation in the screen temperature is measured by an infrared (IR) camera. This technique has proven itself well in installation, adjustment, and evaluation of the characteristics of the MMW transmission lines and sources of CW and long-pulse high-power millimeter waves. The main drawbacks of all thermographic methods of the MMW imaging are their low temporal resolution, which can be 10 ms at best, and low energy flux sensitivity, which is no better than 1 mJ/cm^2. Moreover, distortions, which are caused by heat conduction, can significantly impair the quality of thermographic images of MMW beams. These distortions increase with increasing MMW pulse duration or camera exposure time.

Light emission from microwave-induced gas breakdowns in space and in gas-filled tubes (luminescent lamps, neon indicator lamps, etc.), as well as breakdowns near a dielectric surface coated with a thin layer of metal powder, have been exploited to analyze the field profiles of millimeter waves [24, 25]. However, the microwave-induced breakdown yields rough distorted images of the field pattern of high-power pulsed MM waves. This technique operates only for narrow ranges of MMW intensities and pulse durations. It is impossible to obtain good-quality images by using this technique because of the properties both of the microwave-breakdown plasma and the light emission from such a plasma. No plasma glow is observed when the intensity of millimeter waves is below the breakdown threshold, or the MMW pulse duration is shorter than the time to breakdown. When the breakdown threshold is exceeded, a filamentary array with subwavelength spacing is formed in space if a gas pressure is higher than a few Torr [26]. The electron density in the filaments is higher than the critical plasma density; therefore, millimeter waves are reflected and scattered by the filaments. The boundary of the region occupied by the breakdown plasma propagates towards the MMW beam. The main contribution to the light emission from the breakdown plasma is given by a bound-bound transition between electronic levels of atoms and molecules. The dependences of the intensity of the atomic and molecular emission lines on the MMW intensity are sophisticated and nonlinear. Moreover, the dependences are different for different spectral lines.

The above-specified drawbacks of the conventional techniques for imaging and measurement of the MMW field profiles require their further improvement, as well as the development of alternative techniques. Recently, a technique for imaging of MMWs based on visible continuum (VC) radiation from a slab of the positive column (PC) of a medium-pressure DC discharge in a mixture of cesium vapor and xenon (a Cs-Xe DC discharge) was proposed and developed [27–35]. The idea behind this MMW imaging technique is to use the effect of the increase in intensity of the VC radiation from the PC of a Cs-Xe discharge due to MMW electron heating. By means of this technique, MMW beam profiles are converted into visible images, thus allowing a conventional digital camera with an optical filter to acquire the images. This review is focused on the operating principles, fundamentals, and applications of this new technique for imaging of MMWs. The chapter is organized as follows. Section 2 provides methods for a plasma slab generation using a Cs-Xe DC discharge and its diagnostics. Section 3 focuses on the experimental evaluation of some important characteristics of the technique for MMW imaging, which is based on using visible continuum radiation from the discharge (VCRD) in a mixture of Cs-Xe. We shall call this technique for imaging of MMWs a VCRD technique. In Section 3, experiments on imaging of the field profiles at the output of MMW horn antennas and quasioptical MMW beams are also discussed. The subjects of Section 4 are the fundamentals of the VCRD technique for imaging of MMWs. In particular, it discusses the nature of the visible continuum radiation (VCR) from the PC of a Cs-Xe discharge, the mechanism of microwave-induced variations in the VCR intensity, and the causes of violation of the local relation between the VC emissivity and the MMW intensity. Section 5 reviews applications of the VCRD technique for imaging of output field patterns of short-wavelength MMW vacuum electron tubes and determination of some characteristics of these tubes. Applications of the VCRD technique for real-time imaging and nondestructive testing using MMWs are also the subject of Section 5.

2. Plasma slab generation using a Cs-Xe DC discharge and experimental study of its characteristics

A sealed discharge tube (DT) was used to generate a slab of the positive column of a Cs-Xe DC discharge in the experiments on imaging of MM waves [27]. Figures 1(a) and 1(b) show side and top views of the discharge tube. A hollow rectangular parallelepiped was located at the center of the tube. It was glued from fused quartz plates. Two square quartz windows with 10 × 10 cm^2 apertures and 0.65 cm thick were set at a distance of 2 cm from each other. They were used to form the plasma slab and input the MMW beam into the discharge tube. Two plane anodes and two heated cathodes were sealed in glass cylinders 10 cm in diameter. The glass cylinders were glued to the quartz cell. The distance between the anodes and cathodes was 30 cm. Each pair of electrodes was powered by a separate power supply. Two pairs of thin quartz plates with a size of 5 × 9 cm^2 were installed inside the cylinders parallel to each other at a distance of 2 cm. These two pairs of quartz plates restricted the region in the cylinders occupied by the positive column and increased the length of a homogeneous part of the PC plasma. The longitudinal electric field E in the positive column was determined using the difference in potentials between the electrical probes. The discharge tube was filled with 45 Torr xenon. The tube had side arms in which drops of cesium metal were placed. Cesium was a readily ionized seed. To obtain the required density of cesium vapor, the discharge tube was heated by a hot air flow in an oven (see Figure 1(c)). The DT wall temperature was measured using thermocouples. The oven has one or two quartz windows 20 cm in diameter. The spectrum of visible radiation emitted by the PC of a Cs-Xe discharge was recorded by a monochromator (see Figure 1(c)). A lens focused light from the PC onto the input slit of the monochromator. A photomultiplier tube was attached to the exit slit of the monochromator. A signal from the photomultiplier was recorded by a digital oscilloscope. The data from the oscilloscope were saved to a computer and processed. The relative calibration of the spectroscopic system in a spectral region from 350 to 700 nm was performed with a tungsten ribbon lamp.

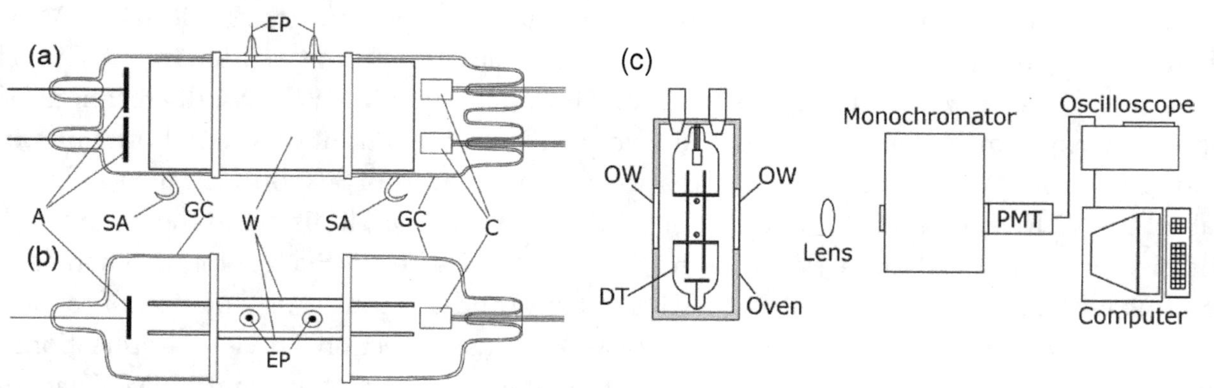

Figure 1. (a) Side and (b) top views of the discharge tube used to generate a slab of the positive column of a Cs-Xe DC discharge; A: anodes; C: cathodes; W: quartz windows; GC: glass cylinder; SA: sidearm; EP: electrical probes. (c) Schematic diagram of the experimental setup (top view) used for a study of emission spectra of light from the PC of a Cs-Xe discharge; DT: discharge tube; OW: oven window; PMT: photomultiplier tube.

The characteristics of the positive column of a Cs-Xe discharge and the spectrum of the light from the PC plasma were studied when there were no incident MM waves [27, 28]. The discharge current J was 1.5 A and the DT wall temperature varied from 80°C to 120°C, which corresponded to the partial pressures of Cs vapor from about 10^{-4} to 2 10^{-3} Torr. For a fixed value of the discharge current, three modes of the positive column were observed depending on the tube temperature [27, 28, 36], specifically, a constricted PC, a wide homogeneous PC, and a PC with a filament. The first PC mode occurred at a high DT wall temperature, i.e., under a relatively high density of Cs vapors. At the tube wall temperature exceeding 110°C the positive column of a Cs-Xe discharge was constricted and occupied only part of the cross section of the discharge tube. For the first PC mode, the DC electric field in the positive column E_0 was less than 0.8 V/cm. Emission lines of cesium atoms dominated in the spectrum of visible light from the constricted PC. As the tube temperature decreased, the width of the positive column and the DC electric field increased. The second mode of the PC was observed for the tube wall temperature ranging from 100°C to 80°C. At these temperatures, the homogeneous positive column filled the entire cross section of the discharge tube. The aperture of the homogeneous plasma slab was approximately equal to the operating aperture of the discharge tube windows 10 × 8 cm² (see Figure 2(a)). The DC electric field in the homogeneous PC increased from 0.9 to 1.7 V/cm when the tube temperature decreased from 100°C to 80°C. If the discharge tube was cooled further, a bright filament 5 mm in diameter appeared. The filament was directed along the discharge current. The rest of the tube cross section was filled with a homogeneous positive column. The positive column with the filament was the third PC mode [27, 36–38]. Xenon was excited and ionized in the filament, and bright xenon atom lines along with the cesium atom lines were observed in its emission spectrum.

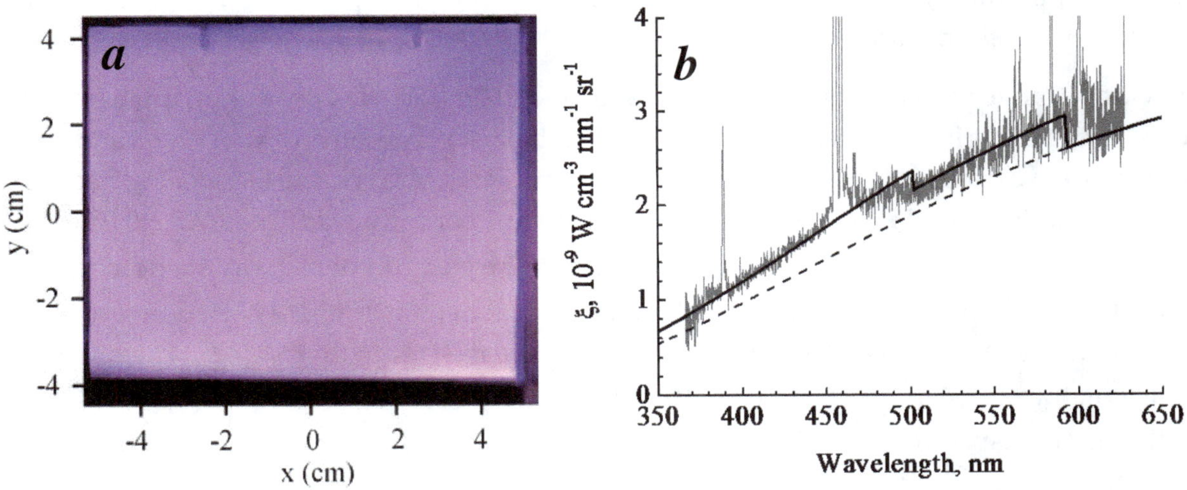

Figure 2. (a) A photo of the homogeneous positive column of a Cs-Xe discharge (the second PC mode). (b) Spectrum of the light emitted from the homogeneous positive column of a Cs-Xe discharge in the visible region.

The decrease in the cesium density with decreasing DT temperature was the main reason for the deconstriction of the PC of a Cs-Xe discharge and the transition from the first to the second PC mode. The electron temperature in the first and second PC modes is too low for a considerable ionization of xenon, and the main ion species in these PC modes are cesium atomic ions Cs^+ [27, 28, 35]. The ionization degree of cesium in the Cs-Xe discharge plasma is high and the number density of positive cesium ions is more than 80% of cesium [35]. Therefore, a decrease in the cesium density with decreasing tube temperature results in an almost proportional decrease in the density of cesium atomic ions. Owing to the plasma quasineutrality, the electron density in the PC is equal to the density of cesium atomic ions; hence, a decrease in the tube temperature also leads to a decrease in the electron density in the PC of the Cs-Xe discharge. For a fixed value of the discharge current, a decrease in the electron number density in the PC plasma leads to an increase in the transverse dimension of the current flow region, i.e., deconstriction of the PC. The positive column ceased to become wider with decreasing tube temperature only when the PC plasma filled entirely the internal cross section of the quartz cell.

Gray thin line in Figure 2(b) shows emission spectrum of the homogeneous positive column (second PC mode) in the visible region for the discharge current density j=0.1 A/cm^2 and DC electric field E_0 = 1.1 V/cm. This spectrum clearly reveals the continuum and the cesium atom emission lines. The electron temperature T_e in the thermal nonequilibrium PC plasma was measured using the spectra of visible continuum radiation from the PC of a Cs-Xe discharge [27, 28, 39–41]. In the homogeneous positive column T_e rose from 0.4 ± 0.05 to 0.5 ± 0.05 eV as the DC electric field increased from 1 to 1.4 V/cm. For the second PC mode, the DC electric field, electron temperature, and density vary in space only weakly, except for the narrow layers near the tube walls: $E_0(x, y) \approx$ const, $T_{e0}(x, y) \approx$ const, and $N_{e0}(x, y) \approx$ const, where the subscript zero denotes the values of the parameters in the absence of MMW exposure. Using the results of E_0 and T_{e0} measurements, the electron number density N_{e0}in the positive column was determined by the equation $N_{e0} = j / e\mu(T_{e0})E_0$ [42], where e is the electron charge, $\mu(T_e)$ is the DC electron mobility, and j is the discharge current density equal to $j = J / S$; here S is the internal cross-sectional area of the quartz cell.

The wide homogeneous plasma slab, i.e., the second PC mode, was used as an imager for MM waves. In the experiments on imaging of MMWs, the discharge current was 1.5 A, and the DT temperature was about 90°C. When there were no incident MMWs, the DC electric field in the PC was equal to E_0= 1.15 ± 0.05 V / cm, the electron temperature was T_{e0} = 0.47 ± 0.03 eV, and the electron density was N_{e0} = (2.7 ± 0.3)•10^{12} cm^{-3}. The electron density in the homogeneous PC of a Cs-Xe discharge was much less than the critical density for MMW frequency band, so reflection of the MM waves from the plasma was weak (less than 1%) [42]. The absorption coefficients $\gamma_0 = \gamma(T_{e0})$ of the PC plasma for Ka and D band MM waves were equal to 0.3 cm^{-1} and 0.01 cm^{-1}, respectively [27, 29].

3. Evaluation of the performance of the VCRD technique

The MMW effects on a slab of the PC of a Cs-Xe discharge have been experimentally studied to develop the VCRD technique for imaging of MMWs and determine its basic characteristics. The model experiments[1] on imaging of the field patterns of horn antennas and quasioptical beams have been performed. Figure 3 shows a layout of the experiments on imaging of the near-field patterns at the output of a MMW horn antenna. The images of the plasma slab were captured by a black and white charged coupled device (CCD) camera. The data from the CCD camera were processed with a computer. A set of optical filters rejecting atomic emission lines and transmitting the continuum in the visible region was placed in front of the camera lens.

1 In the model experiments, the MMW field profiles have been known from calculations or measurements by conventional techniques.

The filter set had a transmissivity of less than 0.1% in the near IR range and in the region from 450 to 460 nm, where the bright lines of the second resonance doublet of cesium occur (see Figure 2(b) and refs. [27, 28]). The contribution of atomic-line intensity into the wavelength-integrated intensity of the transmitted light was a few percent only.

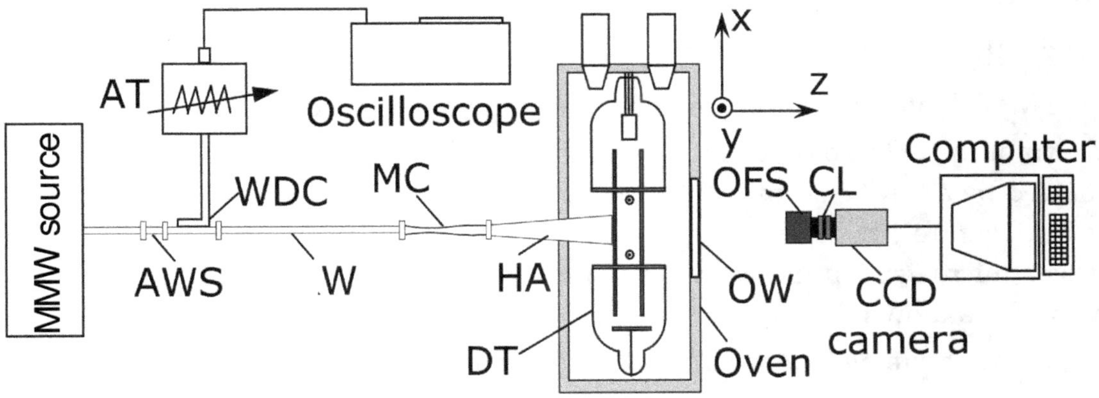

Figure 3. Schematic diagram of the experimental setup (top view) for imaging of the field patterns at the output of a horn antenna; W: waveguide; AWS: waveguide section with MMW absorber; WDC: waveguide directional coupler; AT: MMW attenuator; MC: mode converter; HA: horn antenna; DT: discharge tube; OW: oven window; OFS: optical filter set; CL: camera lens.

To evaluate the microwave-induced variation in the VCR energy flux $\Delta S_C(x, y)$, the background VCR energy flux $S_{C_0}(x, y)$ was subtracted from the VCR energy flux $S_C(x, y)$: $\Delta S_C(x, y) = S_C(x, y) - S_{C_0}(x, y)$. The VCR energy flux $S_C(x, y)$ was obtained using the CCD camera frame simultaneous with the MMW pulse, and the background VCR energy flux $S_{C_0}(x, y)$ was obtained using the CCD camera frame preceding the MMW pulse. When the VC emissivity is time-independent during the camera exposure time τ_e, the VCR energy flux is equal to $S_C(x, y) = \tau_e I(x, y)$, where $I(x, y)$ is the intensity of the visible continuum radiation, and the relative variation in the VCR intensity is equal to the relative variation in the VCR energy flux $\Delta I(x, y) / I_0(x, y) = \Delta S_C(x, y) / S_{C_0}(x, y)$, where $\Delta I(x, y) = I(x, y) - I_0(x, y)$; here $I_0(x, y)$ is the background VCR intensity.

Study of the millimeter wave effect on the homogeneous positive column of a Cs-Xe discharge was performed using conical horn antennas (see Figure 3) excited by the TE_{11}^{O} mode of a circular waveguide [27]. A 35.4-GHz magnetron was used as a source of MM waves. The magnetron supplied coherent radiation with an output power of up to 20 W in the long-pulse mode. The MMW pulse length was up to 100 ms, and the pulse leading edge was less than 0.1 μs. Mode converters were used to transform the TE_{10} mode of a rectangular waveguide into the TE_{11} mode of a circular waveguide [43, 44]. The conical horns had aperture radii from 10 to 30 mm. The MMW electric field polarization in the vicinity of the beam axis was directed along the x coordinate (see Figure 3). The discharge tube window was attached directly to the output of a horn antenna, and the spatial distribution of the MMW intensity was measured in the near-field region. A set of rectangular waveguide sections with MMW absorber was used to decrease the power of millimeter waves that are incident on the PC plasma. By using the waveguide section with a MMW absorber, the MMW power can be decreased up to 15 dB with discrete steps of about 3 dB. The MMW power was also varied smoothly by changing voltage and current supplied to the magnetron. For relative measurements of the radiated power, a small part of the MMW radiation was branched to the calibrated attenuator by a waveguide directional coupler. The output signal from the attenuator was detected using a MMW detector and measured using an oscilloscope.

When the MMW intensity W exceeded the breakdown threshold W_{Br}, a microwave-induced breakdown of the PC slab occurred, and bright thin filaments were observed in the plasma area affected by a MMW beam. The filaments were elongated in the direction of the MMW electric field. Figure 4(a) shows the spatial distribution of the VCR energy flux variation when there was a microwave-induced breakdown of the plasma slab. The threshold value of the microwave breakdown for the Ka-band was equal to about 4 W/cm^2 for E_0= 1.1 V/cm. It decreased to approximately 2 W/cm^2 for E_0= 1.5 V/cm. When $W > W_{Br}$, MMW heating increases the electron temperature so that excitation and ionization of xenon become significant. Ionization of xenon is the major cause of ionization instability of the homogeneous PC and the appearance of filaments with a high electron density, i.e., a microwave breakdown of the plasma slab [35].

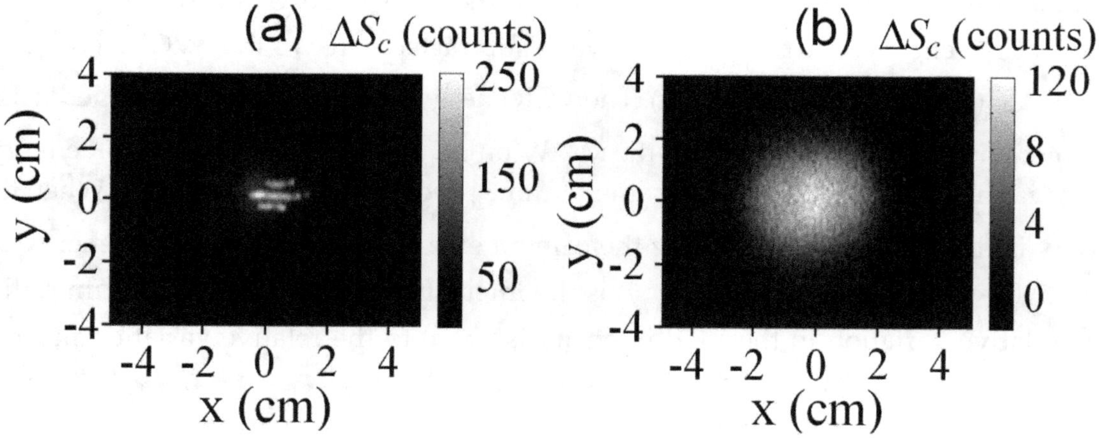

Figure 4. Spatial distribution of the VCR energy flux variation at the output of a conical horn antenna excited by the TE_{11}^{O} mode for the MMW intensities (a) above and (b) below the microwave-induced breakdown threshold of the plasma slab.

Figure 4(b) shows a 2D distribution of the VCR energy flux variation $\Delta S_C(x, y)$ at the output of a conical horn antenna,[2] when the MMW intensity in the field maximum was less than the microwave breakdown threshold. The horn antenna with the taper angle of 5° and output aperture radius R_h = 22.5 mm was excited by the TE_{11}^{O} mode. The CCD camera exposure time was equal to 200 μs. The spatial distribution of the VCR energy flux variation is smooth, and

2 The coordinates are denoted x and y in the reference frame, whose origin coincides with the center of the DT window, and the coordinates are denoted x' and y' in the reference frame, whose origin coincides with the MMW beam center.

it resembles the intensity profile for the TE_{11}^{O} mode (see Figure 4(b)). Using the same conical horn excited by the TE_{11}^{O} mode, the dependences of the relative VCR intensity variation $\Delta I(0, 0) / I_0(0, 0) = \Delta S_C(0, 0) / S_{C_0}(0, 0)$ on the intensity of millimeter waves incident on the plasma slab $W(0, 0)$ was measured. The CCD camera exposure time was 100 and 200 μs. The measured dependence of $\Delta I / I_0$ on W / W_{Br} is shown in Figure 5 by diamonds. In the range of MMW intensities from 0 to W_{Br}, the second-order polynomial function fits the dependence $\Delta I(W) / I_0$ with an accuracy of up to a few percent (black solid line in Figure 5). For the MMW intensity range $0 < W < W_{Br} / 3$, the value of ΔI could be considered approximately directly proportional to the intensity of millimeter waves (with an accuracy of about 10%). If the MMW intensity is below the breakdown level, MMW beam profile can be obtained from the distribution of the VCR intensity variation using the dependence of $\Delta I / I_0$ on W.

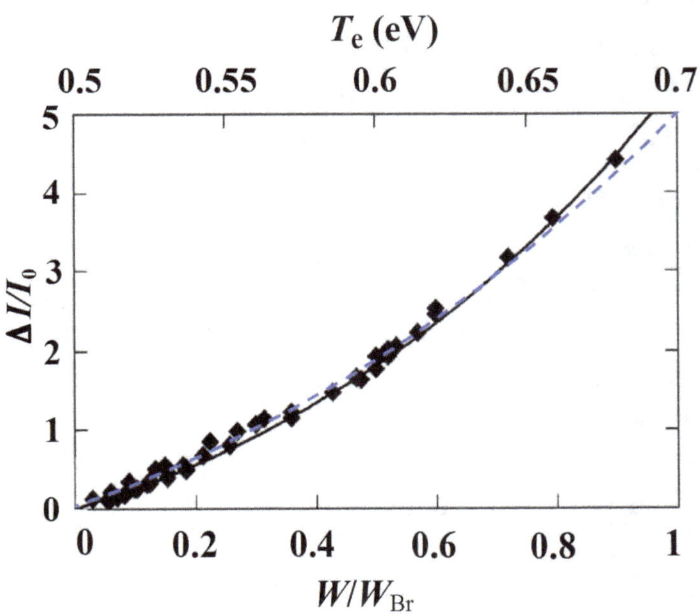

Figure 5. Experimental dependence of the relative VCR intensity variation as a function of the intensity of MM waves incident on the plasma slab (black diamonds), and calculated dependence of the relative VCR intensity variation as a function of the electron temperature (blue dashed line).

The data of the experiment on the imaging MMW intensity profile at the output of the conical horn antenna with an aperture radius of 22.5 mm excited by the TE_{11}^{O} mode (see Figure 4(b)) were compared with the calculated spatial distribution of the EM wave intensity for the TE_{11}^{O} mode of a circular waveguide of the same radius. For $x' = 0$, the calculated EM wave intensity decreases as a function of $|y'|$ without discontinuities from the maximum value W_m at the center of the waveguide to zero near the waveguide wall, i.e., for $y' = \pm 22.5$ mm [43, 44]. The measured dependences of the VCR intensity variation vs. coordinate y' for $x' = 0$ were in good agreement with the calculated spatial distribution of the EM wave intensity $W(x'=0, y')$. For $y' = 0$, the measured dependence of the VCR intensity variation vs. the coordinate x' was different from the calculated spatial distribution of the EM wave intensity for the TE_{11}^{O} mode $W(x', y'=0)$ in the regions near $x' = \pm 22.5$ mm. For $y' = 0$ the sharp edges of the MMW intensity distribution near the waveguide wall were blurry at the image. The blurring effect appeared due to the diffraction of MMWs at the edge of a horn antenna and electron heat conduction (see Section 4.2).

The experiments on imaging of the MMW field profiles at the output of the conical horn antennas excited by the axially symmetric TE_{01}^{O} mode of a circular waveguide were carried out in ref. [27]. The same 35.4 GHz long-pulse magnetron was used in these experiments. The TE_{10} mode of a rectangular waveguide was transformed into the TE_{01}^{O} mode of a circular waveguide by a Marier transducer (see Figure 3). Figure 6(a) shows a 2D distribution of the MM wave-induced variation of the VCR intensity $\Delta I(x, y)$ at the output of a conical horn with a taper angle of 6° and output aperture radius R_h = 28 mm. The MMW intensity at the field maximum was $W_m \approx$ 1 W/cm². The CCD camera exposure time was 100 μs. The obtained pattern of the VC glow corresponds to the TE_{01}^{O} mode with a small admixture of other modes[3] (see Figure 6(a)). Black lines in Figures 6(b) and 6(c) show the dependence of the VCR intensity variation under the MMW effect vs. the coordinates y for $x' = 0$ and x for $y' = 0$, respectively. For the TE_{01}^{O} mode, the dependence of the MMW intensity on the radius $r = \sqrt{(x')^2 + (y')^2}$ is $W(r) \approx W_m \cdot [1.72 J_1(3.832 r / R_h)]^2$ for $r < R_h$, where J_1 is a first-order Bessel function and R_h = is the radius of a circular waveguide or the aperture of a conical horn [43, 44]. Gray lines in Figures 6(b) and 6(c) show the function $W(r)$. The function $W(r)$ fits well with the measured dependences $\Delta I(x, y'=0)$ and $\Delta I(x'=0, y)$. However, the intensity variation of the VC radiation from the plasma region near the axis of the MMW beam $\Delta I(x'=0, y'=0)$ was equal to 15% of the VC intensity variation at the field maximum ΔI_m (see Figures 6(b) and 6(c)), while the MMW intensity at the beam center $W(0, 0)$ is equal to zero. Thin solid line in Figure 6(d) shows the dependence of the VCR intensity variation under the MMW effect vs. the coordinate x' for y' = 0, which was obtained for a conical horn with an aperture radius R_h = 20 mm. For a horn antenna of this aperture size, $\Delta I(0, 0) / \Delta I_m$ was about 25%. When conical horns with smaller aperture radii were used, the relative value of the VCR intensity variation in the beam center $\Delta I(0, 0) / \Delta I_m$ increased, i.e., the central deepening became shallower. Thus, a violation of the local relation between the intensity of the plasma VC radiation and the MMW intensity was observed in these experiments. The causes of the nonlocality of microwave-induced variations in the intensity of the VC radiation from a medium-pressure cesium-xenon DC discharge are discussed in Section 4.2.

An experiment on imaging of the Ka-band quasioptical millimeter-wave beam using the VCRD technique is described in ref. [27]. The quasi-Gaussian MMW beam was focused by a double-

3 An experiment showed that the dependence of the VCR intensity on the azimuthal coordinate (see Figure 6(a)) is caused by the admixture of other waveguide modes excited in a conical horn, but not the spurious plasma effect. In this experiment, a rectangular waveguide, which was twisted through an angle of approximately 90°, was installed before the Marier transducer. The image of the spatial distribution of the MMW intensity was also rotated through an angle 90° in the same direction.

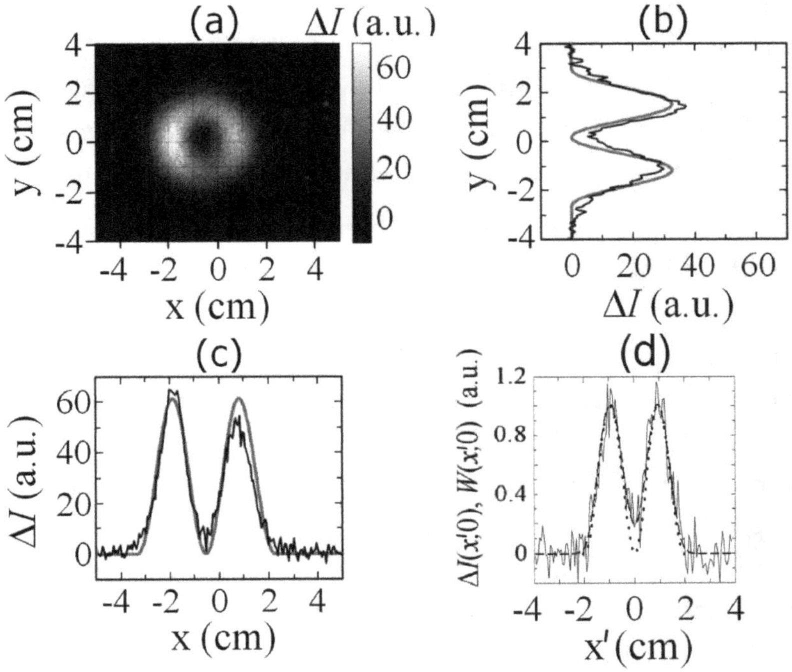

Figure 6. Spatial distributions of the VCR intensity variation at the output of the conical horn antennas excited by the TE^{O}_{01} mode. (a) Two-dimensional distribution of the VCR intensity variation. (b) and (c) Measured dependence of the VCR intensity variation (black lines) and calculated dependences of the MMW intensity variation (gray lines) vs. the coordinates y for $x' = 0$ (b) and x for $y' = 0$ (c) for a horn with the aperture radius R_h = 28 mm. (d) Measured (solid gray line) and calculated (dashed black line) dependences of the VCR intensity variation vs. the coordinates x' for $y' = 0$ for a horn with the aperture radius R_h = 20 mm. Calculated dependence of the MMW intensity variation vs. x for $y' = 0$ (dotted line).

convex Teflon lens 20 cm in diameter and a focal length of 32 cm at the center of the plasma slab. The width of the MMW beam (FWHM) in the focus was approximately 2.5 cm. The distribution of the VCR intensity variations in the focal plane of the lens was compared with the MMW field profile. The MMW intensity profile was measured by a mechanically scanned receiver with calibrated MMW detector. The patterns measured by the two techniques coincided within the limits of the random experimental error. Thus, model experiments on imaging of the field profiles of horn antennas and quasioptical beams have demonstrated that the VCRD technique can be used for good-quality imaging of Ka-band MMW beams.

The response time of the VCR intensity to the variation in the MMW intensity was measured using a photomultiplier tube (PMT) [27]. In these experiments, the photomultiplier tube was set in place of the CCD camera (see Figure 3). The optical filter set rejecting atomic emission lines and transmitting the VC radiation from the PC of a Cs-Xe discharge was mounted before the PMT input window. The temporal resolution of the setup used for recording of the VCR intensity time history was less than 50 ns. After the MMW radiation was switched on, the measured waveform of VCR intensity fit well with the exponential time dependence:

$$\Delta I = \Delta I_{\infty}\left\{1 - \exp\left[-\left(t - t_0\right)/\tau\right]\right\}, \quad (1)$$

where ΔI_{∞} is the VCR intensity variation, t_0 is the time point at which the MMW pulse was switched on, and τ is the characteristic time of VCR intensity variation. The VCR response time was about $\tau = 0.8 \pm 0.1$ μs. This time is approximately equal to the time of the electron temperature variation in the PC plasma (see Section 4.2 and refs. [27, 35]). Thus, it was shown that the VCRD technique for MMW imaging is very rapid and its temporal resolution was less than 1 μs.

In the experiments on imaging of Ka-band MM waves using the VCRD technique, a single-shot signal to noise ratio (SNR) was about one when the MMW intensity was $W \approx 1$ W/cm² and the CCD camera exposure time was set at 10 μs [27]. Thus, in the Ka-band, a single-shot threshold energy flux sensitivity (SNR = 1) of about 10 μJ/cm² was achieved. In the performed experiments, the energy flux sensitivity was restricted by the noise performance of the CCD camera. The actual sensitivity of the VCRD technique is determined by fluctuations of the parameters of the PC of a Cs-Xe discharge, primarily, by the fluctuations of the electron temperature in the plasma [45]. Estimated sensitivity of the VCRD technique is higher by at least an order than the sensitivity achieved in the experiments. Further study is required for more accurate determination of the VCRD technique sensitivity.

The VCRD technique can be used to image the intensity patterns of the EM waves in a wide band from short-wavelength centimeter to submillimeter waves. When the MMW frequency increases, the plasma becomes more transparent for MMWs due to a decrease in the MMW absorption coefficient, and the MMW effect on the plasma parameters decreases [27, 29]. At the xenon pressure of tens of Torr, the electron-Xe atom collision frequency for the momentum transfer was much less than the angular frequency ω of MM waves [27, 29]. In this case, the scaling law for the efficiency of electron heating under the MMW effect is $\Delta T_e \propto W / \omega^2$, where ΔT_e is the electron temperature variation. Therefore, on passing from long-wavelength to short-wavelength millimeter waves, the sensitivity of the technique decreases if the identical systems are used for VC radiation recording. The threshold of the microwave breakdown of a plasma slab is significantly increased in a short MMW band. For example, on passing from Ka-band to D-band the threshold of the microwave breakdown of a plasma slab increases to the value of the order of 100 W/cm², but the energy sensitivity of the VCRD technique deteriorates to the level of hundreds of μJ/cm².[4]

The ability to image the intensity profiles of short-wavelength MM waves using the VCRD technique was demonstrated in refs. [29, 30]. Thus, the wide frequency band of the technique has been confirmed. In paper [29], the intensity profile of the D-band MMW beam was measured at the output of a corrugated conical horn antenna. The arrangement of this model experiment was similar to the arrangement of the experiment on imaging of Ka-band MMWs at the output of conical horn antennas (see Figure 3). A pulsed F- and D-band orotron OR-180 was used as a source of MM waves (see Section 5.1 and ref. [46]). In this experiment, the orotron radiation frequency was 140 GHz. The maximum orotron power was about 0.6 W. The MMW pulse length was 3 ms, and the pulse repetition rate was 10 Hz. The corrugated conical horn was fed by the orotron through an E-band rectangular waveguide and a down taper. The

4 In the D band, a single-shot energy flux sensitivity of about 200 μJ/cm² was demonstrated [29, 30].

output diameter of the corrugated horn was 16 mm, and its length was about 90 mm. The discharge tube window was attached to the output of the horn. The corrugated horn forms a high-quality axially symmetric Gaussian beam [47]. The width (FWHM) of the Gaussian MMW beam in the middle plane of the plasma slab was equal to 7.8 mm. The MMW electric field vector at the output of the horn was directed along the y coordinate. Figure 7(a) shows the 2D distribution of the VCR intensity variation induced by the MMW beam, which was measured at the output of the corrugated conical horn. The solid lines in Figures 7(b) and 7(c) show the VCR intensity variation as functions of the x' and y' coordinates, respectively. The CCD camera exposure time was 2 ms. To improve the SNR, the images of the MMW field pattern (see Figure 7) were averaged using 300 camera frames. The image was broader than the MMW beam by approximately 1 mm, and it had no axial symmetry in the region of the beam tail.[5] The causes of the broadening of the MMW beam images and their axial asymmetry are discussed in Section 4.2. The MMW intensity profile in the focal region of a Teflon conical lens (axicon) was imaged by the VCRD technique in refs. [29, 30]. A 140 GHz Gaussian EM beam, which is radiated from a corrugated conical horn and collimated by a Teflon lens, was incident on the axicon. The corrugated horn was fed from orotron OR-180. The 2D intensity profiles of the focused MMW beam were imaged at various distances from the axicon. It has been shown that the axicon forms a beam with the width (FWHM) of a focal spot of less than 3 mm and a focal length of about 10 cm. The imaging of the profiles of the MMW beams, which were formed by a corrugated horn antenna and an axicon, demonstrates that the VCRD technique can be used for imaging of MMW beams having a width of more than several millimeters.

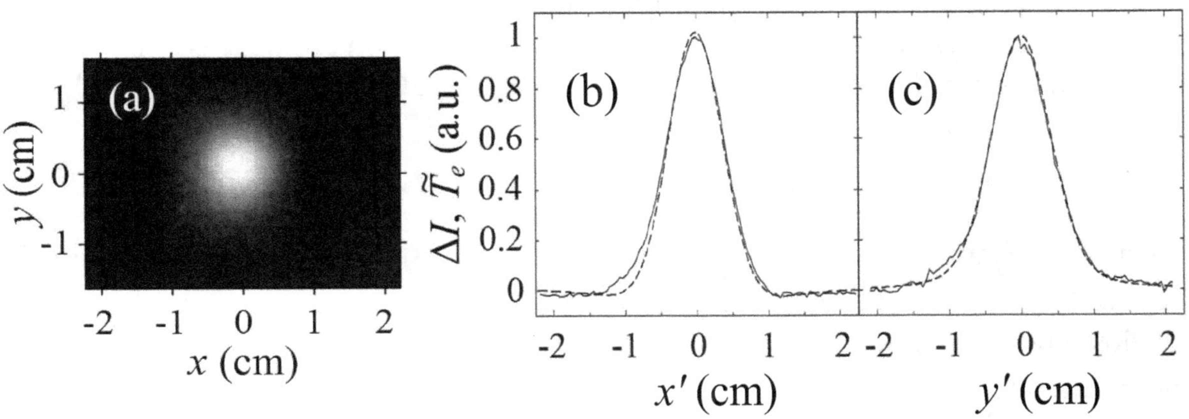

Figure 7. (a) 2D distribution of the VCR intensity variation at the output of a D-band corrugated conical horn. (b) and (c) Measured dependences of the VCR intensity variation ΔI (solid lines) at the output of a corrugated conical horn and calculated electron temperature perturbation $\widetilde{T}_e$ (dashed lines) on the coordinates x' for $y' = 0$ (b) and y' for $x' = 0$ (c).

5 When a section of the D-band rectangular waveguide twisted by an angle of 90° was installed before the corrugated conical horn, the horn rotated around its axis by 90°, and the MMW polarization in the Gaussian beam also changed from direction along the y-axis to direction along the x-axis. However, this rotation does not change the shape of the image of the MMW beam.

4. Fundamentals of the VCRD technique

This section discusses the nature of the visible continuum radiation from the PC of a Cs–Xe discharge and the mechanism responsible for the VCR intensity variation under the MMW effect. The relationship of the spatial distributions of the VC emissivity variation and intensity of MMW radiation incident on the plasma slab is also studied. The factors determining the spatial and temporal resolution of the VCRD technique are discussed.

4.1. Mechanism of the MMW effect on the intensity of the VC radiation from the PC of a Cs-Xe discharge

The nature of the visible continuum radiation from the positive column of a Cs-Xe discharge for plasma parameters that are typical for the MMW imaging experiments was studied in refs. [27, 28]. It was shown that the electron-Xe atom bremsstrahlung continuum (e-Xe BC) and the cesium recombination continuum (RC) give the main contributions to the visible continuum radiation from the PC of a Cs-Xe discharge. The VC emission coefficient $\xi(\lambda)$ is equal to the sum of emission coefficients for the electron-Xe atom bremsstrahlung continuum $\xi^{BC}_{e-Xe}(\lambda)$ [39–41, 48–50] and the cesium recombination continuum $\xi^{RC}(\lambda)$ [39, 51]:

$$\xi(\lambda) = \xi^{BC}_{e-Xe}(\lambda) + \xi^{RC}(\lambda). \tag{2}$$

Visible RC radiation originates from recombination of the electrons and Cs^+ ions into $6P$ and $5D$ electronic states of cesium (blue and yellow cesium RC, respectively) [51]. Thus, the emission coefficient for the recombination continuum of cesium in the visible region is equal to $\xi^{RC}(\lambda)=\xi^{RC}_{6P}(\lambda)+\xi^{RC}_{5D}(\lambda)$, where $\xi^{RC}_{6P}(\lambda)$ and $\xi^{RC}_{5D}(\lambda)$ are the emission coefficients for $6P$- and $5D$-recombination continua of cesium, respectively. The emission coefficients of the blue and yellow cesium recombination continua for $\lambda < hc/E_n$ are equal to [27, 39, 51]

$$\xi^{RC}_{n}(\lambda) = A\lambda^{-3}\sqrt{\varepsilon}\sigma_n(\varepsilon)N_+N_e f_0(\varepsilon), \tag{3}$$

where n means $6P$ or $5D$, E_n is the binding energy of atomic state n, h is the Planck constant, c is the speed of light, λ is the wavelength of the emitted light, A is a constant which is equal to approximately $6\bullet10^{-36}$ $kg^{3/2}m^6s^{-4}sr^{-1}$, ε is the electron energy, $\sigma_n(\varepsilon)$ is the cross section of radiative recombination into the nth electronic state of a cesium atom, N_+ is the number density of the Cs^+ ions ($N_+ \approx N_e$), and $f_0(\varepsilon)$ is an isotropic component of the electron energy distribution function (EEDF). For RC radiation, the relationship between the free electron energy and the light wavelength is $\varepsilon = hc/\lambda - E_n$. For $\lambda > hc/E_n$, $\xi^{RC}_n(\lambda)$ is equal to zero. The emission coefficient of the e-Xe BC is given by [27, 28, 40]

$$\xi^{BC}_{e-Xe}(\lambda) = C\frac{N_e N_{Xe}}{\lambda^2}\int_{\frac{hc}{\lambda}}^{\infty}\sqrt{\varepsilon-\frac{hc}{\lambda}}\left\{\varepsilon\sigma_t\left(\varepsilon-\frac{hc}{\lambda}\right)+\left[\varepsilon-\frac{hc}{\lambda}\right]\sigma_t(\varepsilon)\right\}f_0(\varepsilon)d\varepsilon, \tag{4}$$

where C is a constant which is equal to approximately 0.9 sr^{-1}kg$^{-1/2}$m, N_{Xe} is the Xe atom number density, and $\sigma_t(\varepsilon)$ is the momentum transfer cross section for the electron scattering from a xenon atom. Thick solid line in Figure 2(b) shows the VC emission coefficient $\xi(\lambda)$, which was calculated by Eqs. (2)–(4) for $N_{e0} = 2.5\ 10 \cdot ^{12}$ cm^{-3} and EEDF with the effective electron temperature $T_{e0} = 0.47$ eV [28]. The results of calculations agree well with the measured VC emission spectra. Dashed line in Figure 2(b) shows calculated e-Xe BC emission coefficient for the same plasma parameters. In the visible region, the intensity of e-Xe BC radiation is several times greater than the maximum intensity of the cesium RC radiation.

The intensity of the visible continuum from the PC of a Cs-Xe discharge rises when the millimeter waves incident on the plasma slab or the DC electric field in the PC plasma increases. It follows from Eqs. (3) and (4) that the rise in the VCR intensity can be caused by an increase in the electron number density or by EEDF variation. The electron-density variation induced by the MMW or DC electric field cannot explain the observed increase in the VCR intensity. Firstly, due to a high ionization degree of cesium in the homogeneous positive column, the microwave-induced variation in the electron density is relatively small ($\Delta N_e / N_{e0} << 1$) [35], whereas the effect of watt-scale Ka-band MMWs on the PC plasma increases the VCR intensity significantly ($\Delta\xi / \xi \sim 1$) (see Figure 5). Secondly, the characteristic time of the electron density variation [35, 36] is several orders of magnitude longer than the VCR response time (see Section 3). Thirdly, when the discharge tube temperature was decreased, the electron number density N_{e0} in the homogeneous PC also decreased, whereas the VCR intensity and the DC electric field E_0 increased [28]. Hence, the VCR intensity variation is caused by the EEDF variation resulting from the MMW or DC electron heating. Using Eqs. (2)–(4), the VC emission coefficient were calculated for different electron temperatures in the homogeneous PC plasma. The electron density was assumed constant. Solid lines in Figure 8 show the continuum emission coefficient $\xi(\lambda)$ in the visible region calculated for 45 Torr xenon, $N_e = 2.5 \bullet 10^{12}$ cm^{-3} and the EEDF with the effective electron temperatures $T_e = 0.45$, 0.5, 0.55, and 0.6 eV. Dashed lines in Figure 8 show the e-Xe BC emission coefficient $\xi_{e-Xe}^{BC}(\lambda)$ calculated for the same plasma parameters. In the visible region, the wavelength-integrated emission coefficient for the e-Xe BC radiation exceeds the wavelength-integrated emission coefficient for the cesium recombination continuum by about one order of magnitude. Figure 8 demonstrates that the increase in the VC emissivity is caused by the e-Xe BC emissivity increase due to an additional electron heating and is not caused by an increase in the cesium recombination continuum emissivity as was believed earlier [31]. As the electron temperature rises from $T_e = 0.45$ to 0.6 eV, the intensity of the e-Xe BC in the visible region increases more than fourfold.

Explanation of the phenomenon of the increase in the e-Xe BC radiation intensity in the visible region induced by the electron heating is quite obvious. Bremsstrahlung photons with wavelength λ can be emitted only by electrons with kinetic energies greater than the energy of a light quantum, i.e., $\varepsilon > hc / \lambda$. The visible light wavelengths ranging from 620 nm to 410 nm correspond to photon energies from 2 to 3 eV. The bremsstrahlung photons with such an energy can be emitted by the electrons from the tail of the EEDF ($\varepsilon > 2$–3 eV) because the electron temperature in the homogeneous PC of a Cs-Xe discharge is from 0.4 to 0.8 eV. Thus, the

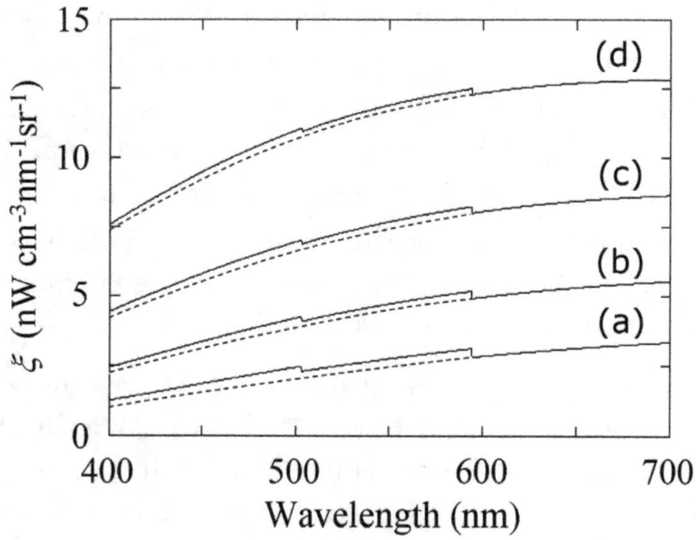

Figure 8. The visible continuum (solid line) and e-Xe bremsstrahlung continuum (dashed line) emission coefficients vs. wavelength for effective electron temperatures (a) 0.45, (b) 0.5, (c) 0.55, and (d) 0.6 eV.

increase in the e-Xe bremsstrahlung radiation intensity in the visible region is caused primarily by an increase in the fast electron number density, which results from the additional electron heating induced by the MMW or DC electric field. The bremsstrahlung cross section for electron-atom collisions increases with increasing electron energy (see Eq. (4) and refs. [40, 50]). This is another cause of the increase in the VC brightness due to DC- or microwave-induced electron heating.

Verification of the previously discussed mechanism of the electromagnetic field effect on the brightness of the visible continuum radiation from the PC of a Cs-Xe discharge was carried out in ref. [28], where the effects of the DC electric field strength and electron density on the intensity of VC radiation were studied experimentally. There was no MMW radiation incident on the plasma. The electron density was varied by changing the discharge current. The DC electric field strength was varied by changing the tube temperature for a fixed value of the discharge current. Experimental data are in a good agreement with the results of calculations done using the above model [28]. In ref. [27], the dependence of the relative intensity variation of the VC radiation from the PC of a Cs-Xe discharge, $\Delta I / I_0$, on the electron temperature was calculated. The best agreement between the results of calculations and the experimental data is achieved when it is assumed that the microwave breakdown of the plasma slab occurs at an electron temperature equal to T_{eBr} = 0.7 eV. Dashed line in Figure 5 shows the calculated dependence $\Delta I(T_e) / I_0$ for T_{eBr} = 0.7 eV.

4.2. Relation between spatial profiles of the VC emissivity and the MMW intensity. Spatial and temporal resolution of the VCRD technique

The causes of violation of the local relation between the visible continuum emissivity and the MMW intensity, as well as the main factors which limit the spatial resolution of the VCRD technique, were explored in ref. [29]. The emissivity of the e-Xe bremsstrahlung continuum depends locally on the effective electron temperature $T_e(x, y, z)$ and electron density $N_e(x, y, z)$ in the PC of a Cs-Xe discharge (see Eqs. (2)–(4)). The electron density variation is relatively small in the PC plasma, and its effect on the brightness of VC radiation is not significant. The main cause of the increase in the VC intensity is the microwave-induced increase in T_e (see Section 4.1). The value of ΔI is directly proportional to the local value of the electron temperature variation ΔT_e when $\Delta T_e < 0.1$ eV. Spatial profiles of the microwave-induced perturbations in the electron temperature and density in the slab of the PC of a Cs-Xe DC discharge were analytically modeled [29]. The model considers the effects of the electron heating by the DC and MMW electric field, the electron energy loss due to electron-atom collisions, electron heat conduction, diffusion, and thermal diffusion of the charged particles. The perturbations of the electron temperature induced by MMW beams which had different shapes were described by the electron-energy balance equation. Perturbation technique was used to find the solutions for this equation. It was sufficient to calculate the first- and second-order terms for the electron temperature perturbation to explain the results of the experiments. For a one-dimensional (1D) MMW beam having a width much smaller than the size of the operating aperture of a plasma slab, the spatial distribution of the electron temperature perturbation of the first order is given by the convolution of the MMW intensity $W(\zeta)$ and the line spread function (LSF) for the VCRD technique $G_e(\zeta)$:

$$\tilde{T}_{e1}(\zeta) = \frac{2\gamma_0 \tau_\varepsilon}{3kN_{e0}} \int_{-\infty}^{+\infty} W(\hat{\zeta}) G_e(\zeta - \hat{\zeta}) d\hat{\zeta}. \tag{5}$$

In Eq. (5), the coordinate ζ is a linear function of x and y, k is the Boltzmann constant, and $\tau_\varepsilon \approx (2P'_0/3k)^{-1}$ is the characteristic time of electron temperature variations due to the electron energy loss in electron-atom collisions, where $P' = \partial(Q/N_e)/\partial T_e$; here, Q is the power density of the electron energy loss due to electron-atom collisions. The line spread function for VCRD technique is in direct proportion to the electron temperature perturbation induced by an infinitely narrow 1D MMW beam with the intensity $W(\zeta) = W_m \delta(\zeta - \zeta_0)$, where $\delta(\zeta)$ is a delta function. In a first-order approximation, LSF is given by the Laplace distribution

$$G_e(\zeta) = \frac{1}{2\Lambda_T} \exp\left(-\frac{|\zeta - \zeta_0|}{\Lambda_T} \right); \tag{6}$$

here $\Lambda_T = (2\tau_\varepsilon \chi_0 / 3k)^{1/2}$, where χ is the electron heat conduction coefficient divided by the electron density. For a xenon pressure of 45 Torr and an electron temperature of 0.5 eV, Λ_T is equal to about 1.5 mm, so the width of the line spread function (FWHM) equals to approximately 2 mm. When the plasma slab is affected by a two-dimensional beam, whose width is significantly smaller than the aperture sizes of the discharge tube, the electron temperature perturbation of the first order is given by a 2D convolution of $W(x, y)$ with a modified Bessel function of the second kind of order zero $K_0(r/\Lambda_T)$,

$$\tilde{T}_{e1}(x,y) = \frac{\gamma_0}{2\pi N_{e0}\chi_0}\int_{-\infty}^{\infty}\int_{-\infty}^{\infty} W(\hat{x},\hat{y})K_0(\xi/\Lambda_T)d\hat{x}d\hat{y}; \tag{7}$$

where $\xi = \sqrt{(x-\hat{x})^2 + (y-\hat{y})^2}$. It follows from Eqs. (5) to (7) that the value of spatial broadening of the MMW beam image is determined by a parameter Λ_T, which depends on the electron heat conduction coefficient and the time of electron energy relaxation due to electron-atom collisions. Therefore, at a first-order approximation, the nonlocality of the microwave-induced variation in the intensity of the VC radiation is caused by the electron heat conduction effect. The electron heat conduction restricts the spatial resolution of the VCRD technique. Using Eq. (7), one can find that the plasma imager with aperture dimensions of about 10 cm has an equivalent number of effective sensor elements of the order of one thousand.

The calculations of the microwave-induced electron temperature perturbation based on the above-described model were compared with the data of the experiments on imaging of 34.5GHz MMW field profiles at the output of conical horn antennas with different aperture radii, which were excited by the TE_{01}^{O} mode (see Section 3). The dashed line in Figure 6(d) shows the calculated dependence of the first-order perturbation of the electron temperature on transverse coordinate r for $R_h = 20$ mm. The calculated dependence fits well with the measured spatial distributions of the VC brightness variations. Figure 7 allows one to compare the data of the experiment on imaging of a D-band MMW beam at the output of the corrugated conical horn and the calculated profiles of the electron temperature perturbation induced by a Gaussian MMW beam with a width of 7.8 mm [29]. The dashed lines in Figures 7(b) and 7(c) show the calculated dependences $\tilde{T}_e = \tilde{T}_{e1} + \tilde{T}_{e2}$ on coordinates x and y, respectively, wher $\tilde{T}_{e2}$ is the second-order perturbation of the electron temperature. The calculations fit well with the experimental data. Figures 7(b) and 7(c) show that the calculated distributions of the electron temperature perturbation, as well as the measured variations in the VCR intensity, are not axisymmetric. Considering the second-order perturbation of the electron temperature, one can explain the violation of the axial symmetry in the images of axisymmetric beams. The difference in the VCR intensity profile along the x and y coordinates is due to the MM wave-induced spatial inhomogeneity of the electron heating by the DC electric field [29].

When a smooth and wide (i.e., wider than 2 cm) MMW beam affects the plasma slab, the influence of the electron heat conduction and other manifestations of nonlocality can be neglected in a first-order approximation. In this case, the perturbation in the electron temperature induced by a long pulse of MMW radiation with a sharp leading edge is given by [27]

$$\tilde{T}_e = \frac{2}{3}\frac{\gamma_0 W\tau_\varepsilon}{kN_{e0}}\left[1\text{-}\exp\left(-\left(t-t_0\right)/\tau_\varepsilon\right)\right]. \tag{8}$$

The time history of the electron temperature perturbation, calculated using Eq. (8), coincides well with the measured time history of the VCR intensity variation. Therefore, the response time τ of the VC radiation is approximately equal to the characteristic time τ_{ε} of electron temperature variations due to the electron energy loss in electron-atom collisions.

5. Applications of the VCRD technique

The study of the performance and fundamentals of the VCRD technique, which was reviewed above, shows that this technique has some advantages over conventional techniques for MM imaging. In contrast to the imaging techniques, which use the receiving antenna arrays, a continuous medium with a small MMW reflection coefficient is exploited in the VCRD technique as an imager. Hence, it is possible to perform imaging of MMW beam profiles in the near-field region via the VCRD technique. The advantages of the VCRD technique over the thermographic techniques include a much higher energy sensitivity and a microsecond temporal resolution. These and some other advantages of the VCRD technique give hope for its applications for determination of the characteristics of output radiation from moderate-power MMW sources as well as for imaging and nondestructive testing with MM waves. At the same time, the VCRD technique has some peculiarities; therefore, experimental verification of the possibility of its use for solving applied problems is required. In this regard, Section 5 is focused on demonstration of two application capabilities of the VCRD technique. Subsection 5.1 discusses the experiments, where this technique is successfully used to measure the parameters of MM waves at the output of new-designed electron tubes. These studies are very relevant because short-wavelength millimeter waves are supposed to be widely used in new advanced radar, communication, diagnostic, and spectroscopy systems. However, in these frequency bands, the power of most types of modern CW and pulsed sources of EM waves is insufficient for use of thermographic imaging techniques. Due to its higher sensitivity, the VCRD technique can be successfully used for measuring the output characteristics of such MM wave sources. Subsection 5.2 provides an overview of experiments on the use of the VCRD technique for real-time imaging and nondestructive testing in the millimeter-wave band. Time-resolved imaging and NDT using MM waves are required for many applications [1–9]. Millimeter waves can penetrate through many opaque dielectric materials, such as paper, wood, fabric, ceramics, polymers, semiconductors, and others. This fact gives MMW imaging and NDT systems some advantage over optical and infrared systems, even though the spatial resolution of MMW systems is much worse than that of optical and IR systems. MMW imaging also has some advantages over THz imaging: first, atmospheric absorption and scattering is lower in the millimeter-wave region, and second, many types of conventional MMW sources and components are widely available. In contrast to EM radiation of the centimeter-wave band, which is also used for imaging and nondestructive testing, millimeter waves have a better spatial resolution. X-ray imaging also has some disadvantages compared to MMW imaging: first, the X-ray is ionizing and requires the implementation of stringent safety rules, and second, some materials, e.g., water, produce only low-contrast images in the X-ray region.

5.1. Imaging of the beam profiles at the output of a D-band orotron and a W-band gyrotron

Orotrons are vacuum electron sources of coherent EM radiation of the short-wavelength band of the MMW range. They are used in the microwave and NMR spectroscopy, for MMW frequency standards. A pulsed orotron OR-180 has been designed and manufactured by a collaboration of Russian organizations [46]. The orotron frequency was tuned electromechanically in the F and D bands. Output coupling of MMW radiation from the orotron cavity occurred through an oversized E-band rectangular waveguide. The design of the orotron assumes that the TE_{10} mode of a rectangular waveguide will be excited predominantly in its output waveguide. Verification of this assumption and determination of the field profiles at the output of the orotron when it was tuned to the different frequencies were important steps towards the comprehension of the operating principle of this device and the improvement of its performance. For the solution of these problems, the VCRD technique for imaging of MM waves has been successfully used [30]. The scheme of this experiment is similar to that shown in Figure 3. Millimeter waves were generated by the orotron OR-180. The orotron pulse duration was 3 ms. A pyramidal horn antenna with aperture dimensions 14 × 14 mm^2 and 21 cm long was installed at the output of the E-band waveguide. The MMW electric field vector at the output of the horn was directed along the y coordinate. The output of the horn was placed very close to the discharge tube window. Near-field intensity profiles at the output of a pyramidal horn antenna were imaged when the frequencies orotron radiation were 130, 140, and 150 GHz. The maximum orotron output power was about 0.3, 0.6, and 0.2 W for the frequencies 130, 140, and 150 GHz, respectively. Thin red lines in Figure 9 show the dependences of the VCR intensity variations on x for $y' = 0$ and on y for $x' = 0$, which were measured at the output of a horn antenna when the orotron frequency was 130 GHz (see Figures 9(a) and 9(b)) and 150 GHz (see Figures 9(c) and 9(d)). The VCR intensity profiles, which are shown in Figure 9, were averaged using 200 CCD camera frames recorded when MMW pulses were incident on the plasma. Figure 9 illustrates that the tuning the orotron frequency changes the intensity profile of the output MMW beam. In the frequency range 130–150 GHz in the E-band rectangular waveguide, besides the fundamental mode $TE_{10}^{\square}$, several higher modes can also be excited [43, 44]. By comparing the measured spatial distributions of VCR intensity variation with the profiles of the microwave-induced electron temperature perturbation, which were calculated using Eq. (7), it is possible to analyze the mode content of the orotron output radiation. Thick blue lines in Figures 9(a) and 9(b) show distributions of the first-order perturbation of the electron temperature on the coordinate x for $y' = 0$ and on y for $x' = 0$, respectively. These dependences were calculated assuming that in the orotron waveguide excited only one fundamental waveguide mode, $TE_{10}^{\square}$. For an orotron frequency of 130 GHz, the profiles of the VCR intensity are approximated well by these calculated distributions of the electron temperature perturbation. The measured relative value of the VCR intensity was higher than calculated electron temperature perturbation at the "tails" of distribution on y for $x' = 0$ due to diffraction of the MM waves at the edges of a horn antenna, as well as the MM wave-induced nonuniformity of the electron heating by a DC electric field (see Section 4.2 and ref. [29]). Therefore, only fundamental waveguide mode is excited for an orotron radiation frequency of 130 GHz. The same situation occurred when the orotron frequency was 140 GHz. For orotron frequencies of 130 GHz and 140 GHz, the admixture of higher modes does not exceed 10%. When the orotron was tuned to 150 GHz, multimode excitation of its output waveguide was observed. This is confirmed, particularly, by strong asymmetry of the distribution of the VCR intensity along the y coordinate (see Figure 9(d)) and its broadening on x (see Figure 9(c)). The results of the experiments on imaging of the MMW intensity profile for an orotron frequency of 150 GHz can be explained assuming that in the output waveguide, a $TE_{11}^{\square}$ mode is excited in addition to the fundamental mode $TE_{10}^{\square}$. Thick blue lines in Figures 9(c) and 9(d) show distributions of the first-order perturbation of the electron temperature on the x coordinate for $y' = 0$ (see Figure 9(c)) and on y for $x' = 0$ (see Figure 9(d)). The distributions were calculated assuming that the electric field amplitude ratio for the $TE_{10}^{\square}$ and $TE_{11}^{\square}$ modes is equal to E^{TE10}/E^{TE11} =3:2 and the phase difference between them is $\pi/4$. Figures 9(c) and 9(d) show that good agreement between the results of calculation and the experimental data occurs for such an amplitude ratio and a phase difference for the two modes. At an orotron frequency of 150 GHz, the power of the spurious mode was about 40% of the power of the fundamental mode.

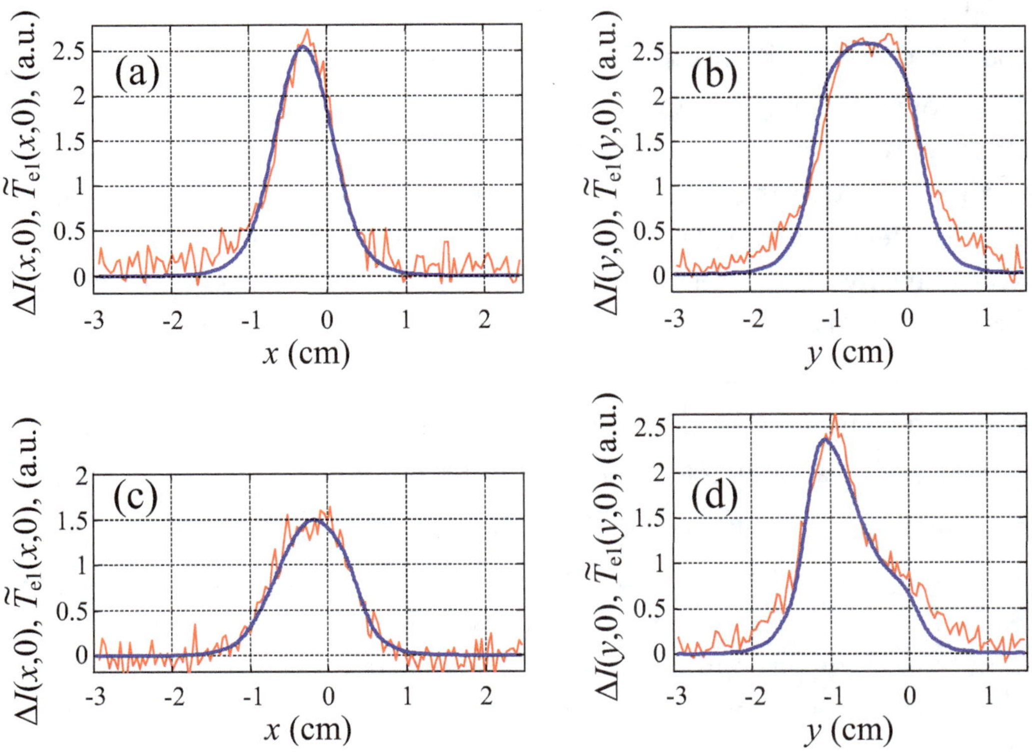

Figure 9. The dependences of the VCR intensity variations (thin red lines) on x for $y' = 0$ ((a) and (c)) and on y for $x' = 0$ ((b) and (d)), which were obtained at the orotron frequency 130 GHz ((a) and (b)) and 150 GHz ((c) and (d)). The first-order perturbation of the electron temperature (thick blue lines) vs. x for $y' = 0$ ((a) and (c)) and vs. y for $x' = 0$ ((c) and (d)).

The VCRD technique was applied for imaging of the field profiles at the output of a 110 GHz kilowatt-scale gyrotron with a pulsed magnetic field [32]. These experiments are of interest for the improvement of this type of gyrotrons and for their successful exploitation in various applications. The scheme of this experiment is similar to that shown in Figure 3. The millimeter waves were output from the gyrotron cavity without mode conversion via a circular hollow metal waveguide 16 mm in diameter and about 200 mm long. A circular hollow metal waveguide 16 mm in diameter and 640 mm long was attached to the gyrotron waveguide. A conical horn antenna 140 mm long and having an aperture diameter of 40 mm was set at the output of the circular waveguide. The gyrotron power can be varied by changing the electron beam voltage and current. The pulse duration of the gyrotron was 60 μs. The CCD camera exposure time was 64 μs. The gyrotron was in the stage of debugging and improvements at the time of the experiments on imaging of its output field profiles. In accordance with the gyrotron design, its operating mode was the rotating mode TE_{32}^{O} of a circular waveguide. In the first series of experiments, the length of the gyrotron cavity was by 1.5 times longer than optimal for its high-efficiency operation. The maximum output power of the gyrotron with that long cavity was about 3 kW. Figure 10(a) shows a 2D distribution of CVR intensity variation under the MMW effect, which was measured at the output of a conical horn for a gyrotron output power P = 1 kW. For such a power, the MMW intensity in the field maxima did not exceed the microwave breakdown threshold of the PC of a Cs-Xe discharge. The VCR image corresponds to the TE_{32}^{O} mode pattern. The standing- wave pattern in the azimuthal direction occurred due to the presence of an admixture of the counter-rotating TE_{32}^{O} mode. Figure 10(b) shows the image of VCR intensity variation recorded at the output of a horn antenna for the gyrotron power $P = 3$ kW. For such a power, the intensity of millimeter waves at their maxima exceeded the breakdown threshold of the plasma slab. Due to the strong increase of VCR intensity in the plasma breakdown areas, the image of the spatial distribution of the MMW intensity was distorted, i.e., it was much more contrasting than in reality. 2D distributions of the MMW intensity were calculated for different power ratios of the counter-rotating TE_{32}^{O} modes of a circular waveguide. The best agreement between experimental data (Figure 10(a)) and computed distribution was obtained for a power of the spurious counter-rotating mode equal to about 0.5% of the power of the fundamental mode TE_{32}^{O}. Figure 10(c) shows the calculated 2D distribution of MMW intensity for TE_{32}^{O} modes with such a power of the counter-rotating mode. The darker areas in Figure 10(c) correspond to the higher-intensity MM waves. In the second series of experiments, a new cavity was installed in the gyrotron. The length of this cavity was optimal for the high-efficiency gyrotron operation. As a result, the maximum gyrotron power increased up to about 10 kW. Figure 10(d) shows 2D distribution of the VCR intensity variation induced by MMW radiation from gyrotron with the optimal cavity. This pattern was recorded for reduced gyrotron power of about 1 kW. At such a gyrotron power, the MMW intensity did not exceed the breakdown threshold of the plasma. The image of MMW intensity profile (see Figure 10(d)) corresponds to the profile of the TE_{32}^{O} mode, which rotates in one direction only. The admixture of the counter-rotating TE_{32}^{O} mode at the output of the gyrotron with a new cavity was less than 0.1% in power.

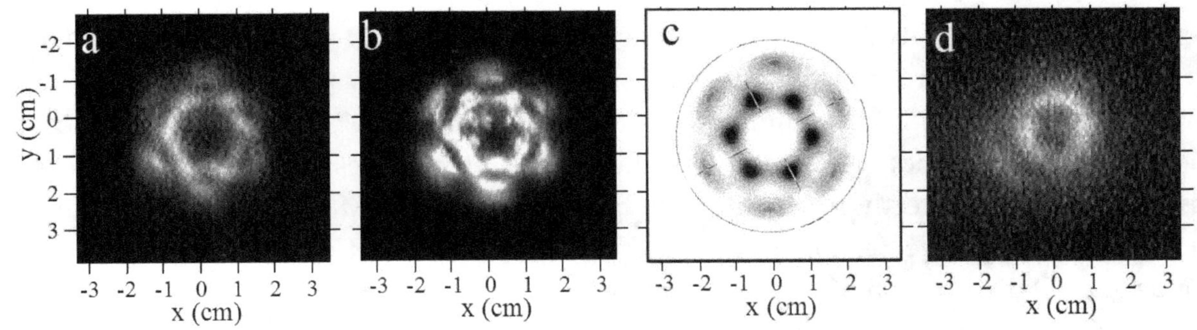

Figure 10. (a), (b), and (d) 2D distributions of the VCR intensity variation at the output of a 110 GHz gyrotron. (c) Calculated 2D distribution of MMW intensity for the TE^{O}_{32} mode.

Thus, the VCRD technique has been successfully applied for determining of the fundamental waveguide modes of moderate-power MMW electron tubes and evaluating the relative power of some spurious modes.

5.2. Applications of the VCRD technique for imaging and nondestructive testing with MM waves

The VCRD technique has been used for active near-field MMW imaging and NDT via shadow projection method [33, 34]. In active-mode MMW imaging and NDT systems MM wave sources are used for object illumination [1–8]. A slab of the positive column of a Cs-Xe DC discharge has been used as a rapid imager for millimeter waves. In the near-field shadow projection method of MMW imaging, the objects are located close to the plasma imager. This is a drawback of the method in comparison with the quasioptical camera-mode method of MMW imaging, in which the object is located far away from the receiver [4, 6, 8, 13]. Realization of the active camera-mode method of imaging based on the VCRD technique is also feasible with using short-wavelength MM waves for object illumination. However, in the camera-mode method distortions of the images occur due to the diffraction and aberrations by lenses and mirrors. Diffraction and aberration effects are very significant when long-wavelength MM waves are used for object illumination. In this case, the use of the shadow projection method may provide a better image quality.

Figure 11 shows schematic of the experimental setup (top view), which was used for active near-field shadow projection MMW imaging using the VCRD technique [34]. A 35.4 GHz magnetron was used as a MMW source. The parameters of the magnetron were specified in Section 3. The MMW pulse length was about 10 ms. The repetition rate of MMW pulses was 12.5 Hz, i.e., half of the frame rate of the CCD camera. Camera frames, which were simultaneous with the MMW pulses, alternated with frames with the background VC radiation from the plasma slab. The CCD camera exposure time was 1 ms. The MMW beam was radiated by a pyramidal horn antenna with a length of 50 cm and aperture dimensions 6 × 8 cm^2. A plane-convex Teflon lens collimated the MMW beam. The polarization of the MMW beam was directed along the x axis (see Figure 11). A test object was placed close to the window of the discharge tube. The width (FWHM) of the quasi-Gaussian beam in the object plane was about

8 cm. The MMW intensity in the center of the beam was approximately 0.3 W/cm^2. Time-averaged MMW intensity in the center of the beam was about 30 mW/cm^2. The MMW dose can be reduced by an order of magnitude by shortening the MMW pulse duration to the CCD camera exposure time. For this millimeter wave intensity, the MMW thermal effect on the majority of objects is not significant.[6] For the MMW intensity less than 1 W/cm^2, the VCR intensity $\Delta I(x, y)$ is directly proportional to the intensity of the MMW radiation incident on the plasma $W(x, y)$ (see Section 3); hence, the image of the VCR intensity variations corresponds to a 2D distribution of the MMW intensity behind the object. The effect of nonlocality of the MM wave-induced variation of the VCR intensity slightly blurs the MMW images.

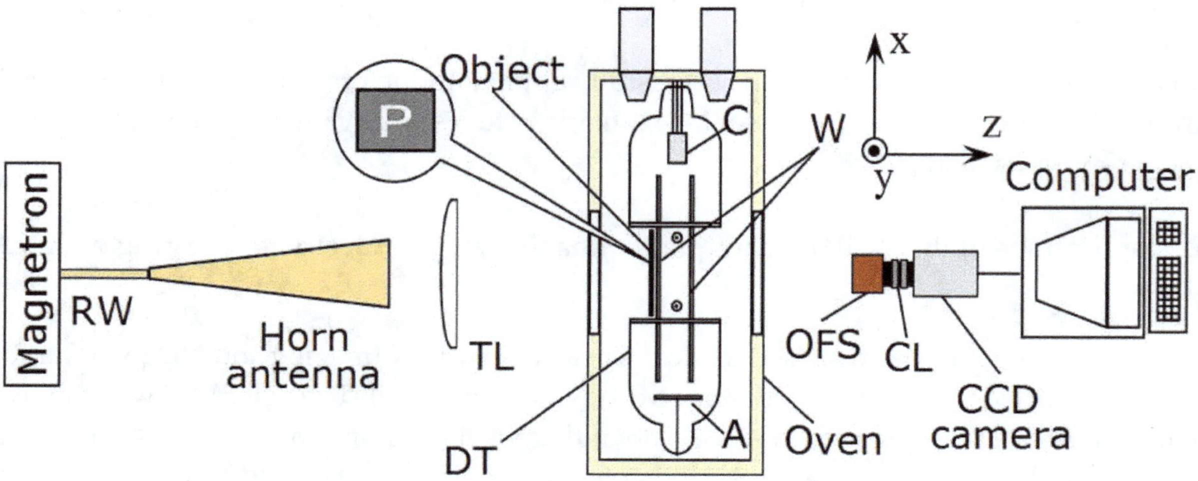

Figure 11. Schematic of the experimental setup (top view); RW: rectangular waveguide; TL: Teflon lens; DT: discharge tube; A: anode; C: cathode; W: quartz windows; OFS: optical filter set; CL: camera lens. Inset, front view of the test object placed before the tube window.

The near-field MMW images of the amplitude objects, which are made of materials that are opaque for MMWs, and the phase objects, which are transparent to the MMW radiation, have been obtained using the VCRD technique. At first, static amplitude and phase objects were imaged. The letters and numbers from IRMMW-THZ 08, an acronym for the name of the International conference, were used as the amplitude test objects. They were cut in the aluminum foil. The foils were glued on cardboard sheets. The letters and numbers were 50 mm high. The width of the rectangular strips transparent for MMWs was 12 mm. The images of the VCR intensity variation under the effect of the MM waves transmitted through the slit objects are shown in Figure 12. The single-shot signal-to-noise ratio for the images was about 20:1. The noise level of the images was given by the noise performance of the CCD camera. The obtained images were distorted by the diffraction because the plasma slab was located in the

6 The VCRD technique is well suited for inspection and MMW imaging of the inanimate objects. The time-averaged MMW intensity of 3mW/cm^2 is several times larger than the maximum MMW dose, which is permitted for general public use in the case of uniform MMW irradiation the whole human body [52]. However, this value is an order of magnitude less than the maximum permitted dose for local MMW irradiation of a part of the human body. Therefore, the VCRD technique cannot be used for entire-body personnel screening, but its applications for security screening and medical imaging with irradiation of a part of human body may be feasible.

Fresnel region. Nevertheless, all of the letters and numbers can easily be recognized. Figure 13(a) shows the image of letter E cut out of aluminum foil and glued on a cardboard sheet. The letter was 50 mm high and 45 mm wide. The width of the foil strips was 10 mm. The shape of this object, which is opaque for MM waves, can be easily determined from the image.

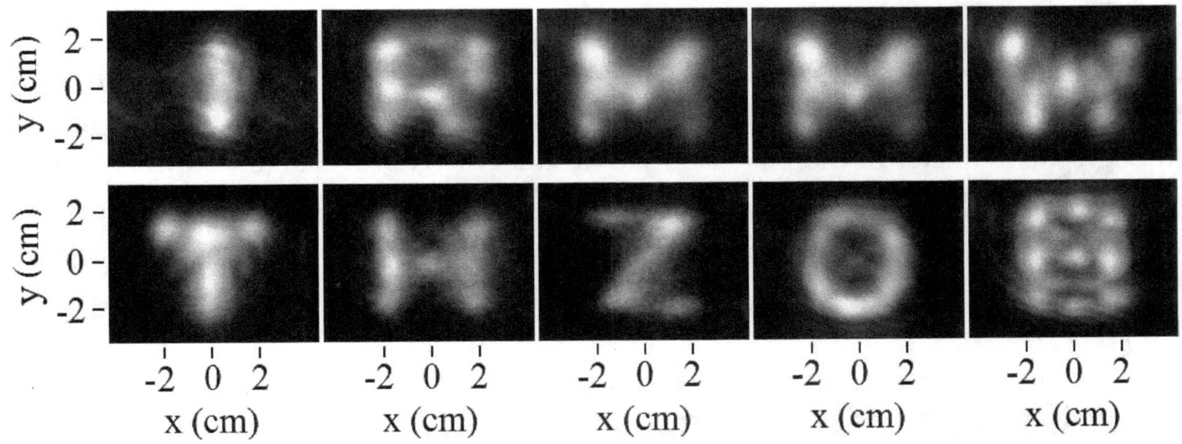

Figure 12. MMW transmission images of the letters and numbers from the acronym IRMMW-THZ 08 cut in aluminum foils.

Due to the peculiarities of the electromagnetic wave diffraction at the edge of transparent and opaque objects, the images of phase objects can have a higher quality as compared to the images of amplitude objects. The EM wave transmitted through a transparent dielectric slab with the thickness L_C acquires an additional phase shift $\Delta\varphi_M = 2\pi(\sqrt{\varepsilon_M} - 1)L_C/\lambda_M$, where ε_M is the permittivity of the material from which the object is made, λ_M is the wavelength of the MMW radiation. If the additional phase shift of the EM wave transmitted through the object differs from the multiple of 2π, then the diffraction gives rise to a narrow minimum of the EM wave intensity in the near-field region behind the edge [53, 54]. The minimum is located between two maxima of intensity, i.e., an edge contrast enhancement occurs in the image of a phase object. Exploiting the VCRD technique, a shadow projection MMW image of a rectangular parallelepiped from Teflon ($\varepsilon_M \approx 2$) was obtained. The length, width, and thickness of the parallelepiped were 180, 28, and 6.8 mm, respectively. In the experiment, the long edges of the parallelepiped were directed parallel to the y axis. When the MMW beam transmitted through Teflon parallelepiped, it acquired an additional phase shift which was equal to about 0.64π. Figure 13(b) shows the 2D MMW image of the parallelepiped. Gray line in Figure 13(c) shows the dependence of the VCR intensity variation on x for $y = 0$. The MMW intensity distribution in the near-field of the Teflon plate 28 mm width was calculated using the theory of Fresnel diffraction from a transparent thin plate [53, 54]. The black line in Figure 13(c) shows the dependence of the integral of the MMW intensity over the plasma slab thickness W_S vs. the coordinate x calculated for such a plate. Calculated dependence $W_S(x)$ is close to the experimental distribution of the VCR intensity variation. Images of more complex dielectric objects are shown in Figures 14 and 15. Figure 14(a) is a drawing of a rectangular Teflon plate with are given in millimeters. Figure 14(b) shows the MMW image of this Teflon plate. The plate edges and both grooves are clearly seen in the image given in Figure 14(b). Perforating circular holes with a diameter of 6 mm are invisible. Figure 15(a) shows the MMW transmission image of a fish-shaped pumice-stone. The height, width, and thickness of the pumice-stone fish were 95, 58, and 18 mm, respectively. In the MMW images of the fish, as well as the Teflon plates, an edge contrast enhancement occurs. Figure 15(b) shows the MMW transmission image of three Septogal pellets (made by JADRAN, Croatia). The pellets were disks 16 mm in diameter and about 3.5 mm thick. The pellets were arranged as a trefoil and placed inside an envelope. Although the diameter of the pills is only twice the MMW wavelength, they are fairly well seen on the image. The experiments have demonstrated that the VCRD technique can be used for a real-time security screening, including the screening of drugs hidden in envelopes.

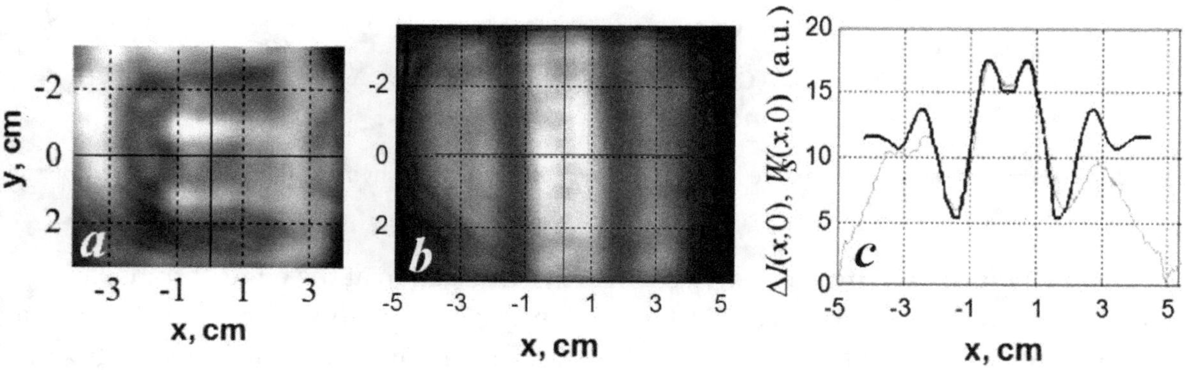

Figure 13. (a) MMW image of the letter E cut out of aluminum foil. (b) 2D MMW image of a rectangular parallelepiped from Teflon. (c) The dependences of the VCR intensity variation (gray line) and the MMW intensity integrated over the plasma slab thickness (black line) on x for $y = 0$.

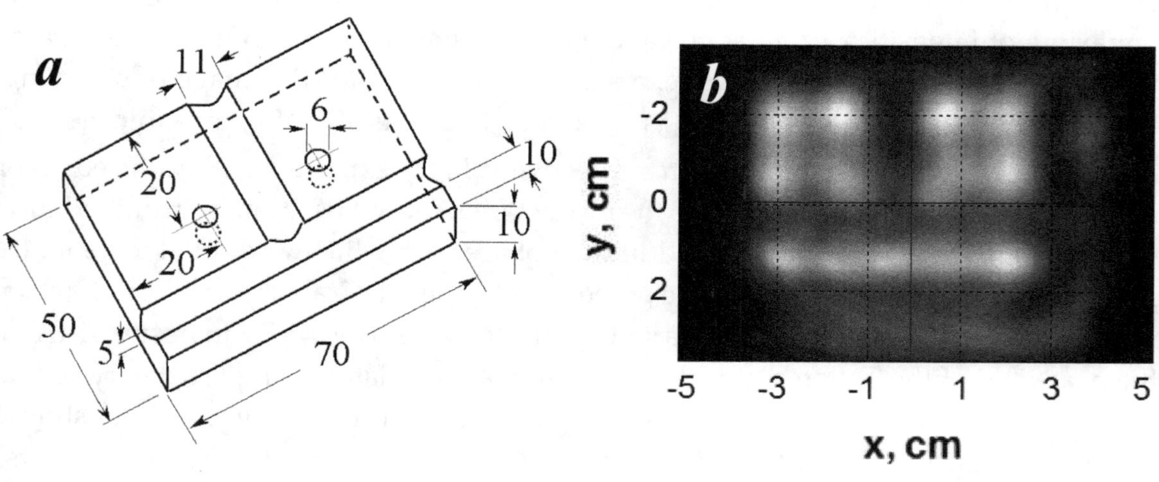

Figure 14. (a) A drawing of the rectangular plate with two intersecting grooves. (b) MMW transmission image of the Teflon plate.

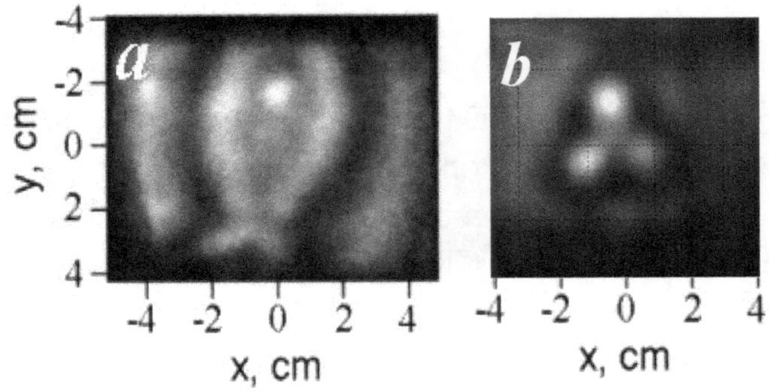

Figure 15. (a) MMW transmission image of a pumice-stone in the form of a fish. (b) MMW transmission image of three Septogal pellets.

The VCRD technique was used for time-resolved record of the MMW images of moving objects and registration of transient processes [34]. In particular, real-time shadow projection images of damped pendulum oscillation were obtained. The pendulum was a Teflon ring suspended by a thread about 15 cm long. The outer and inner ring diameters were 25 mm and 13 mm, respectively, and the ring was 8 mm thick. The ring affected the MMW beam as a three-zone phase filter which focused the MMW beam in the near field in a focal spot about 7 mm in diameter. The decay time of the pendulum oscillation was about 4 s, and the oscillation period was approximately 0.8 s. Figures 16(a) –16(e) show a sequence of MMW images of the pendulum motion, which was recorded at equal time intervals during half of the first period of pendulum oscillations. The process, by which a cylindrical glass tube filled with water became empty, was also imaged (see Figures 16(f) –16(j)). The inner and outer tube diameters were 6 and 8 mm, respectively. The tube filled with water was opaque for MMWs (see Figure 16(f)). After opening a valve which was located at the foot of the glass tube (outside the aperture), the water began to flow out of the tube. Figures 16(g) –16(i) show how the glass tube becomes transparent for MMWs as it is emptied. The water poured out of the tube for about 1 s. The MMW transmission image of the empty tube is shown in Figure 16(j). The tube wall is well defined in Figure 16(j) because of the considerable difference in the refractive indices of glass and air.

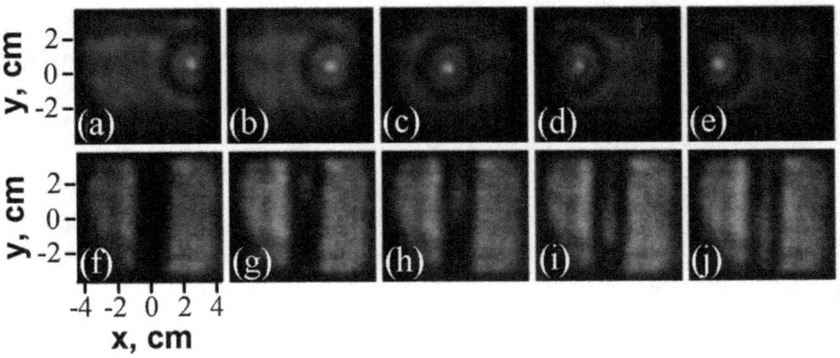

Figure 16. (a)–(e) Real-time MMW transmission images of pendulum motion obtained during one-half of the oscillation period. (f)–(j) MMW transmission movie showing how a glass tube filled with water was emptied.

6. Conclusions

The chapter presents a review of the VCRD technique for imaging and measurement of the 2D spatial profiles of MMW intensity based on the use of visible continuum radiation from the positive column of a Cs-Xe DC discharge. The design of the discharge tube and the experimental setup that were used to create a CW homogeneous slab of thermal nonequilibrium discharge in a mixture of medium pressure xenon and cesium vapor were described. Results of the study of the basic parameters of the positive column of a Cs-Xe discharge were presented. Three modes of the positive column of a Cs-Xe discharge were observed depending on the tube temperature for a fixed value of the discharge current: a constricted PC, a spatially homogeneous PC, and a PC with a filament. For imaging of MM waves, a wide homogeneous PC plasma slab was used. The MMW effects on the plasma slab have been experimentally studied. It was shown that brightness of the VC radiation from the PC plasma increases by several times when the MMW intensity changes from zero to the threshold of the microwave-induced plasma breakdown. The model experiments on imaging of the field patterns of horn antennas and quasioptical beams have been performed using the VCRD technique. These experiments demonstrate that the VCRD technique can be used for a good-quality time-resolved imaging of MMW beams with a width of about 10 mm or more. The temporal resolution of the technique is about 0.8 μs. It was shown that the VCRD technique is wideband and can be used for imaging of the intensity profiles of the EM waves in the entire MMW band. Energy flux sensitivities of about 10 μJ/cm^2 in the Ka band and about 200 μJ/cm^2 in the D band have been demonstrated.

The fundamentals of the VCRD technique for imaging of MMWs have been discussed. The studies of the nature of the visible continuum radiation from the PC of a Cs-Xe discharge and mechanism of microwave-induced variations in the VCR intensity were reviewed. It is shown that the electron-xenon atom bremsstrahlung is the dominant component of the visible continuum emitted by the homogeneous PC of a Cs-Xe discharge. The increase in the e-Xe bremsstrahlung intensity in the visible region is caused by an increase in the number density of electrons with the energy more than 2 eV, which is a result of the additional electron heating induced by the MM waves. The profiles of the microwave-induced variation in the intensity of the VC radiation from the plasma slab were analytically modeled. It was proved that the nonlocality of the microwave-induced variations in the VC radiation intensity, as well as the spatial resolution of the VCRD technique of MMW imaging, are primarily determined by the influence of the electron heat conduction. The line spread function for the VCRD technique has been calculated. It was found that its width is about 2 mm. It was shown that the main cause of the axial asymmetry of images of the axisymmetric MMW beams is the MM wave-induced spatial inhomogeneity of the electron heating by a DC electric field.

The final part of the paper reviews some applications of the CVRD technique for MMW imaging. The experiments on imaging of the output field pattern of a D-band watt-scale orotron and a W-band kilowatt-scale gyrotron were described. Operating modes for these vacuum tubes were identified and the power of spurious modes were determined. Applications of the VCRD technique for real-time imaging and nondestructive testing using MM waves are also the subject of the final part of the review. Near-field shadow projection images of the objects which are opaque and transparent for MM waves have been obtained using pulsed watt-scale MM waves for object illumination. Near video frame rate millimeter-wave shadowgraphy has been demonstrated. It was shown that this technique can be used for a single-shot security screening, including the screening of drugs hidden in envelopes, and for a real-time imaging of the transient processes.

Acknowledgements

The author acknowledges A. E. Fedotov, M. Yu. Glyavin, V. V. Golovanov, A. G. Luchinin, A. G. Spivakov, S. E. Stukachev, A. I. Tsvetkov, and V. V. Zelenogorsky for their contributions to this work. The author is also grateful to N. A. Bogatov, I. L. Epstein, Yu. A. Lebedev, V. I. Malygin, and V. E. Semenov for helpful discussions.

Author details

Mikhail S. Gitlin*

Address all correspondence to: gitlin@appl.sci-nnov.ru

Institute of Applied Physics, Russian Academy of Sciences, Nizhny Novgorod, Russia

References

[1] Bolomey J.-C. Recent European developments in active microwave imaging for industrial, scientific, and medical applications. IEEE Trans. Microwave Theory Tech. 1989;37:2109–2117. DOI: 10.1109/22.44129

[2] Kharkovsky S., Zoughi R. Microwave and millimeter wave nondestructive testing and evaluation—Overview and recent advances. IEEE Instrum. Meas. Mag. 2007;10:26–38. DOI: 10.1109/MIM.2007.364985

[3] Gomez-Maqueda I., Almorox-Gonzalez P., Callejero-Andres C., Burgos-Garcia M. A millimeter-wave imager using an illuminating source. IEEE Microwave Mag. 2013;14:132–138. DOI: 10.1109/MMM.2013.2248652

[4] Sato M., Mizuno K. Millimeter-wave imaging sensor. In: Minin I, editor. Microwave and Millimeter Wave Technologies from Photonic Bandgap Devices to Antenna and Applications. InTech; 2010. p. 331–349. ISBN: 978-953-7619-66-4

[5] Ahmed S. S., Schiessl A., Gumbmann F. F., Tiebout M., Methfessel S., Schmidt L.-P. Advanced microwave imaging. IEEE Microwave Mag. 2012;13:26–43. DOI: 10.1109/MMM.2012.2205772

[6] Tang A., Gu Q. J., Chang M.-C. F. CMOS receivers for active and passive mm-wave imaging. IEEE Communications Magazine. 2011;49:190–198. DOI: 10.1109/MCOM.2011.6035835

[7] Sheen D. M., McMakin D. L., Hall T. E. Three-dimensional millimeter-wave imaging for concealed weapon detection. IEEE Trans. Microwave Theory Tech. 2001;49:1581–1592. DOI: 10.1109/22.942570

[8] Watabe K., Shimizu K., Yoneyama M., Mizuno K. Millimeter-wave active imaging using neural networks for signal processing. IEEE Trans. Microwave Theory Tech. 2003;51:1512–1515. DOI: 10.1109/TMTT.2003.810132

[9] Greeney N. S., Scales J. A. Dielectric microscopy with submillimeter resolution. Appl. Phys. Lett. 2007;91:222909. DOI: 10.1063/1.2818674

[10] Ozhegov R., Gorshkov K., Vachtomin Y., Smirnov K., Finkel M., Goltsman G., Kiselev O., Kinev N., Filippenko L., Koshelets V. Terahertz imaging system based on superconducting heterodyne integrated receiver. In: Corsi C., Sizov F, editors. THz and Security Applications, Detectors, Sources and Associated Electronics for THz Applications. Netherlands: Springer; 2014. p. 113-125. DOI: 10.1007/978-94-017-8828-1_6

[11] Kopeika N. S., Fartiat N. H. Video detection of millimeter waves with glow-discharge tubes. IEEE Trans. on Electron Devices. 1975;22:534–548. DOI: 10.1109/T-ED.1975.18175

[12] Abramovich A., Kopeika N. S., Rozban D. Design of inexpensive diffraction limited focal plane arrays for millimeter wavelength and terahertz radiation using glow discharge detector pixels. J. Appl. Phys. 2008;104:033302-1-033302-4. DOI: 10.1063/1.2963714

[13] Brooker G., Johnson D. G. Low-cost millimeter wave imaging using a commercial plasma display. IEEE Sensors J. 2015;15:3557-3564. DOI: 10.1109/JSEN.2015.2393306

[14] Hisatake S., Nguyen Pham H. H., Nagatsuma T. Visualization of the spatial-temporal evolution of continuous electromagnetic waves in the terahertz range based on photonics technology. Optica. 2014;1:365–371. DOI: 10.1364/OPTICA.1.000365

[15] Tsuchiya M., Kanno A., Sasagawa K., Shiozawa T. Image and/or movie analyses of 100-GHz traveling waves on the basis of real-time observation with a live electrooptic imaging camera. IEEE Trans. Microwave Theory Tech. 2009;57:3373-3379. DOI: 10.1109/TMTT.2009.2033890

[16] Bazhulin A. P., Vinogradov E. A., Irisova N. A., Fridman S. A. Obtaining a visible image of radio emission in the millimeter band. JETP Lett. 1968;8:160–162.

[17] Cherkassky V. S., Gerasimov V. V., Ivanov G. M., Knyazev B. A., Kulipanov G. N., Lukyanchikov L. A., Merzhievsky L. A., Vinokurov N. A. Techniques for introscopy of condense matter in terahertz spectral region. Nuclear Instruments and Methods in Physics Research A. 2007;575:63–67. DOI: 10.1016/j.nima.2007.01.025

[18] Chang T. H., Yu C. F., Fan C. T. Polarization-controllable TE21 mode converter. Rev. Sci. Insrum. 2005;76:074703-1-074703-6. DOI: 10.1063/1.1942528

[19] Junchang L., Yanmei W. An indirect algorithm of Fresnel diffraction. Optics Communications. 2009;282:455–458. DOI: j.optcom.2008.10.060

[20] Thumm M. K., Kasparek W. Passive high-power microwave components. IEEE Trans. on Plasma Science. 2002;30:755–786. DOI: 10.1109/TPS.2002.801653

[21] Malygin V. I., Paveljev A. V. Determination of the mode content in spurious microwave radiation of the gyrotron with a straight axisymmetric output. Int. J. Infrared Millimeter Waves. 1999;20:33–56. DOI: 10.1023/A:1021747516720

[22] Jawla S., Hogge J-P., Alberti S., Goodman T., Piosczyk B., Rzesnicki T. Infrared measurements of the RF output of 170-GHz/2-MW coaxial cavity gyrotron and its phase retrieval analysis. IEEE Trans. on Plasma Sci. 2009;37:414-424. DOI: 10.1109/TPS. 2008.2011488

[23] Idehara T., Kosuga T., Agusu L., Ogawa I., Takahashi H., Smith M. E., Dupree R. Gyrotron FU CW VII for 300 MHz and 600 MHz DNP-NMR spectroscopy. J. Infrared Millim. Terahz Waves. 2010;31:763–774. DOI: 10.1007/s10762-010-9637-9

[24] Gold S. H., Fliflet A. W., Manheimer W. M., McCowan R. B., Lee R. C., Granatstein V. L., Hardesty D. L., Kinkead A. K., Sucy M. High peak power Ka-band gyrotron oscillator experiments with slotted and unslotted cavities. IEEE Trans. on Plasma Sci. 1988;16:142–148. DOI: 10.1109/27.3806

[25] Bratman V. L., Denisov G. G., Ofitserov M. M., Korovin S. D., Polevin S. D., Rostov V. V. Millimeter-wave HF relativistic electron oscillators. IEEE Trans. on Plasma Sci. 1987;15:2–15. DOI: 10.1109/TPS.1987.4316655

[26] Cook A., Shapiro M., Temkin R. Pressure dependence of plasma structure in microwave gas breakdown at 110 GHz. Appl. Phys. Letters. 2010;97:11504-1-011504-3. DOI: 10.1063/1.3462320

[27] Gitlin M. S., Golovanov V. V., Spivakov A. G., Tsvetkov A. I., Zelenogorskiy V. V. Time-resolved imaging of millimeter waves using visible continuum from the positive column of a Cs-Xe DC discharge. J. Appl. Phys. 2010;107:063301-1-063301-11. DOI: 10.1063/1.3327218

[28] Gitlin M. S., Spivakov A. G. Mechanism of the electric field effect on the intensity of visible continuum emission from the positive column of gas discharge in a cesium vapor-xenon mixture. Tech. Phys. Lett. 2007;33:205-208. DOI: 10.1134/ S1063785007030078

[29] Gitlin M. S., Fedotov A. E., Stukachev S. E., Tsvetkov A. I. Nonlocality of microwave-induced variations in the intensity of the visible continuum from a medium-pressure cesium-xenon dc discharge. Physics of Plasmas. 2012;19:033508-1-033508-11. DOI: 10.1063/1.3692077

[30] Fedotov A. E., Gitlin M. S., Golovanov V. V., Perminov A. O., Stukachev S. E. Imaging of short millimeter waves using the visible continuum emitted by the Cs-Xe DC discharge. In: Proc. Sixth Int. Kharkov Symposium on Physics and Engineering of Microwaves, Millimeter and Submillimeter Waves and Workshop on Terahertz Technologies; Kharkov. Kharkov, Ukraine: IRE NASU; 2007. 263–265.

[31] Gitlin M. S., Zelenogorsky V. V., Perminov A. O. Microwave beam imaging by recombination continuum from the positive column of gas discharge in a cesium vapor–xenon mixture. Tech. Phys. Letters. 2002;28:445–447. DOI: 10.1134/1.1490955

[32] Gitlin M. S., Glyavin M. Yu., Luchinin A. G., Zelenogorsky V. V. Imaging the output field pattern of a 110 GHz gyrotron with pulsed magnetic field using recombination continuum emitted by a slab of the Cs-Xe DC discharge. IEEE Trans. on Plasma Science. 2005;33:380-381. DOI: 10.1109/TPS.2005.845294

[33] Gitlin M. S., Golovanov V. V., Tsvetkov A. I. Real-time shadow projection millimeter-wave imaging using visible continuum from a slab of the Cs-Xe DC discharge. IEEE Trans. on Plasma Sci. 2008;36:1398–1399. DOI: 10.1109/TPS.2008.920901

[34] Gitlin M. S., Tsvetkov A. I. Real-time millimeter-wave shadowgraphy using the visible continuum from a slab of the Cs-Xe DC discharge. Appl. Phys. Lett. 2009;94:234102-1-234102-3. DOI: 10.1063/1.3152285

[35] Gitlin M. S., Epstein I. L., Lebedev Yu. A. Modeling of the positive column of a medium pressure Cs-Xe dc discharge affected by a millimeter wave pulse. J. Phys. D: Appl. Phys. 2013;46:415208-1-415208-11. DOI: 10.1088/0022-3727/46/41/415208

[36] Bogatov N. A., Gitlin M. S., Dikan D. A., Luchinin G. A. Cs-Xe dc gas discharge as a fast highly nonlinear volumetric medium for microwaves. Phys. Rev. Lett. 1997;79:2819–2822. DOI: 10.1103/PhysRevLett.79.2819

[37] Wazink J. H., Polman J. Cesium depletion in the positive column of Cs-Ar discharge. J. Appl. Phys. 1969;40:2403–2408. DOI: 10.1063/1.1658005

[38] van Tongeren H. Positive column of the Cs-Ar law-pressure discharge. J. Appl. Phys. 1974;45:89–96. DOI: 10.1063/1.1663024

[39] Batenin V. M., Chinnov V. F. Electron bremsstrahlung in the field of argon or helium atoms. Sov. Phys. JETP. 1972;34:30–33.

[40] Golubovskii Yu. B., Kagan Yu. M., Komarova L. L. Diagnostics of the discharge in argon at medium pressures by bremsstrahlung continuum. Opt. Spectrosc. (USSR). 1972;33:646–648.

[41] Park J., Henins I., Herrmann H. W., Selwyn G. S. Neutral bremsstrahlung measurement in an atmospheric-pressure radio frequency discharge. Phys. Plasmas. 2000;7:3141–3144. DOI: S1070-664X(00)03808-8

[42] Raizer Y. P. Gas Discharge Physics. Berlin: Springer-Verlag; 1991. 449 p.

[43] Harvey A. F. Microwave Engineering. London and New York: Academic Press; 1963. 1313 p.

[44] Marcuvitz N. Waveguide Handbook. London: P. Peregrinus; 1986. 434 p.

[45] Camparo J., Fathi G. Effects of rf power on electron density and temperature, neutral temperature, and Te fluctuations in an inductively coupled plasma. J. Appl. Phys. 2009;105: 103302-1-103302-9. DOI: 10.1063/1.3126488

[46] Bratman V. L., Dumesh B. S., Fedotov A. E., Makhalov P. B., Movshevich B. Z., Rusin F. S. Terahertz orotrons and oromultipliers. IEEE Trans. Plasma Sci. 2010;38:1466–1471. DOI: 10.1109/TPS.2010.2041367

[47] Clarricoats P. J. B., Olver A. D. Corrugated horns for microwave antennas. London: P. Peregrinus; 1984. 231 p.

[48] Rutscher A., Pfau S. On the origin of visible continuum radiation in rare gas glow discharges. Physica C. 1976;81:395–402. DOI: 10.1016/0378-4363(76)90078-4

[49] de Regt J. M., van Dijk J., van der Mullen J. A. M., Schram D. C. Components of continuum radiation in an inductively coupled plasma. J. Phys. D: Appl. Phys. 1995;28:40–46. DOI: 10.1088/0022-3727/28/1/008

[50] Burm K. T. A. L. Continuum radiation in a high pressure argon–mercury lamp. Plasma Sources Sci. Technol. 2004;13:387–394. DOI: 10.1088/0963-0252/13/3/004

[51] Agnew L., Reichelt W. H. Identification of the ionic species in a cesium plasma diode. J. Appl. Phys. 1968;39:3149–3155. DOI: 10.1063/1.1656749

[52] Wu T., Rappaport T. S., Collins C. M. Safe for generations to come. IEEE Microwave Mag. 2015;16:65–84. DOI: 10.1109/MMM.2014.2377587

[53] Anokhov S. P. Plane wave diffraction by a perfectly transparent half-plane. J. Opt. Soc. Am. A. 2007;24:2493–2498. DOI: 10.1364/JOSAA.24.002493

[54] Amiri M., Tavassoly M. T. Fresnel diffraction from 1D and 2D phase steps in reflection and transmission modes. Optics Communications. 2007;272:349–361. DOI: 10.1016/j.optcom.2006.11.048

4

Electrical Discharge in Water Treatment Technology for Micropollutant Decomposition

Patrick Vanraes, Anton Y. Nikiforov and Christophe Leys

Abstract

Hazardous micropollutants are increasingly detected worldwide in wastewater treatment plant effluent. As this indicates, their removal is insufficient by means of conventional modern water treatment techniques. In the search for a cost-effective solution, advanced oxidation processes have recently gained more attention since they are the most effective available techniques to decompose biorecalcitrant organics. As a main drawback, however, their energy costs are high up to now, preventing their implementation on large scale. For the specific case of water treatment by means of electrical discharge, further optimization is a complex task due to the wide variety in reactor design and materials, discharge types, and operational parameters. In this chapter, an extended overview is given on plasma reactor types, based on their design and materials. Influence of design and materials on energy efficiency is investigated, as well as the influence of operational parameters. The collected data can be used for the optimization of existing reactor types and for development of novel reactors.

Keywords: electrode configuration, electrohydraulic discharge, energy yield, organic degradation efficiency, dielectric barrier discharge

1. Introduction

In this introductory section, the current status and limitations of both conventional and advanced water treatment systems are explained in detail, with a focus on their performance on micropollutant removal. This context is necessary to understand the role that plasma technology can play in the future challenges of water purification, which is the main topic of this chapter. Further, it opens a clear perspective on research that still needs to be done in order to turn the plasma treatment of water into a mature technology.

With ongoing improvement of chemical analytical methods, a wide spectrum of compounds and their transformation products are increasingly detected in water bodies and sewage sludge. Many of these compounds occur in low concentrations in the range of microgram to nanogram per liter. They are therefore called micropollutants. Among these are food additives, industrial chemicals, pesticides, pharmaceuticals, and personal care products. Even with such low concentrations, various environmental effects have been observed. Continuous release of antibiotics in the environment leads, for example, to increasing resistance of microorganisms [1]. Chronic exposure of aquatic life to endocrine disruptive compounds causes feminization, masculinization, and immunomodulating effects in fish and frogs, which has tremendous effects on the ecosystem [2, 3]. Additionally, there is growing concern for the direct acute and chronic effects of micropollutants on human health and safety [4–8]. Although more insight is gained on the impact of individual micropollutants, their synergistic, additive, and antagonistic effects are still vastly unknown. Some countries or regions have adopted regulations for a small number of compounds. Nevertheless, effluent limitation guidelines and standards do not exist for most micropollutants.

The primary source of many micropollutants in aquatic systems is the effluent of conventional wastewater treatment plants. An example of such treatment plant is schematically shown in Figure 1. The treatment process occurs in sequential steps:

- Primary treatment consists of a series of mechanical and/or physicochemical treatment steps, which remove solids, oil, and fat. This step is usually common to all wastewater treatment plants [9]. During pretreatment, large debris is eliminated, as it can damage or clog pumps and sewage lines in later steps. Next, abrasive particles are withdrawn, such as grit, sand, oil, and fat. As a third step, a primary clarifier removes the last part of solid material before the wastewater is send to the secondary treatment unit. With this solid material, as much as possible biodegradable material is removed as well in order to significantly reduce oxygen demand and therefore costs in the later biological treatment unit. Overall, primary treatment removes almost half of the suspended solids in raw wastewater [10]. Sometimes, pH adjustment for neutralization is added as a last step of primary treatment to make the effluent suitable for secondary treatment [11].

- Secondary treatment usually consists of biological conversion of dissolved and colloidal organic compounds into stabilized, low energy compounds, by means of a diversified group of microorganisms in presence of oxygen. Additionally, new cells of biomass are generated during this biological treatment step. The mixture of microorganisms with inorganic and organic particles contained in the suspended solids is referred to as activated sludge. Secondary treatment can differ substantially between wastewater treatment plants, as it depends on the nature of the activated sludge [9].

- Tertiary treatment, also called water polishing, consists of any additional treatment that is carried out to make the effluent quality more suitable for discharge in the receiving environment. Chlorination is the most common method of wastewater disinfection due to its low costs. Nevertheless, it can lead to formation of chlorinated organic compounds that may be carcinogenic or harmful to the environment. Therefore, also ultraviolet disinfection is increasingly finding application in wastewater treatment plants [12]. Additional pollutant removal can be achieved with sand filtration, biological nutrient removal, adsorption on activated carbon, membrane processes, and advanced oxidation processes [9, 12]. However, many conventional wastewater treatment plants have no or limited tertiary treatment, due to the high costs associated with most of these methods.

As should be noted, organic pollution that enters a wastewater treatment plant has three outlets: effluent water, excess sludge, and exhaust gas in the form of CO_2 or volatile organic compounds. Excess sludge disposal is complex since it contains a high concentration of harmful substances and only a small part of it consists of solid matter. Sludge treatment is required to reduce the water and hazardous organic content and to render the processed solids suitable for reuse or final disposal. Therefore, sludge is processed in sequential steps of thickening, digestion, and dewatering before disposal [10].

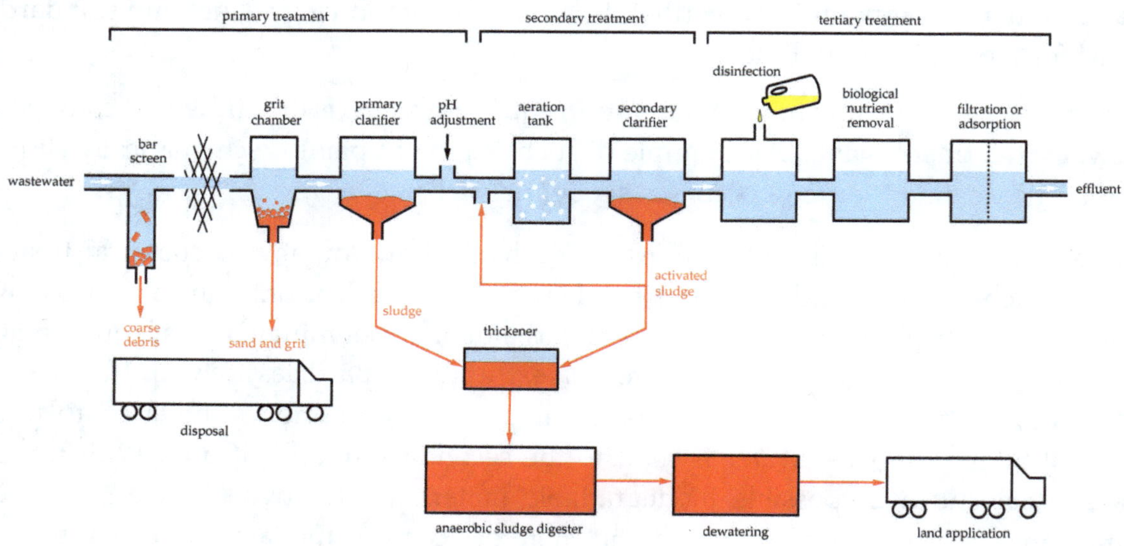

Figure 1. Simplified scheme of a conventional wastewater treatment plant, adapted from Flores Alsina and Benedetti et al. [10, 13].

Up to now, conventional treatment plants cannot sufficiently remove micropollutants, as shown by many studies worldwide [14]. Several options are available to improve the elimination of these contaminants, including:

- Prevention by application of products without micropollutants or only with micropollutants which are easily removed
- Reassessment and optimization of current treatment processes
- Pretreatment of hospital and industrial effluents
- End-of-pipe upgrading of wastewater treatment plants

Preventive measures will always be limited by the increased demand for industrial, pharmaceutical, and personal care products. Moreover, highly stable micropollutants such as the pesticide atrazine are detected even several years after the discontinuation of their use. Hence, more effective water treatment processes are required. Existing processes in water treatment plants can be optimized by increased sludge ages and hydraulic retention times in conjunction with nutrient removal stages and the varying redox conditions associated with them. Temperature and pH control can also enhance micropollutant removal. According to the reviews by Jones et al. and Liu et al., this is expected to be the most economically feasible approach to increase overall water treatment plant performance [15, 16]. While such measures indeed lead to more effective removal of micropollution in general, they often have limited or negligible effect on several specific persistent micropollutants, as shown in the extended review by Luo et al. [14].

The removal difference among different compounds can be partly ascribed to micropollutant properties. For example, Henry's law constant as a measure of compound volatility gives an indication how easily a contaminant will be removed by aeration and convection. In activated sludge processes, the solid–water distribution coefficient K_d has been proposed as a relative accurate indicator of sorption behavior [17, 18]. It is defined as the partition of a compound between the sludge and the water phase and takes into account both hydrophobicity and acidity of the molecule. The biodegradation of micropollutants is more complex to predict, but compound structure is an important indicator. Highly branched side chains and sulfate, halogen, or electron withdrawing functional groups generally make a contaminant less biodegradable. Also, saturated or polycyclic compounds show high resistance to biodegradation [19, 20].

In order to remove such persistent compounds, an additional secondary or tertiary treatment step for wastewater is required. As a cost-effective alternative, the treatment of hospital and industrial effluent can solve the problem at its source. For these purposes, several advanced treatment techniques have been proposed in recent years. Each removal option has its own limitations and benefits in removing trace contaminants. A complete review on this topic was conducted by Luo et al. [14]. Advanced biological treatment techniques are activated sludge, membrane bioreactors, and attached growth technology. As in the case of conventional biological treatment, these methods are commonly unable to remove polar persistent micropollutants. Coagulation–flocculation processes yield ineffective elimination of most micropollutants. In contrast, electrochemical separation treatment methods such as electrocoagulation and internal microelectrolysis give very good results, as discussed by Sirés and Brillas [21]. Activated carbon, nanofiltration, and reverse osmosis are generally also very effective for removal of trace contaminants. However, they are associated with high energy costs and high financial costs. Moreover, all these separation methods have the additional problem of toxic concentrated residue disposal, as they only remove the compounds without further decomposition into less toxic by-products. Advanced oxidation techniques, on the other hand, are able to oxidize toxic compounds to smaller molecules, ideally with full mineralization to CO_2 and H_2O [22]. Nevertheless, their weakest point is as well their high energy demand. Additionally, they can be hazardous to the environment, by their unwanted production of CO_2, long-living oxidants, and potentially toxic oxidation products. Therefore, research needs to focus on optimizing existing systems in order to overcome these problems.

Optimal water treatment schemes will eventually be decided upon, achieving effluent limitations set by national or international environmental regulations at a reasonable cost. In their extended reviews, Jones et al. and Liu et al. conclude that it seems unpractical for activated carbon, nanofiltration, reverse osmosis, and advanced oxidation techniques to be widely used in conventional wastewater treatment [15, 16]. However, with more and more shortages of drinking water all over the world, the recycling of wastewater treatment plant effluents as a drinking water source seems just a question of time. For such purpose, these technologies may be advantageous due to their high removal efficiency. Moreover, with the recent developments in energy harvesting in wastewater treatment [23, 24], one needs to consider the possibility to make wastewater treatment plants self-sustainable in power consumption, even with additional implementation of advanced treatment. As mentioned above, the treatment of hospital and industrial effluent can be a cost-effective approach as well. This last approach has gained a lot of attention over the past several years. For an overview of research done in this field for the case of hospitals, the reader is referred to the reviews of Verlicchi et al. [25, 26].

Comninellis et al. propose in their perspective article a strategy for wastewater treatment, depending on the water's total organic content, biodegradability, toxicity, and other physicochemical requirements, such as transparency [27]. According to this strategy, the use of cost-effective biological treatment is only advised when total organic content is high enough. If such wastewater is not biodegradable, advanced oxidation can be used as pretreatment step, to enhance biodegradability, and reduce toxicity. One needs to consider, however, that presence of oxidant scavengers can sabotage the oxidation efficiency. Therefore, the used oxidation technology needs to be geared toward the wastewater under treatment. In the case that total organic content is low, wastewater possesses little metabolic value for the microorganisms. Then, advanced oxidation technology that effectively mineralizes the targeted pollution can be applied as a one-step complete treatment method. Alternatively, separation treatment can be applied prior to advanced oxidation, where pollutants are transferred from the liquid to another phase and subsequently posttreated.

Following this line of thought, advanced oxidation techniques take a promising place in the quest for micropollutant removal, as they appear the most effective methods for the decomposition of biorecalcitrant organics. However, in this time of a growing energy crisis and concerns over global warming, removal efficiency should not be the only objective. Sustainable development on the whole must to be considered. Therefore, the main objective of research on advanced oxidation technology should be optimization in terms of energy cost and effluent toxicity as well as its compatibility with biological treatment.

Examples of advanced oxidation processes are ozonation, hydrogen peroxide addition, chlorination, Fenton process, UV irradiation, radiolysis, microwave treatment, subcritical wet air oxidation, electrochemical oxidation, homogeneous and heterogeneous catalytic oxidation, ultrasonication, and combinations thereof, such as peroxonation, photocatalysis, and electro-Fenton process.

One type of oxidation method is typically insufficient for micropollutant removal, while a combination of oxidation methods with each other or with other advanced treatment techniques leads to significant improvement up to complete removal, as concluded in many reviews, such as for antibiotics [28], for pharmaceuticals [29], for UV-based processes [30], and for the general case [31, 32]. Subcritical wet air oxidation is not feasible for micropollutant removal [33].

Oturan and Aaron conclude from their review that chemical methods such as Fenton's process and peroxonation perform worse than photochemical, sonochemical, and electrochemical advanced oxidation techniques [32]. For the latter three technologies, they further conclude the following:

- Among the photochemical methods, photo-Fenton process and heterogeneous photocatalysis possessed in most cases a better efficiency than H_2O_2 and O_3 photolysis. Moreover, the solar photo-Fenton process and solar photocatalysis reduce energy consumption even further. The heterogeneous catalyst TiO_2 has many advantages, as it is chemically highly stable, biologically inert, very easy to produce, inexpensive, and possessing an energy gap comparable to that of solar photons.

- The sonochemical combination of Fenton's process with ultrasonication has effective results at lab scale, but application at the industrial level in real time is needed to demonstrate its economic and commercial feasibility.

- Electrochemical methods have the advantage of minimizing or eliminating the use of chemical reagent. Anode oxidation and electro-Fenton have very good performance when a boron-doped diamond anode is used. They can be combined with other advanced oxidation methods for further improvement of efficiency. Examples of such combinations are photoelectro-Fenton, solar photoelectro-Fenton, sonoelectro-Fenton, and peroxyelectrocoagulation. One needs to consider, however, that boron-doped diamond anodes are very expensive [21].

In conclusion, a synergetic combination of multiple oxidants and oxidation mechanisms is recommended for efficient micropollutant decomposition. Therefore, water treatment by means of plasma discharge takes an interesting and promising place among the advanced oxidation techniques, as it can generate a wide spectrum of oxidative species and processes in proximity of the solution under treatment, including shock waves, pyrolysis, and UV radiation. The hydroxyl radical OH is often named as the most important oxidant, due to its high standard oxidation potential of 2.85 V and its unselective nature in organic decomposition. Further, plasma in contact with liquid can generate significant amounts of O_3 and H_2O_2. These two oxidants are frequently used in other advanced oxidation methods and lead together to the peroxone process. Other important reactive oxygen plasma species include the oxygen radical O, the hydroperoxyl radical HO_2, and the superoxide anion O_2^-. When electrical discharge occurs in air, also reactive nitrogen species such as the nitrogen radical, the nitric oxide radical NO, and the peroxynitrite anion $ONOO^-$ will play an important role. Oxidants can either enter the liquid phase through transfer from the gas phase or be formed directly in the liquid phase at the plasma–water interface by interaction of plasma species with water or dissolved molecules. Radical hydrogen H, a powerful reducing agent, is directly formed in aqueous phase by the electron collision with water molecules. Plasma discharge also leads to aqueous electrons, which even have a stronger reduction potential. A more detailed overview of oxidative plasma species and processes without nitrogen-containing oxidants was conducted by Joshi and Thagard [34]. Recent insights on the chemistry of plasma-generated aqueous peroxynitrite and its importance in water treatment are given by Brisset and Hnatiuc and Lukes et al. [35, 36].

The presence of different oxidants reduces the selectivity of an oxidation method. Direct oxidation of organics by ozone is, for example, very selective. This is illustrated with the reaction rate constants listed by Jin et al., Sudhakaran and Amy, and Von Gunten [37–39], which range from 10^{-5} to 3.8×10^{7} $M^{-1}s^{-1}$. Hydroxyl, on the other hand, is considered unselective in the decomposition of organics, with reaction rate constants from 2.2×10^{7} to 1.8×10^{10} M^{-1} s^{-1}, as listed in the same 3 references and [40]. Hence, organic oxidation by means of plasma is expected to be rather unselective as well. Nevertheless, the chemistry behind plasma treatment is very complex due to the interactions between the various reactive species in the gaseous phase, in the aqueous phase and at their interface. Moreover, this chemistry is strongly dependent on the used electrode configuration and material, discharge regime, applied voltage waveform, water properties, and feed gas. Therefore, the optimization of plasma reactors for water treatment in terms of energy costs and effluent toxicity is a complicated task, which still requires more research effort and insight. On the positive side, plasma-based water treatment has already shown itself as a versatile technology, which can find application in the treatment of biological, organic, and inorganic contamination, after sufficient optimization has been reached. As an additional advantage, its flexible design allows it to be easily combined with other advanced treatment techniques. Such combinations can lead to interesting synergetic effects and further optimization.

As pointed out already by Sillanpää et al. [41], no information has been reported in scientific literature on the treatment cost of advanced oxidation processes. To our knowledge, plasma technology has only been compared with other water treatment methods in the review of Sirés and Brillas [21]. According to their review, the main drawback of plasma technology would be its high energy requirement. However, the authors did not validate this claim with quantitative data. Comprehensive quantitative assessment is needed to compare different techniques better from both economic and technical points of view, as Luo et al. conclude in their review [14]. For the specific case of plasma technology for water treatment, only 3 reviews have extensively compared different reactors in their energy efficiency for organic decomposition. Malik was the first to do this for dye degradation [42]. Recently, Bruggeman and Locke and Jiang et al. have made a comparison for mostly phenolic compounds [43, 44]. Their findings will be discussed in next sections. Nevertheless, the main comparative parameter in these reviews is energy yield G_{50} for 50% pollutant removal, expressed in g/kWh, which is only used in literature on plasma treatment. Moreover, G_{50} is strongly dependent on the initial pollutant concentration. It is therefore not fit for comparison with other advanced treatment techniques.

In this chapter, several plasma technologies are discussed that can be applied in water treatment. First, different approaches for reactor classification will be analyzed in detail. In Section 3, an overview will be given of different plasma reactor types, where classification is based on reactor design and reactor materials. The influence of multiple working parameters on reactor energy efficiency is discussed in Section 4. Section 5 gives a summary with future prospects and concluding remarks on the research that is still required in this field.

2. Approaches for reactor classification

Plasma reactors for water treatment can be classified in many ways, depending on several criteria. Such classifications have already been made in a few reviews. One popular approach starts from 2 or 3 main plasma–water phase distributions and subdivides reactors further based on their electrode configuration, as in Bruggeman and Leys and Locke et al. [45, 46]. Here, we will adapt this approach with 6 plasma–water phase distributions, where electrical discharge is generated

1. directly in the water bulk,
2. directly in the water bulk with externally applied bubbles (bubble discharge reactors),
3. in gas phase over water bulk or film,
4. in gas phase with water drops or mist,
5. as a combination of the previous types, or
6. not in direct contact with the solution under treatment.

Accordingly, these reactor types are called (1) electrohydraulic discharge, (2) bubble discharge, (3) gas phase discharge, (4) spray discharge, (5) hybrid, and (6) remote discharge reactors. The basic idea behind this classification is that the total plasma–water interface surface is an important, determining parameter for a reactor's energy efficiency. A larger interface surface is expected to cause higher pollutant degradation efficiency, in agreement with the reviews of Malik, Jiang et al., and Bruggeman and Locke [42–44]. Interface surface can be enlarged by generating plasma in bubbles, by spraying the solution through the active plasma zone, and by making the solution flow as a thin film along the discharge. It can be further enlarged by extending the plasma volume, which has led to many possible choices of electrode configurations and geometries. Additionally, dielectric barriers and porous layers are often introduced in the setup to avoid unwanted energy losses to Joule heating of water and spark formation, while enhancing the local electric field for easier breakdown. In Section 3, this approach for reactor classification will be studied in more detail.

Another popular approach classifies reactors based on the used discharge regime and applied voltage waveform. The following discharge regimes should be considered:

a. Corona and streamer discharge

b. Glow discharge

c. Dielectric barrier discharge (DBD)

d. Arc discharge

Microdischarge and Townsend discharge are not included in this list, as they are not used for water treatment plasma reactors to our knowledge, except in ozone generation. Spark discharge is to be understood as a transient form of arc discharge and therefore falls in the fourth category. All four discharge regimes can be formed in the gas phase or the liquid phase, although we only found one very short and recent report on submerged DBD discharge without bubbles for water disinfection [47]. In the gas phase, corona discharge needs lowest power input and arc discharge the highest. Underwater discharge requires additional energy for plasma onset by cavitation, but it has the advantage of a large plasma–liquid contact surface. Locke et al. and Jiang et al. suggested in their reviews to apply low-energy plasma such as corona and glow discharge for the treatment of water with low contaminant concentration. On the other hand, high-energetic arc discharge might be more effective for high pollutant concentration [44, 46].

Depending on the voltage waveform used, the following types of discharge can be distinguished:

i. DC discharge

ii. AC (low frequency) discharge

iii. Radio frequency discharge

iv. Microwave discharge

v. Monopolar pulsed discharge

vi. Bipolar pulsed discharge

Variations are possible, such as periodically interrupted AC discharge to avoid excessive heating of the plasma gas [48]. Multiple voltage-related parameters can influence reactor energy efficiency, such as voltage amplitude and polarity, sinusoidal or pulse frequency, and pulse rise time and width. Their influence will be discussed in more detail in Section 4. By combining different voltage waveforms with the different discharge regimes in either the liquid or the gas phase, a list of discharge types can be obtained. Malik, for example, identified 27 distinct reactor types in his literature study, using this approach [42]. Locke et al. made a classification into 7 types [46]:

- Pulsed corona and corona-like electrohydraulic discharge
- Pulsed spark electrohydraulic discharge
- Pulsed arc electrohydraulic discharge
- Pulsed power electrohydraulic discharge
- Gas phase glow discharge
- Gas phase pulsed corona discharge
- Hybrid gas–liquid electrical discharge

This classification partly overlaps with the 6 types of plasma–water phase distribution mentioned above, where bubble discharge is included in hybrid gas–liquid discharge. Jiang et al. add 4 more types [44]:

- DC pulseless corona electrohydraulic discharge
- Dielectric barrier discharge
- Gas phase gliding arc discharge
- DC arc discharge torch

While the first 3 types generate plasma in or in contact with the liquid, the last type introduces the water under treatment directly in the torch, where it subsequently gets vaporized and becomes the plasma forming gas. It can therefore be classified as a subtype of spray discharge reactors. For more summarized information on the basic chemistry and physics of the discharge types in this list, the reader is referred to [43, 46].

The list illustrates how the 6 reactor types based on plasma–water phase distribution can be further split into subtypes. This has been done in a more extensive way by Locke and Shih, who identified more than 30 subtypes during their comparative study for the reactor energy efficiency of H_2O_2 production [49]. While this method has proven to be useful for their purpose, it disregards the influence of reactor design, materials, and working parameters. As Locke and Shih point out, studying this influence is a challenging task, but it will reveal valuable information on reactor optimization.

A small number of reviews have been made on plasma reactor energy efficiency for organic decomposition, which mostly focused on the influence of plasma–water phase distribution, discharge type, and a few working parameters. Nevertheless, no comprehensive overview is found in literature up to now on reactor design and reactor materials. The next section will deal with this topic. While the section is not meant to be fully comprehensive, we want to give a broader overview of reactor types reported in literature than has been done in prior reviews. Such overview is not only useful to get easier and faster understanding of reactor operation and development but also serves as source of inspiration for future reactor designs. Moreover, it reveals the close relationship between electrode configuration and discharge type. Both popular and more exotic, unique designs will be discussed, with extra focus on several general working parameters in Section 4. This review also includes plasma reactors that have been used for inorganic removal or biological treatment, as these reactors are also valuable candidates for aqueous organic decomposition.

3. Overview of reactor types

3.1. Electrohydraulic discharge reactors

Electrohydraulic discharge reactors have been studied for many years due to their importance in electrical transmission processes and their potential for water treatment. From theoretical point of view, they are attractive for water treatment due to the relatively high ratio of plasma–water contact surface to plasma volume and proximity of plasma to the water surface. Moreover, they generate shock waves that can aid in organic decomposition. Nevertheless, electrohydraulic discharge reactors are usually less efficient than other reactor types for water decontamination [42, 50]. This is likely due to the additional input energy required for cavitation, i.e., gas phase formation during discharge onset.

Most commonly reported types of electrohydraulic discharge are pulsed arc and pulsed corona discharge. To our knowledge, all arc electrohydraulic discharge reactors reported in literature have pulsed input power. Mostly, a rod-to-rod electrode configuration is used (Figure 2a), but there are also reports on a reactor with a grounded L-shaped stationary electrode and a vibrating rod electrode (Figure 2b) [51, 52]. As learned from personal communication with Dr. Naum Parkansky, the vibrating electrode has the purpose to facilitate electrical breakdown and mix the treated solution. Apart from organic degradation, pulsed arc electrohydraulic discharge is also gaining more recent attention for biological treatment [53, 54]. Often, refractory metal such as tantalum, titanium, tungsten, or a corresponding alloy is selected as electrode material, as it needs to be sufficiently resistant to corrosion and shock waves. Tungsten has proven to be less corrosive than titanium and titanium alloy [53]. Particles that eroded from titanium electrodes enhanced methylene blue decomposition during aging in one study. The authors explained this enhancement with titanium peroxide formation from interaction of H_2O_2 with the particle surface [55]. In contrast, particles that eroded from low carbon steel electrodes diminished the decomposition of the same pollutant, possibly through the catalytic decomposition of H_2O_2 and scavenging of OH radicals [52]. While an increasing energy efficiency is reported for the decomposition of 2,4,6-trinitrotoluene [56] and methyl-tert-butyl ether [57] when the electrode gap is reduced, the opposite effect has been observed for atrazine degradation in another study [58]. Further investigation is needed to understand this apparent contradiction. In a small comparative study, Hoang et al. found the energy efficiency of 4-chlorophenol decomposition of this type of reactor to be one order of magnitude lower than degradation efficiency with UV, UV/H_2O_2, and O_3 systems [59].

Submerged pin-to-pin electrode configuration, as in Figure 2c, is uncommon in scientific literature. Such reactor has been used with high frequency bipolar pulsed power with reduced voltage and low pulse energy by Potocký et al. [61], to lower temperature loading of the electrodes and to clean both electrode tips continuously from any possible adjacent products. With addition of a high inductance in series with the discharge, transition from glow type to arc type discharge can be suppressed [62]. For sufficiently large interelectrode distance and low voltage amplitude, unbridged nonarc discharge was observed. For closer electrodes and higher voltage, bridged arc discharge was obtained, which, according to the authors, was initiated with spark formation at both electrodes [63].

A pin-to-plate electrode configuration (Figure 2d) is often used in electrohydraulic discharge reactors, either for DC glow [64, 65] or pulsed corona [66–68] discharge with positive polarity. The pin curvature radius determines the local electric field strength and is therefore an important parameter that influences discharge initiation [69]. For pulsed corona discharge, the anode pin material has been reported to lead to catalytic effects for organic decomposition.

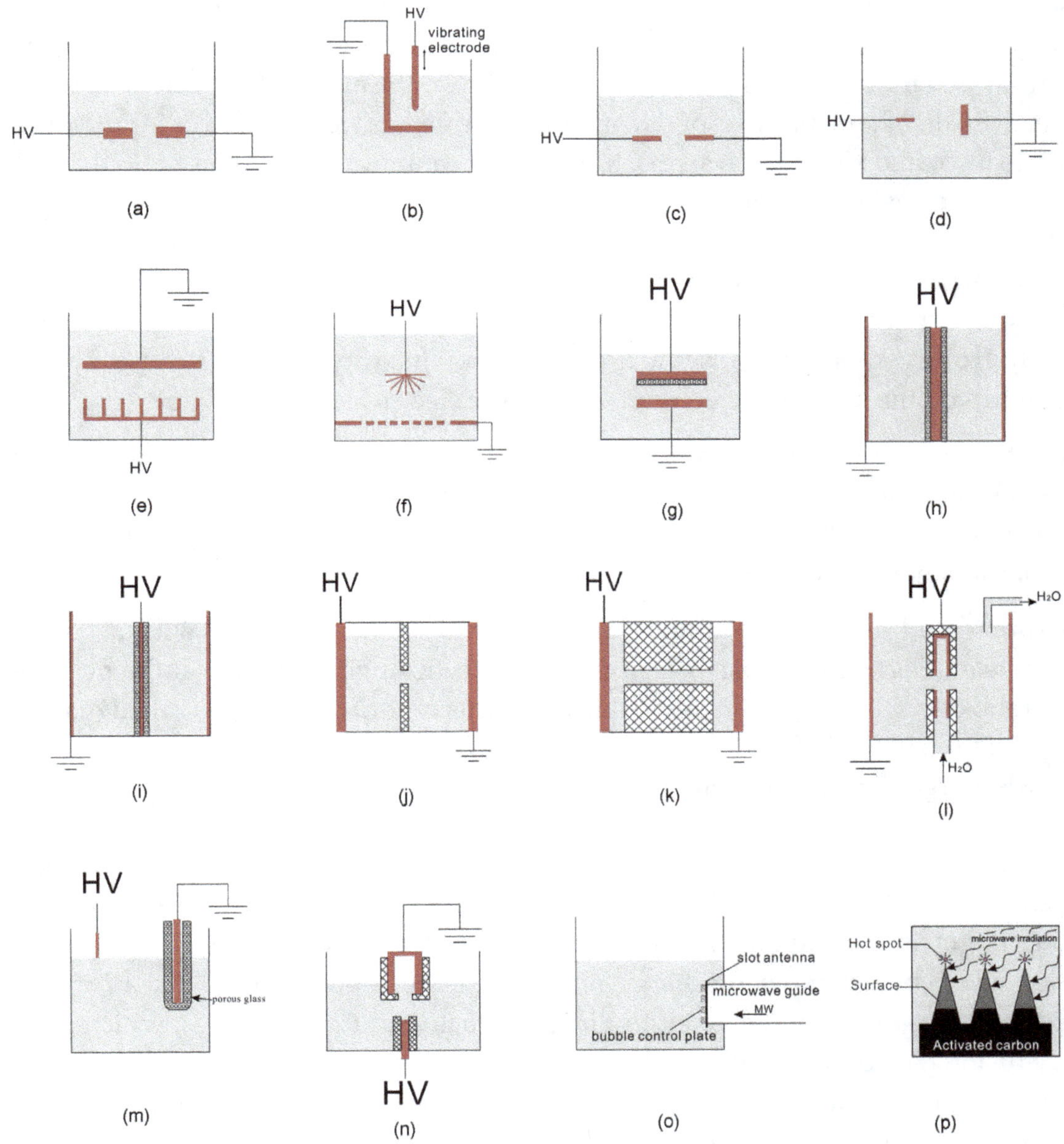

Figure 2. Types of electrohydraulic discharge reactors: (a) pulsed arc, (b) pulsed arc with vibrating electrode, (c) pin-to-pin, (d) pin-to-plate, (e) multi-pin-to-plate, (f) brush-to-plate, (g) plate-to-plate with porous ceramic coating, (h) coaxial rod-to-cylinder with ceramic coating on rod, (i) coaxial wire-to-cylinder with ceramic coating on wire, (j) diaphragm discharge, (k) capillary discharge, (l) coaxial diaphragm discharge reactor from Šunka et al. [60] with perforations in tubular electrode covered by polyethylene layer, (m) contact glow discharge electrolysis, (n) RF-discharge in cavitation bubble on electrode, (o) microwave bubble plasma from waveguide with antenna slot, and (p) "hot spot" plasma formation on activated carbon surface under influence of microwave irradiation.

Platinum enhances pollutant degradation as compared to a NiCr, but only in combination with certain electrolytes. In the case of ferrous salts, this is due to reduction of Fe^{3+} to Fe^{2+} by erosion particles from the platinum electrode [66, 67]. Erosion of tungsten, on the other hand, causes catalysis of oxidation by plasma-generated H_2O_2 [70].

To achieve a higher plasma volume, the pin electrode can be replaced by a multipin electrode [71], a brush electrode [72], or a plate electrode coated with a thin ceramic layer [73], respectively, shown in Figures 2e–g. The concentration of predischarge in the pores of the ceramic layer enhances the electric field strength on the electrode surface [68]. Due to inhomogeneities such as entrapped microbubbles inside the ceramic layer, the electric field can be locally even higher. As a result, a large number of streamers can be generated with lower input voltage as compared to uncoated electrode systems. Moreover, a ceramic coating does not only facilitate an upscale of the system but also serves as a support for a suitable catalyst of the plasma chemical reactions. Also, certain ceramic materials can enhance organic decomposition by catalytic effects [74]. Usually, positive high-voltage pulses are applied to the coated electrode. In some cases, negative high voltage is used to avoid arc formation, as this can damage the coating [73]. However, bipolar pulses are advised since monopolar pulses cause a polarized charge buildup on the ceramic, which can quench the electrical discharge [75].

To enlarge the plasma volume even further, coaxial geometry has been used, where either a coated rod [75] or wire [68, 74] high-voltage electrode is located at the symmetry axis of the grounded cylindrical electrode (Figures 2h–i). Analog to the curvature of pin electrodes, the diameter of the inner electrode is an important system parameter.

When a submerged anode and cathode are separated from each other with a perforated dielectric barrier, electrohydraulic discharge will occur through cavitation at the perforation. For larger ratio of perforation diameter to thickness, this type of discharge is called diaphragm discharge (Figure 2j), while for lower values, the term capillary discharge is used (Figure 2k). DC glow discharge [76, 77] and pulsed corona discharge [78] are commonly used for this reactor, but also AC power input is possible. The strongly inhomogeneous electric field during plasma onset has a similar structure to the one in pin-to-plate geometry. Therefore, similar plasma features can be expected. Energy efficiency for dye and phenol degradation with a single diaphragm is the same as in the case of a pin-to-plate electrode system [46]. Also, similarities with contact glow discharge electrolysis are reported [77]. As an important difference, diaphragm discharge is not in direct contact with the electrodes, which prevents electrode erosion [76]. Sunka et al. developed a coaxial reactor where a polyethylene covered tubular anode with perforations was placed inside a cylindrical cathode (Figure 2l) [60]. The generated plasma was reported to be similar as well.

Another common type of electrohydraulic discharge is contact glow discharge electrolysis. As depicted in Figure 2m, a pointed anode is placed with its tip in the water surface. It is separated from the submerged cathode by means of a sintered glass barrier. In such reactor, glow discharge is generated at the anode tip in a vapor layer surrounded by water. Plasma volume can be increased by increasing the anode number. Stainless steel performed better as anode material than platinum for Acid Orange 7 decoloration [79].

Electrohydraulic discharge can also be generated with RF or microwave power, but such reactor types are less common. Figure 2n shows a reactor where plasma is generated in a cavitation bubble on the tip of an RF electrode [80]. Producing cavitation bubbles by means of microwave power is more complicated. Therefore, a slot antenna can be placed in between the liquid and a microwave guide, as illustrated in Figure 2o. The electric field intensity can be enhanced by installing a quartz plate with holes, a so-called bubble control plate, on the slot antenna. Ishijima et al. reported an increase of methylene blue decomposition efficiency with a factor of 20 after installation of the bubble control plate and tripling the amount of slot antennas [81]. Another way to produce underwater plasma with microwave power is by adding a microwave-absorbing material with high surface area, such as activated carbon, to the solution under treatment. Under influence of microwave irradiation, delocalized π-electrons on the activated carbon surface gain enough energy to jump out of the surface and generate confined plasmas (Figure 2p), also called hot spots, which are known to increase organic decomposition efficiency [82].

3.2. Bubble discharge reactors

Since electrohydraulic discharge generally has low energy efficiency due to the difficulty of initiating discharge directly in the water phase, a lot of attention has gone to enhancing efficiency by discharge formation in externally applied bubbles. Bubbling has the additional advantage of mixing the solution. Moreover, discharge initiation in the gas phase minimizes electrode erosion, which lengthens the lifetime of the system. Bubbling gas through the discharge region greatly increases radical density in the plasma, as for example observed by Sun et al. [69] for O_2 and Ar bubbles. Obviously, feed gas plays a determining role. Yasuoka et al. measured highest efficiency for Ar bubbles, in which plasma spread extensively along the inner surface, while lowest efficiency was obtained with He plasma, which has smallest plasma–water contact surface [83]. The importance of the working gas in plasma reactors in general will be further discussed in Section 4.

A common method is to pump gas upward through a nozzle anode, located underneath a grounded electrode [69], as shown in Figure 3a. Often, the nozzle electrode is placed inside a dielectric tube up to its tip, to avoid any energy leakage toward the water. Alternatively, a pin anode is sometimes placed inside a dielectric nozzle which transports the feed gas (Figure 3b) [86]. The pin tip can be placed below or above the nozzle extremity. Many variations can be encountered in literature, such as a pin anode inside a perforation in a dielectric plate (Figure 3c) [87] and different nozzle orientations (Figures 3d–e) [88, 89]. All choices in nozzle or perforation material, shape, dimensions, and orientation determine the bubble shape during formation and its position after detachment, which significantly influences the electric field in the interelectrode region and therefore the plasma characteristics. This complicates comparison of different reactors. Another option is to place the high-voltage electrode underneath the perforated dielectric plate, as shown by Yasuoka et al., Sato et al., and Yamatake et al. [83, 90, 91] (Figures 3f–h), where the ring-shaped grounded electrode is located around the bubble. Also, here, electrode geometry and position influence the electric field. Yasuoka et al. found their single hole reactor (Figure 3h) to be more energy efficient than advanced oxidation with photochemical persulfate, photocatalyst heteropoly acid, photodegradation, and ultrasonic cavitation for the decomposition of 2 surface active compounds [83]. By increasing the number of nozzles or holes, energy efficiency can be enhanced. In a study by Sato et al. [90], a reactor with a single hole (Figure 3f) was compared to a reactor with 9 holes (Figure 3g). Discharge power deposited per hole was lower in the reactor with 9 holes, which seemed to minimize self-quenching of OH radicals, resulting in higher efficiency. Following this line of thought, a multibubble system as in Figure 3i with a high-voltage mesh in the gas phase attached to a

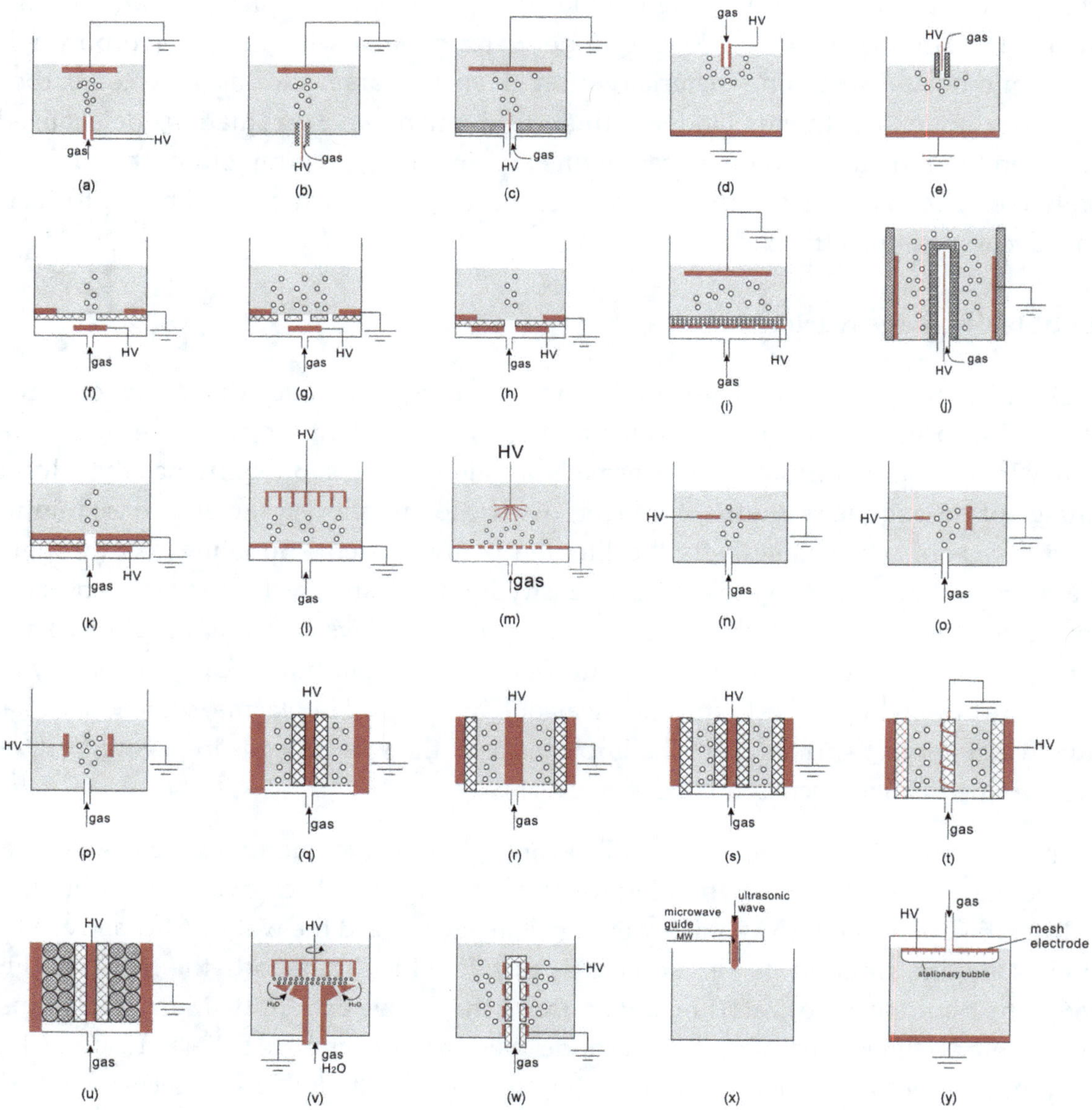

Figure 3. Types of bubble discharge reactors: (a) upward nozzle electrode-to-plate, (b) upward nozzle containing needle electrode-to-plate, (c) hole containing needle electrode-to-plate, (d) downward nozzle electrode-to-plate, (e) downward nozzle containing needle electrode-to-plate, (f–h) hole above electrode and inside circular ground electrode, (i) multibubble discharge on porous ceramic with high-voltage mesh, (j) multibubble discharge on porous ceramic tube surrounding a high-voltage wire electrode, (k) bubble discharge on hole in grounded electrode, (l–m) bubbles rising toward high-voltage electrode, (n–p) bubbles rising in between electrodes, (q) DBD reactor with inner barrier, (r) DBD reactor with outer barrier, (s) DBD reactor with double barrier, (t) DBD reactor with spiral electrode from Aoki et al. [84], (u) glass bead packed-bed DBD reactor, (v) coaxial arc discharge reactor with rotating multipin electrode from Johnson et al. [85], (w) multielectrode slipping surface discharge reactor, (x) microwave discharge in ultrasonic cavitation bubble, and (y) stationary bubble under high-voltage mesh.

porous ceramic seems a promising alternative [92]. Similarly, bubbling gas through a porous ceramic tube containing a high-voltage wire electrode [93] (Figure 3j) gives a large plasma–water contact surface as well. Increasing gas flow rate had no influence in this reactor on decomposition efficiency of phenol, while energy efficiency was enhanced for Acid Orange II removal.

Another situation is found when bubbles are formed on the grounded electrode, as shown by Yamatake et al. and Nikiforov [91, 94] (Figure 3k). Reactor from Figure 3h has more aggressive reaction with water and improved durability of electrode in comparison to the reactor from Figure 3k, as concluded by Yamatake et al. [91]. Nevertheless, one needs to take into account the applied voltage type. While for most of the reactors above positive pulsed corona discharge is used, less commonly AC voltage [94] or positive DC voltage [89, 91] is applied, leading to significantly different phenomena. As Yamatake has shown, plasma can be stably generated in the reactor from Figure 3h even without gas flow since O_2 gas is generated from electrolysis by application of positive DC voltage, while this is not the case for the reactor from Figure 3k. The microdischarge channel makes plasma more stable than that in the study of Kurahashi et al. [89], where the same phenomenon has been studied. Therefore, the system of Figure 3k requires higher gas flow due to the short lifetime of the oxygen radical, while in the reactor of Figure 3h decomposition did not depend on flow rate due to direct reaction with water.

Gas can also be bubbled from the ground electrode toward a multipin or brush high-voltage electrode, as shown by Chen et al. and Wang et al. [72, 95] (Figures 3l–m), or in between 2 sideways positioned electrodes, as shown by Lee et al., Miichi et al., and Vanraes et al. [96–98] (Figures 3n–p). The discharge systems are often very similar to the ones described in Section 3.1 without bubbles. For a plate-to-plate configuration, pulsed streamer discharge mostly occurred inside bubbles adjacent to the electrodes [97].

A last common type of bubble discharge reactor has a coaxial DBD geometry. The cylindrical dielectric barrier can either be placed at the inner [99] or at the outer electrode [91], or at both [91] (Figures 3q–s). For a single barrier, a reticulate electrode can be placed in contact with the water in order to enhance the local electric field [99]. With double barrier, unwanted erosion of the electrodes can be avoided. In the study of Yamatake et al. [91], a double-barrier reactor was found to have higher decomposition efficiency of acetic acid in comparison to a single barrier reactor with similar dimensions, which is most likely caused by a difference in energy density. Monopolar pulsed, bipolar pulsed, and AC high-voltage are most commonly applied to the inner electrode. In the study of Aoki et al. [84], however, the reactor from Figure 3t was used to generate RF glowlike plasma in bubbles. The grounded spiral electrode impeded the motion of the bubbles in the small gap between the electrodes, which increases the probability of discharge in the bubbles. Nevertheless, most of the input energy seems to be dissipated as heat in this system, resulting in low energy efficiency. Bubble movement can also be impeded by adding obstacles in the water bulk. For the single barrier reactor of [100] (Figure 3u), the addition of spherical glass beads significantly enhanced the energy efficiency of indigo carmine decomposition. Porous ceramic sphere gave worse performance than glass beads, but better one than inert conductive fragments. As an interesting research question, it is still unclear whether bubbles in contact with the electrodes in these DBD systems give rise to a better efficiency than freely rising bubbles or not.

A more exotic and patented reactor design is investigated by Johnson et al. [85] (Figure 3v). In this reactor, the high-voltage electrode is a pin array that can rotate at speeds up to 2500 rpm. Oxygen is pumped through the stationary electrode and nebulized to form a bubble mist between the electrodes. DC voltage is applied to the system, generating arc discharge. Rotating the electrode distributes erosion particles evenly on the stationary electrode, preventing pitting and unwanted changes in the relative distance between the electrodes. Moreover, it reduces mass transfer limitations that are apparent in pin-to-plate reactors. Additionally, it lowers the inception voltage by abating the effective distance between the pin electrodes and the stationary electrode. Energy efficiency for methyl tert-butyl ether decomposition increased with an increase in spin rate. The system was found to be more energy efficient than many corona-based technologies but still requires further optimization.

Anpilov et al. developed the multielectrode slipping surface spark discharge system depicted in Figure 3w [101]. A series of cylindrical electrodes are mounted on the outside of a dielectric tube. One of the extreme electrodes is grounded, and high-voltage pulses are applied to the other extreme electrode. All other electrodes are on floating potential. To increase system efficiency, the outer surface of the electrodes is coated with a thin insulating layer. Gas is pumped into the electrode gaps through drilled holes in the dielectric tube. When a high-voltage pulse is applied, plasma discharge occurs initially in the first interelectrode gap adjoining the high-voltage electrode. The breakdown of this gap quickly transports the high-voltage potential to the next electrode, which leads to the breakdown of the second gap. The process repeats itself until the grounded electrode has been reached. This way, the discharge load on each electrode can be kept low, enhancing erosion resistance and increasing the system lifetime. The effectiveness of the system has been demonstrated for disinfection of biological wastewater, methane conversion, and aqueous organic waste decomposition [102].

Apart from bubble cavitation by electrical discharge, the classical gas pumping method, and gas formation by electrolysis, bubbles can also be generated with ultrasonic cavitation. Since microwave power cannot easily generate cavitation bubbles, a combination of ultrasound and microwave discharge can be an attractive method. An example is given by Horikoshi et al. [103] (Figure 3x). Stationary bubbles, however, are not commonly used in bubble discharge reactors, unless for diagnostic purposes. An example is given by Yamabe et al. [104], with the reactor of Figure 3y. This reactor was used to investigate plasma formation and propagation along the gas–water interface, which is a common feature for bubble discharge reactors and which makes it, therefore, different in nature than most gas phase discharge reactors.

3.3. Gas phase discharge reactors

Electrical discharge in the gas phase is usually more energy efficient for organic degradation than discharge in the liquid phase [42, 43, 50]. In this section, we will distinguish 4 subgroups of gas phase discharge reactors: corona and glow discharge over a horizontal water surface (Figure 4), DBD over a horizontal water surface (Figure 5), falling water film reactors (Figure 6), and arc discharge over a water surface (Figure 7).

3.3.1. *Corona and glow discharge over water surface*

The most standard version of a discharge over water surface has a pin-to-water configuration with a grounded water electrode, as depicted in Figure 4a. The type of discharge produced in this reactor, corona, glow, or transient glow-to-spark, depends on the applied voltage, pin curvature, interelectrode distance, and voltage polarity [105, 106]. For the application of water treatment, both positive and negative DC and monopolar pulsed voltage have been reported. AC input power is less common but has been used as well [107]. Plasma volume can be increased by replacing the high-voltage pin electrode with a multipin [108], a brush [72], or a horizontal wire [109] (Figures 4b–d). In the study of Miyazaki et al. [108] with a multipin electrode, the energy efficiency for a certain amount of phenol decomposition was independent of the type of discharge, the voltage amplitude, the polarity of the applied voltage, and the amount of pin electrodes.

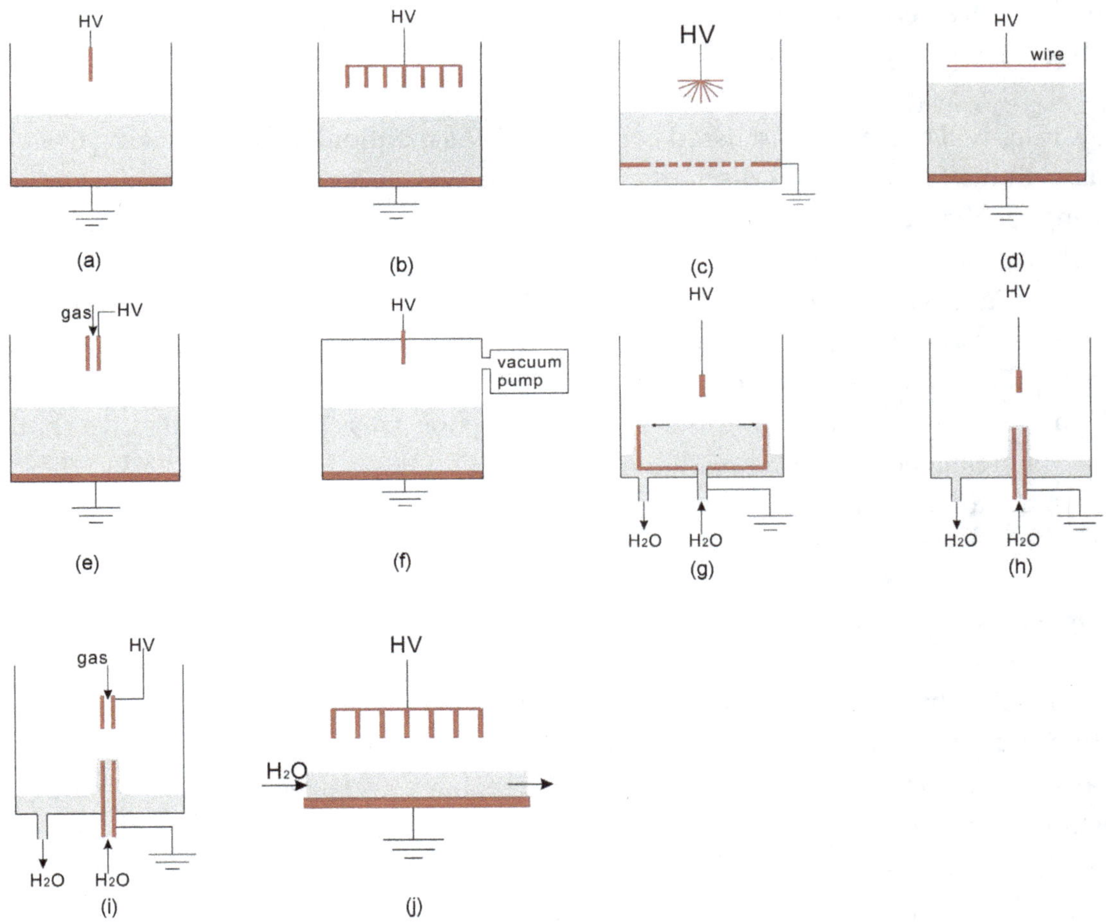

Figure 4. Types of reactors with corona or glow discharge over water surface: (a) pin-to-water, (b) multi-pin-to-water, (c) brush-to-water, (d) wire-to-water, (e) nozzle electrode-to-water, (f) low-pressure pin-to-water glow discharge reactor, (g) pin above radial water flow, (h) pin above flowing liquid electrode, (i) miniature microjet above flowing liquid electrode, and (j) multipin above water flow.

According to Dors et al., atmospheric pressure glow discharge in air produces gaseous nitrogen oxides, leading to formation of undesirable aqueous nitrates and nitrites, while DC positive corona produces ozone in air without any traces of nitrogen oxides [110]. The energy efficiency of phenol oxidation in their system, depicted in Figure 4e, was comparable to the results obtained in pulsed corona discharge systems. Sharma et al., however, compared their low-pressure negative DC glow discharge reactor (Figure 4f) with bench scale data of atmospheric pressure corona discharge and concluded that the power cost for pentachlorophenol decomposition was lower for their system [111]. Additionally, the operating cost of their reactor was found to be comparable with power cost of UV-based advanced oxidation technologies. Yet it is unclear how feasible such low-pressure system is for applications with large water volume or continuous water flow.

It is important to note that dimensions and movement of the water phase in this type of reactors can influence energy efficiency significantly. In the system of Sharma et al., for example, stirring rate increased the rate of pentachlorophenol removal [111]. Water movement also plays an important role in radial flow reactors (Figure 4g). In the reactors from Jamróz et al. [112] with small sized flowing liquid cathode (Figures 4h–i), degradation efficiency depended strongly on the water flow rate, while the Ar flow rate from the miniature flow Ar microjet (Figure 4i) gave negligible effect on methyl red decomposition. The importance of water and gas flow rates will be further discussed in Section 4.3. As mentioned above, making the solution flow as a thin film along the discharge is another way to enhance the oxidation process. Promising results from a pilot-scale system with negative pulsed corona from multiple carbon fiber cathodes above a flowing water film (Figure 4j) have been published by Even-Ezra et al., Gerrity et al., and Mizrahi and Litaor [113–115]. The system was similar or more efficient than a pilot-scale UV/H_2O_2 advanced oxidation process and achieved similar energy efficiency to those reported in the literature for other advanced oxidation processes [113, 114]. Moreover, the plasma pilot system with additional ozone injection was more cost-effective than three other commercialized advanced oxidation systems (O_3/H_2O_2, O_3/UV, and O_3/H_2O_2/UV) [115]. However, the total capital costs and reliability of large-scale gas discharge reactors for water treatment are still relatively unclear.

3.3.2. DBD over water surface

DBD over horizontal water surface is most commonly powered with AC voltage, but occasionally pulsed high voltage has been used as well. Often, glass is used as dielectric barrier, especially quartz glass, while Al_2O_3 ceramic barriers have also been reported less frequently. Interestingly, reactors with DBD over water often have energy efficiency for organic decomposition that increases with input power [116–118]. The most standard reactor design for DBD in the gas phase over a water surface is shown in Figure 5a. In the study of Hu et al. [116], energy efficiency in such reactor was found to increase by decreasing the distance between the dielectric barrier electrode and the water surface. This can be explained with a decrease in plasma volume and thus in unused plasma reactions far from the water surface. A water batch can also be placed in between two dielectric barriers to avoid erosion of one of the electrodes, as in the study of Hijosa-Valsero et al. [119] (Figure 5b). A more exotic way of bringing DBD over a water surface is shown in Figure 5c, where water is kept on floating potential. However, this device has not been applied for organic degradation and is instead used for biomedical applications [120]. In another less common design, the water can be located above the dielectric barrier, with an uncovered high-voltage electrode positioned above the water surface, as in the wire-to-water reactor of Marotta et al. [121] (Figure 5d). In this reactor, the energy efficiency of phenol decomposition was 3.2 times higher with stainless steel wires as compared to Ni/Cr wires. However, it is unclear whether this effect should be attributed to the larger diameter of the Ni/Cr wires or their possible inhibiting effect on ozone or other oxidants. The material of the ground electrode in contact with water in many DBD reactors is, on the other hand, clearly important. In the study of Lesage et al. [122], the use of a stainless steel substrate resulted in a better decomposition efficiency in comparison to the use of brass. This is explained with corrosion of the brass substrate under influence of nitrate, leading to formation of aqueous nitrite, which scavenges OH radicals and thus inhibits the degradation process. Stainless steel, as a more inert metal, does not have this effect.

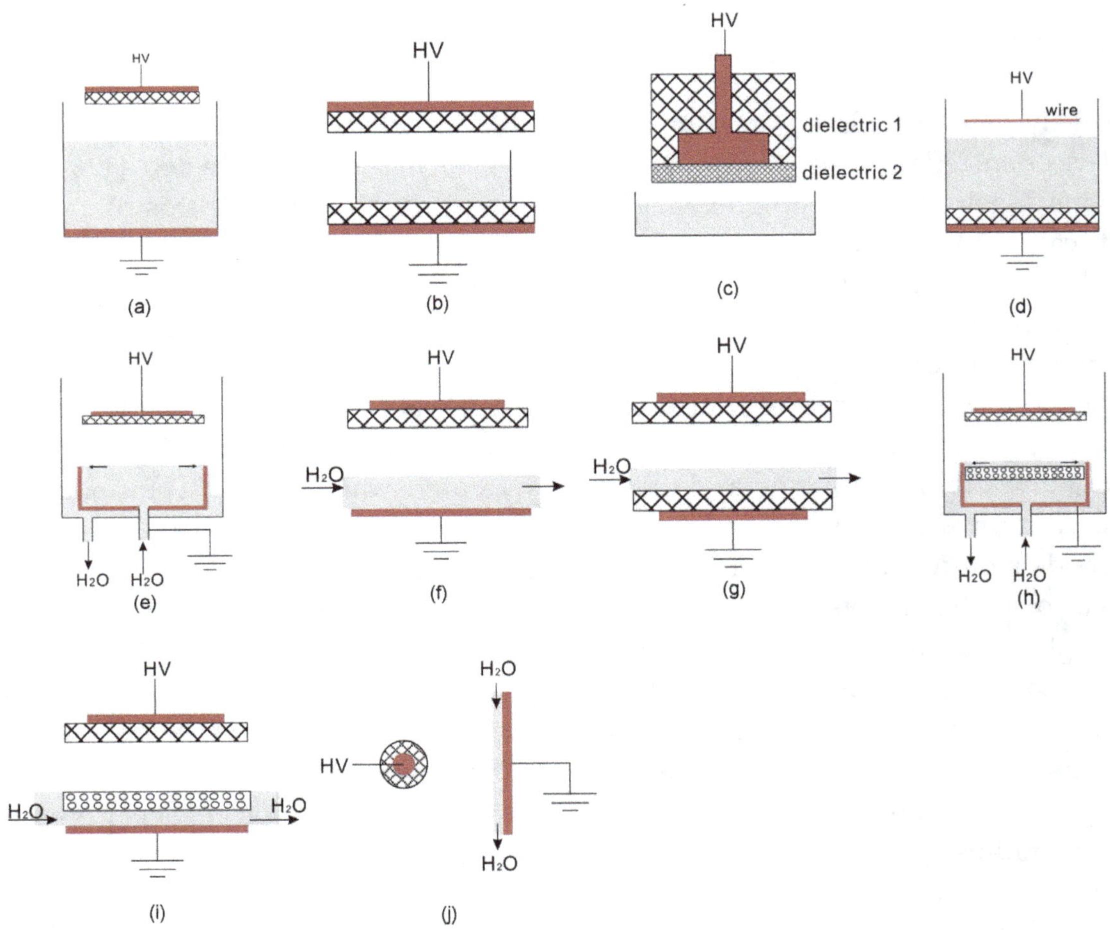

Figure 5. Types of reactors with DBD over water surface: (a) DBD-to-water, (b) water batch in between DBD, (c) DBD above floating electrode, (d) wire-to-water DBD, (e) DBD above radial water flow, (f) DBD above water flow, (g) water flow in between DBD, (h) DBD over radial water flow on porous ceramic, (i) DBD over water flow on porous ceramic, and (j) DBD rod to falling water film.

Also for DBD-based systems, the movement of water influences the degradation efficiency. Reactors with radial flow [117] or flowing water films [118, 123] are investigated in literature for organic decomposition (Figures 5e–g). Both situations have also been reported with incorporation of a porous ceramic segment in the zone between electrodes (Figures 5h–i) [124, 125]. This porous segment serves as guide for the flowing water. It allows the water to remain undisturbed by the electrical discharge due to hydrophilic force. This enables a reduction of the discharge gap and subsequently an increase in the intensity, stability, homogeneity, and efficiency of the discharge. Under such a configuration, a transition from filamentary mode to semi-homogeneous mode of the plasma discharge can be realized [125]. Moreover, such ceramic can be effectively used as substrate for photocatalysts, where both substrate and catalyst remain unchanged after use [124].

An exotic reactor type where a rod high-voltage electrode with dielectric cover is placed next to a falling water film has been investigated by Lesage et al. [122, 123] (Figure 5j). The reactor was found to be significantly more efficient than a gliding arc (see Section 3.3.4) over the same falling water film, partly due to less corrosion of the brass substrate in contact with the water.

3.3.3. *Coaxial reactors with falling water film*

A relatively common falling water film reactor that does not use DBD is shown in Figures 6a–b and is often referred to with the term wetted-wall reactor. Either a rod [126] or more commonly a wire [127–130] high-voltage electrode is placed along the axis of a grounded cylindrical electrode. The falling water film flows along the inner wall of the cylinder electrode where it comes in contact with streamer or corona discharge. Mostly, positive pulsed power is applied on the inner electrode, but also negative DC [127] has been reported. Usually, corona discharge is formed in such reactors for all voltage waveforms, while spark discharge is undesired due to excessive energy dissipation to Joule heating. To prevent spark formation, it is necessary to have the entire inner wall area covered by the water flow [130]. The choice of gas flow direction is important, as concluded by Faungnawakij et al. [131] for negative DC corona. Experiments showed a downward airflow to be more effective than upward airflow for acetaldehyde degradation. In the study of Sano et al. [127], the energy efficiency of a wetted-wall reactor with negative DC voltage applied to the inner wire was calculated to be 3 to 4 times higher than in a wire-to-water corona reactor (Figure 4d) over flowing water with negative DC voltage. For the same reactor, energy efficiency was found to be highest for conditions of a smooth water surface, i.e., for a minimal water flow rate where the flow entirely covers the anode inner wall and for an optimal current, which does not disturb the flow by strong ion wind. There is an optimal wall radius, where decomposition efficiency is maximal. For higher radius, many plasma-generated short-lived radicals cannot reach the water film in time, while for smaller radius, the plasma–water contact surface decreases [127]. Sato et al. compared four kinds of coaxial reactors with falling water film and positive pulsed power. They found phenol most energy efficiently removed with the configuration of Figure 6b [128]. Sealing such reactor seems beneficial for energy efficiency, due to better confinement of the produced ozone [129].

Most falling water film reactors generate plasma by DBD, either with AC or monopolar pulsed high voltage. Pulsed DBD in coaxial configuration using O_2 is considered as one of the most efficient electrical discharge systems evaluated because of the large surface area and small electrode distance [50]. Several configurations are possible, but in the most common design, the water film flows over the surface of an inner stainless steel rod electrode placed inside a glass cylindrical vessel which acts as dielectric barrier. The outer electrode can be a metal mesh or a metal painted layer which is located around the vessel. Four versions of this reactor design

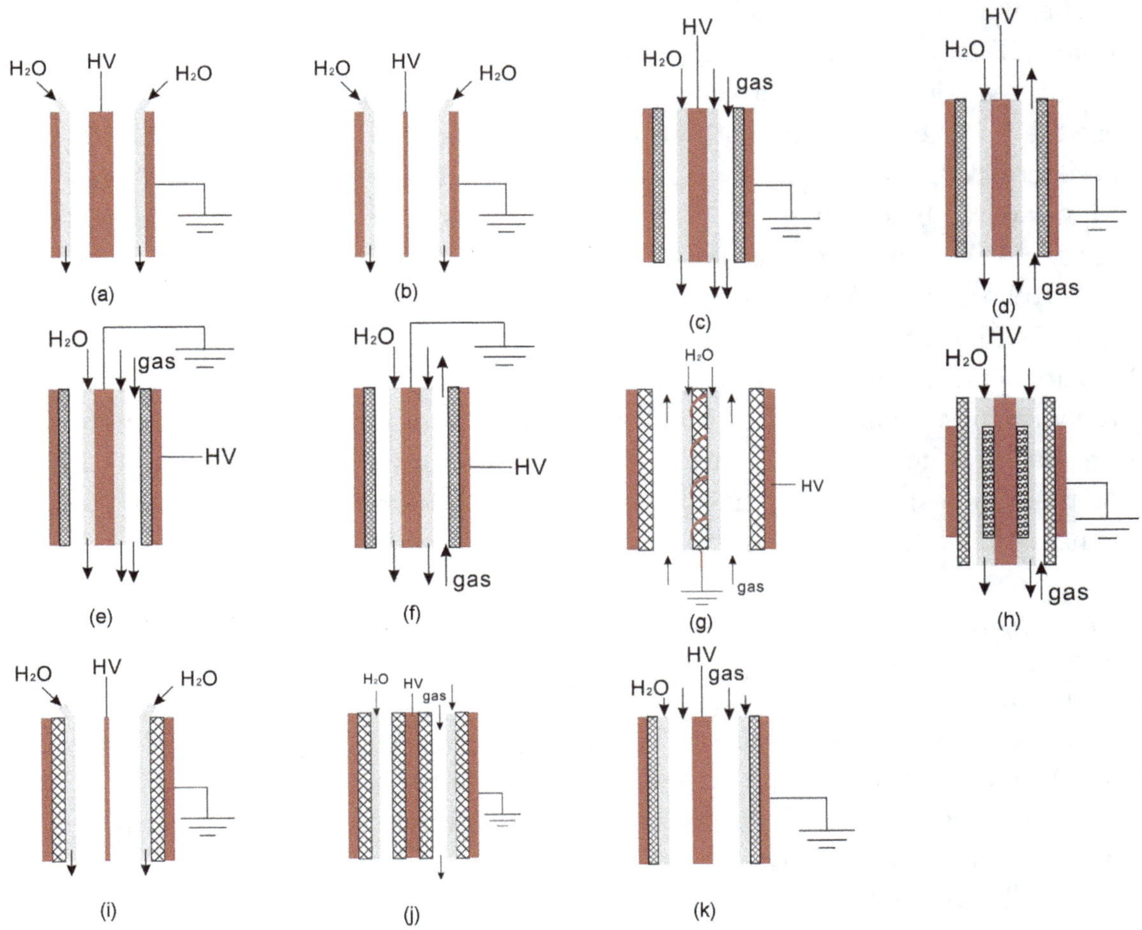

Figure 6. Types of coaxial reactors with falling water film: (a) wetted-wall reactor with rod electrode, (b) wetted-wall reactor with wire electrode, (c–f) 4 variations of falling water film DBD reactor with outer barrier, (g) falling water film DBD reactor with inner spiral electrode, (h) falling water film on glass fiber fabric in DBD reactor with outer barrier, (i) wetted-wall DBD reactor with wire electrode, (j) wetted-wall DBD reactor with double barrier, and (k) configuration of coaxial whirlpool reactor of [132].

are found in literature, where the inner electrode is either grounded or connected to the high voltage and with an upward or downward gas flow (Figures 6c–f). With the reactor of Figure 6f, energy efficiencies of micropollutant decomposition are about one order of magnitude higher than for a water batch in between a DBD reactor (Figure 5b) [119]. A slightly different configuration was used by Ognier et al. [133], where a tungsten wire that was rolled around a dielectric rod served as inner grounded electrode (Figure 6g). In this study, volatile aqueous compounds were treated. The more volatile the compound was, as expressed with the Henry's law constant, the more efficiently it was removed. Therefore, degradation processes of pollutants in the gas phase should be considered in plasma reactors, depending on the volatility of the compound. In the study of Bubnov et al. [134], the inner electrode of a coaxial DBD reactor was covered with a 1-mm-thick porous hydrophilic glass-fiber fabric (Figure 6h). This fabric allows a more homogeneous water flow and higher water retention time. Moreover, it can function as substrate for catalysts, such as Cu and Ni compounds, which enhanced the decomposition efficiency in the research of Bubnov et al. [134].

Less frequently, the water film is chosen to flow along the inner wall of the dielectric barrier. Morimoto et al., for example, investigated the effect of placing a dielectric barrier inside the wetted-wall reactor of Figure 6b, as shown in Figure 6i. Addition of the barrier allows to decrease the interelectrode gap without formation of spark discharge, which is expected to increase energy efficiency. With the application of positive nanosecond pulsed high voltage on the inner wire, the treatment efficiency of the DBD system was found to be, surprisingly, less energy efficient for indigo carmine decomposition as compared to the normal wetted-wall reactor. Another type of wetted-wall DBD reactor with falling water film is investigated by Rong et al. [135], for a reactor with double dielectric barrier (Figure 6j). A more exotic type of DBD reactor with modified water-gas mixing is reported by Chen et al. [132]. The system is powered with high frequency bipolar tailored voltage pulses. It has a configuration as shown in Figure 6k, but water and air are introduced in the reactor with high flow rate of 5 L/min and 100 to 200 L/min, respectively, causing a whirlpool. For this reactor, gas flow rate was shown to have negligible effect on decomposition efficiency of methyl orange.

3.3.4. Arc discharge over water surface

Gliding arc discharge above a water surface (Figure 7a) is a popular approach for water treatment with plasma. In this reactor type, two diverging electrodes are placed above a water solution. An electric arc forms at the shortest electrode gap and glides along the electrode's axis under influence of a gas flow directed toward the water surface. The arc length increases on moving and its temperature decreases, turning the arc from thermal plasma into quenched plasma while breaking into a plume. A new arc then forms at the narrowest gap and the cycle continues. Unfortunately, many publications on this type of reactor are unclear about the use of AC or DC voltage, but AC power is definitely a common choice. Important research in this field has focused on enlarging the plasma treated water surface with adjustments in design. One possibility is to use a couple of controlled electrodes in between the electrode gap to facilitate breakdown, increasing current intensity and allowing a larger interelectrode distance [136]. Another option is to use three main electrodes supplied by two power sources, as proposed by Burlica et al. [137] (Figure 7b). Both approaches have shown to increase reactor efficiency. Gliding arc discharge can also be used for the treatment of falling water films, as shown by Lesage et al. [123] (Figure 7c). Arc discharge with an active water electrode is less commonly researched (Figure 7d). According to Janca et al. [138], the energy efficiency of such system was found to be strongly dependent on the type of discharge produced, such as arc or gliding arc. For more detailed information on water treatment by means of gliding arc, the reader is referred to the review by Brisset et al. [136].

3.4. Spray discharge reactors

3.4.1. Low-energy spray discharge reactors

Although spray discharge reactors have received more attention in recent years due to their high reported energy efficiencies [42–44], still more research is required to characterize and optimize them. One of the most common spray reactors has a wire-to-cylinder geometry

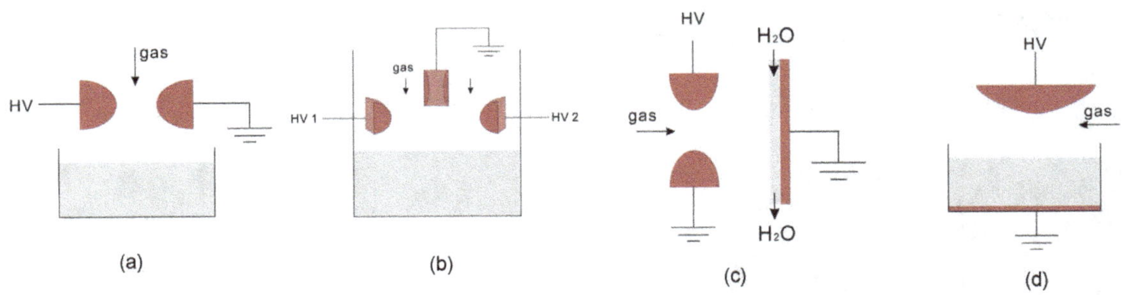

Figure 7. Types of gliding arc discharge reactors over water surface: (a) standard configuration, (b) configuration with extra main electrode, (c) gliding arc discharge to falling water film, and (d) gliding arc discharge with active water electrode.

(Figure 8a), similar to wetted-wall reactors. This reactor type is always operating with positive pulsed corona or streamer discharge, according to our literature review. In the study of Kobayashi et al. [139], different spraying nozzles were used to investigate the influence of water location on the energy efficiency of indigo carmine decomposition. As the results showed, spraying water as droplets into the discharge area is more effective than making it flow as a water film on the inner reactor wall. Moreover, droplets that were sprayed close to the reactor wall underwent 1.5 times faster decolorization than droplets near the wire electrode. Energy efficiency was found to be independent of droplet size for same water flow rate [140].

Sugai et al. adjusted this reactor by addition of packed-bed of pellets (Figure 8b) or fluorocarbon wires (Figure 8c) in order to increase the droplet retention time in the discharge space [141]. The packed pellets were hollow polyethylene balls with 14 holes per ball to increase discharge. The fluorocarbon wires were woven as insulation grids in the outer cylindrical electrode. Addition of the pellets decreased the energy efficiency significantly, due to the narrowing of the discharge space. The fluorocarbon wires, however, kept the discharge space unaltered and increase the energy efficiency with 2–10%.

In the study of An et al. [142], a similar electrode configuration was used as in Figure 8a, but with tooth wheels assembled on the inner electrode (Figure 8d). Here, droplets were not created by spraying, but due to condensation of steam under influence of up-flowing air. Alternatively, positive pulsed corona can also be generated in a spray reactor around wire anodes in parallel with two grounded plate cathodes [143], as depicted in Figure 8e. In another approach, water is sprayed in between rod electrodes that are each surrounded with a dielectric barrier (Figure 8f). Monopolar pulsed voltage of both polarities is reported in literature [144, 145]. In the study of Wang et al. [146], grounded water was sprayed from dielectric nozzles in proximity of a high-voltage dielectric barrier plate electrode (Figure 8g). When AC high voltage was applied to the plate, the by electrostatic induction generated electrostatic force pulled the water droplets to the glass dielectric layer. The energy efficiency of indigo carmine decomposition depended on both voltage amplitude and air gap distance, for which optimal values were found. Another interesting reactor type is based on the electrospray process to simultaneously generate and treat water spray under influence of high voltage, as shown in Figure

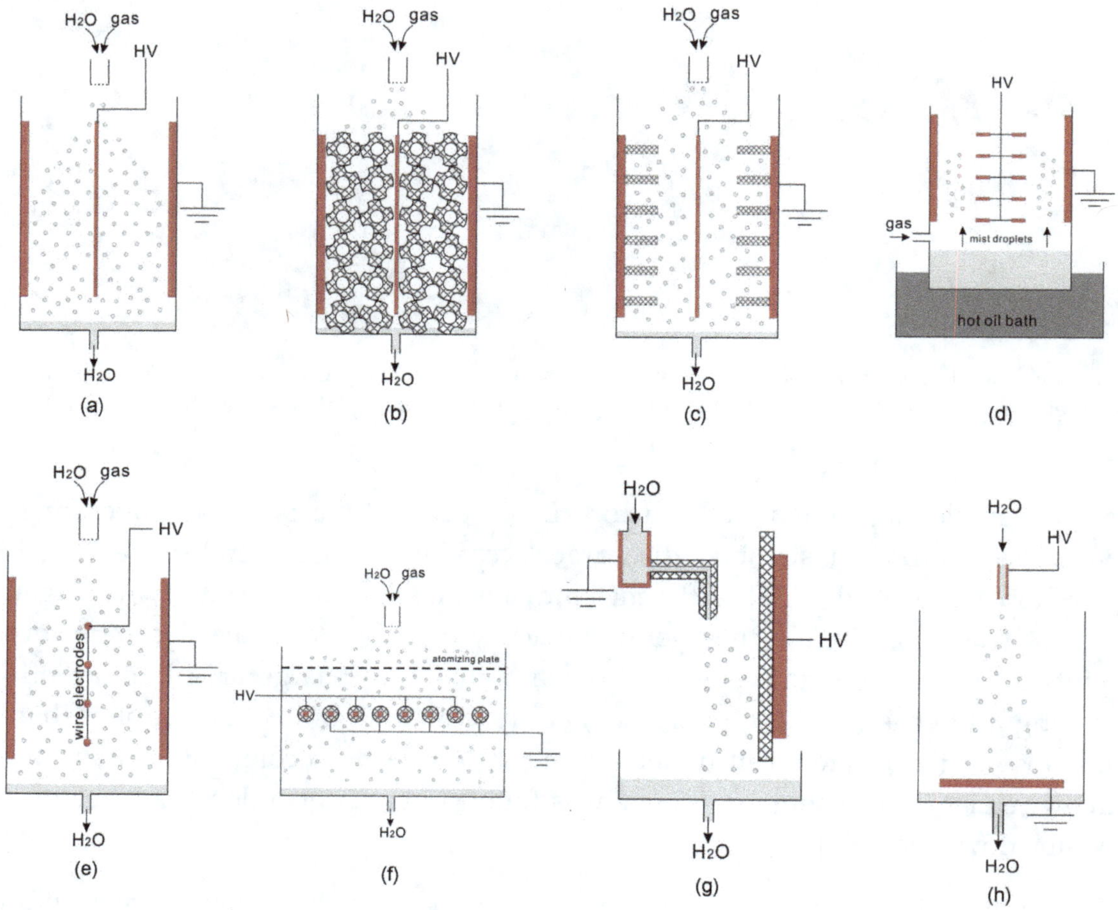

Figure 8. Types of low-energy spray discharge reactors: (a–c) wire-to-cylinder corona or streamer reactor without or with packed-bed of pellets or fluorocarbon wires for increased droplet retention, (d) mist droplets in tooth wheel-to-cylinder corona reactor, (e) multi-wire-to-plate corona reactor, (f) spray through DBD rods, (g) spray from grounded electrode to DBD plate, and (h) electrospray reactor.

8h. In the study of Elsawah et al. [147], water is treated that way with positive pulsed corona electrospray.

3.4.2. *Spray arc reactors*

Water spray can also be introduced in gliding arc discharge for treatment, as depicted in Figure 9a. This method is found to be more energy efficient than gliding arc discharge over a water surface as discussed in Section 3.3.4 [136, 137]. The decomposition of 4-chlorophenol in such spray reactor became more energy efficient with increasing gas–water mixing rate [148]. Efficiency is higher with electrodes from stainless steel than for aluminum or brass electrodes. Also here, extending the plasma volume by the use of a controlled electrode couple or an extra main electrode enhances energy efficiency (Figure 9b) [137]. The process can also be optimized by introducing the water with a flat spraying nozzle perpendicular to the gas flow to improve the contact with the plasma (Figure 9c) [149]. For more detailed information on water treatment by means of gliding arc spray reactors, the reader is referred to the review by Brisset et al. [136]. Water can be treated as well with a DC plasma torch, where it is usually directly introduced into the torch as plasma forming gas. A discussion on this treatment method is provided by Brisset et al. [44].

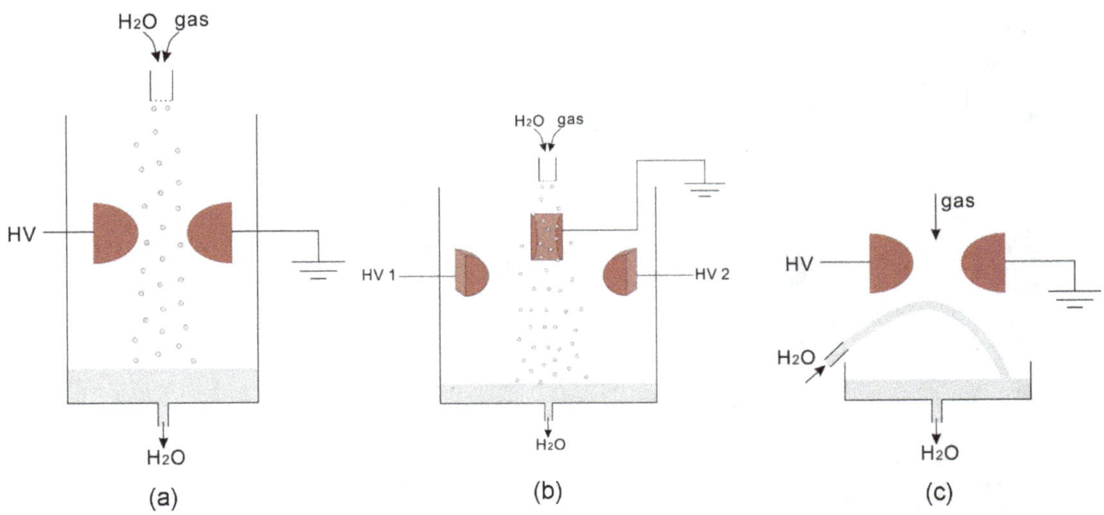

Figure 9. Types of spray gliding arc reactors: (a) standard configuration, (b) configuration with extra main electrode, and (c) water jet under angle through gliding arc discharge.

3.5. Hybrid reactors

Electrohydraulic discharge and gas phase discharge can simultaneously be generated when a high-voltage electrode is placed in the water phase with a grounded electrode above the water surface in the gas phase. Mostly, positive pulsed corona is generated in these systems reported in literature. Several electrode configurations are possible, such as an underwater pin to plate in gas [150] (Figure 10a) and an underwater pin to multipin in gas [151] (Figure 10b). In some reactors, a second high-voltage electrode is placed in the gas phase, either powered by the same high-voltage source [151] (Figure 10c) or by a second one [152] (Figure 10d). Also here, energy efficiency can be enhanced by discharge formation in externally applied bubbles, leading to hybrid reactors that combine bubble discharge with gas phase discharge. One example is given by Ren et al. [153], where bubbles are formed on high-voltage nozzle electrodes located underneath a plate electrode in the gas phase (Figure 10e). Hybrid reactors with discharge in both water and gas phase are sometimes proposed for the treatment of gaseous and aqueous pollutants simultaneously, as in the case of volatile pollutants [44, 46]. Their energy efficiency for organic decomposition is moderately higher than the one of hydraulic discharge reactors [42, 43].

Another type of hybrid reactor combines the treatment of a falling water film and droplets. This situation naturally occurs by spraying water from a shower nozzle at an angle. In the study of Kobayashi et al. [139], a wetted-wall hybrid reactor with inner wire anode was used, where water solutions from the falling film and from the droplets were collected separately (Figure 10f). Indigo carmine was decomposed 0.57 times faster in the droplets than in the water

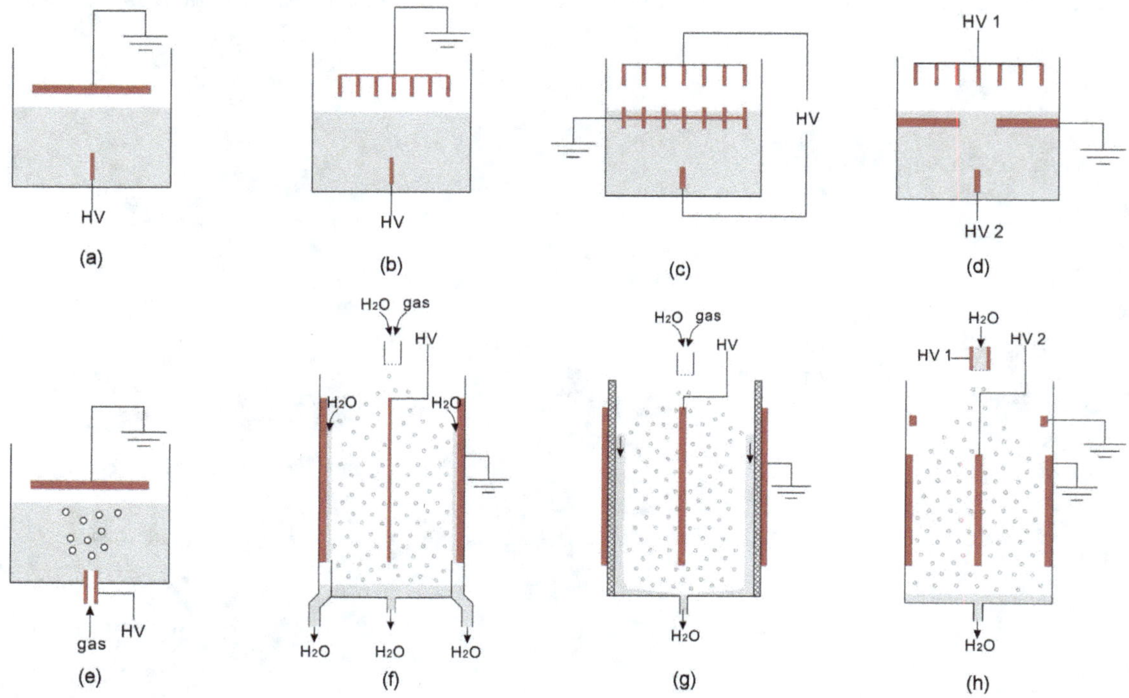

Figure 10. Types of hybrid reactors: (a, b) underwater high-voltage pin to plate or multipin in gas, (c) reactor from Lukes et al. [151], (d) reactor from Lukes et al. [152], (e) bubble discharge on underwater nozzle electrode to plate in gas, (f) wetted-wall spray corona reactor with inner wire electrode, (g) wetted-wall spray DBD reactor with outer barrier, and (h) electrospray through wire-to-mesh corona reactor.

film, as could be expected from the higher energy efficiency of spray discharge reactors as compared to falling water film reactors. In the study of Nakagawa et al. [154], the water film was formed on the dielectric barrier which separated the inner rod anode and the grounded mesh surrounding the barrier (Figure 10g). For low water flow rate, the energy efficiency of rhodamine B decomposition was similar for 3 of such reactors with different inner diameter and barrier thickness. For higher flow rates, however, one reactor performed significantly better. This reactor had larger inner diameter and equal barrier thickness compared to one of the other reactors.

A special case is found when droplets from an electrospray are treated a second time with plasma discharge. In the study of Njatawidjaja et al. [155], electrostatically atomized droplets passed through pulsed corona discharge in between a wire-to-mesh electrode configuration (Figure 10h). In both parts of the reactor, positive polarity performed better than negative one for the decomposition of Chicago sky blue dye. In the electrospray, positive DC voltage produced a larger number of finer droplets with a wider spray angle than in the case of negative voltage.

3.6. Remote discharge reactors

The concept of remote discharge reactors for water treatment is not new. An early example is ozonation, where ozone is generated by means of plasma discharge in gas phase and subsequently transported toward the solution under treatment. More recently, electrical discharge reactors have been developed where remotely generated plasma gas is bubbled through the solution. As main difference, the plasma gas does not only contain ozone, but also other reactive species, such as H_2O_2 and OH. In the study of Tang et al. [156], this is accomplished by using humid air as feed gas of a DBD gas phase reactor (Figure 11a). With corona discharge in dry air, positive and negative ions can be generated in addition to ozone for water treatment, as in the air ionization device reported by Wohlers et al. [157]. Yamatake et al. compared the bubble discharge reactor from Figure 3k with ground electrode in contact with water to a reactor where plasma is generated separately from the water and subsequently bubbled through the solution (Figure 11b) [158]. Both reactors had identical electrode configuration, used oxygen as feed gas, and were powered by the same positive DC voltage source. Acetic acid was not decomposed in the remote discharge reactor, while decomposition was significant in the bubble discharge reactor. This difference was explained with the production of oxygen radicals, which acted as main oxidant during bubble discharge but had too short lifetime to reach the water in the remote discharge reactor.

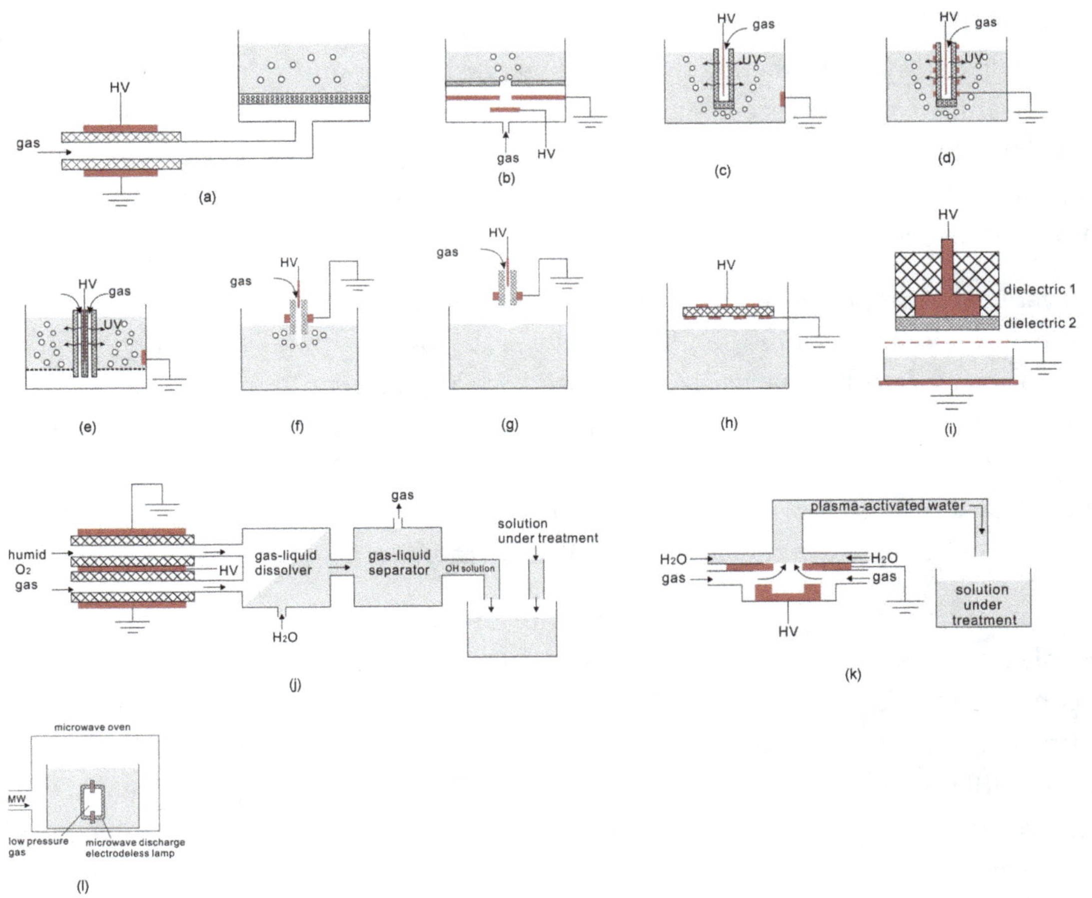

Figure 11. Types of remote discharge reactors: (a, b) plasma gas bubbling reactors, (c–e) plasma gas bubbling reactors with UV irradiation through quartz barrier, (f) underwater DBD plasma jet, (g) DBD plasma jet over water surface, (h) metal strips on quartz disc for DBD above water surface, (i) setup from Dobrynin et al. [159] with removable grounded mesh, (j) reactor from Zhang et al. [160] for production of hydroxyl radical solution, (k) gliding arc reactor from Kim et al. [161] for plasma-activated water production, and (l) reactor with tungsten-triggered microwave discharge electrodeless lamp.

Plasma gas bubbling can be combined with UV irradiation by generating plasma inside a quartz tube submerged in the solution. Several variations are reported in literature, where the high-voltage electrode inside the tube is a screw [162], a spiral wire [163], a rod covered by dielectric barrier [164], or a glass tube filled with NaCl solution [165]. A metal mesh surrounding the tube [162] or a metal rod in contact with the water [165] serves as ground electrode (Figures 11c–d). DBD plasma is usually generated inside the tube with AC power. The working gas is pumped in the tube and subsequently bubbled through the solution by means of a gas diffuser or a series of air distribution needles (Figure 11e) [164]. Comparison with other oxidation techniques has indicated that such reactors may be competitive technology to other plasma systems such as the hybrid reactor by Nakagawa et al. [154] (Figure 10g) and to photocatalytic oxidation [165, 166].

A system closely related to these plasma gas bubbling reactors is a submerged DBD plasma jet for water treatment. In the study of Foster et al. [167], a nanosecond pulsed DBD plasma jet is investigated for oxidation of aqueous organic pollutants (Figure 11f). Its decomposition efficiency was higher than the one reported in literature for glow discharge and pulsed corona discharge [167]. The energy efficiency of methylene blue decomposition significantly dropped with increasing treatment time. As the results suggest, using multiple plasma jets powered in parallel can improve the process significantly. The plasma jet can also be placed above the water surface (Figure 11g), as is done for the production of plasma-activated water by Ma et al. [168]. Nevertheless, application of non-thermal plasma jets for water treatment is relatively uncommon for now.

A few systems are reported in literature where plasma is generated above a water surface without direct contact with the solution. In the study of Olszewski et al. [169], copper strips adhered to both sides of a quartz disc were used as electrode configuration for methyl orange decomposition with AC DBD plasma (Figure 11h). In the study of Dobrynin et al. [159], inactivation of spores was investigated by means of the DBD electrode system of Figure 5c above a removable mesh for reference experiments (Figure 11i). Experiments with and without the mesh were compared to reveal the role of UV irradiation.

Recently, a new approach is gaining popularity in which water treated with plasma, often termed plasma-activated water, is added to the solution under treatment. Plasma-activated water contains several long-living oxidants that are able to inactivate biological organisms. According to Zhang et al., long-living aqueous hydroxyl radicals were produced in their setup by mixing DBD-treated humid O_2 gas with water (Figure 11j) [160]. This hydroxyl radical solution was sprayed in a sea enclosure to effectively inactivate red tide organisms. Plasma-activated water can be produced by means of any of the reactors mentioned above, but gliding arc discharge is most commonly used. Figure 11k depicts the gliding arc reactor used by Kim et al. [161] for plasma-activated water generation utilizing a vortex flow with two circular disk electrodes. Introducing the produced water–air mixture in the solution under treatment through microbubble generators significantly enhanced the process. Up to now, plasma-activated water is used for disinfection of water, while its effects on organic contamination are still largely unknown.

In principle, plasma technology is also frequently used in water treatment technology solely as UV source in plasma lamps or excimer lamps. An upcoming new technology is the microwave discharge electrodeless lamp, which self-ignites under influence of microwave power. As an example, Figure 111 schematically shows a tungsten-triggered microwave discharge electrodeless lamp proposed by Horikoshi et al. [170] for low microwave power levels. The tungsten wire was embedded in a synthetic quartz tube attached to the lamp system to act as a trigger. In that manner, the high microwave power usually required for self-ignition in aqueous medium is avoided.

4. Influence of working parameters on energy efficiency

In addition to reactor design and materials, there are several other factors that influence reactor energy efficiency, which need to be considered for reactor optimization. Roughly, these factors can be split into two groups: working parameters determining reactor operation and solution parameters. Since solution parameters, such as water temperature, pH, conductivity, and water matrix, are hard to control, especially for large volume of influent water, they are not discussed in this chapter. For a discussion on their influence, the reader is referred to [44]. In contrast, operational parameters such as applied voltage characteristics, working gas, and flow rates of gas and solution are easier to adjust. Therefore, they deserve additional attention for further reactor optimization. In this section, we will shortly review the importance of voltage waveform, working gas, and gas and water flow rates. Their influence on reactor efficiency for organic decomposition will be illustrated with examples from literature.

For a given reactor, energy efficiency depends on voltage-related parameters. In the pulsed bubble discharge reactor of [92] and the positive pulsed streamer discharge in wetted-wall reactor of [129], energy efficiency increased for rising voltage amplitude. In a negative pulsed DBD falling water film reactor [171] and a positive pulsed corona electrospray reactor [147], however, increasing voltage amplitude reduced the energy efficiency. In the gas phase DBD reactor of [169], the interruption period of pulse-modulated AC voltage had no effect on methyl orange degradation, while lowering its duty cycle from 100% to 25% increased energy efficiency 2.11 times. The authors explained the latter effect with additional dye degradation during plasma off time under influence of long living reactive species such as O_3 and H_2O_2. Sinusoidal voltage frequency is an important parameter, as it can lead to different plasma phenomena, which explains the distinction of AC, radio frequency, and microwave discharge. Nevertheless, very limited information is available in literature on the dependence of the energy efficiency of the sinusoidal frequency for a given reactor design. In the study of Lesage et al. [122], no change in energy efficiency was observed for 4-chlorobenzoic acid decomposition with voltage frequency increase from 500 Hz to 2000 Hz for AC powered DBD over moving water film. Increasing AC frequency from 1.5 kHz to 15.6 kHz kept the energy efficiency of a coaxial falling water film DBD reactor in the same order of magnitude as well, in spite of the additional heating that resulted from the higher frequency [172]. In the case of pulsed discharge, pulse properties such as rise time and width are expected to be important. For pulsed positive corona discharge in humid O_2/N_2 atmosphere, the energy efficiency of radical and excited species production increases with decreasing pulse width [173]. This is in agreement with the higher efficiency of indigo carmine decomposition for shorter pulse width observed by Sugai et al. [174] for a positive pulsed streamer discharge spray reactor. A faster pulse rise rate generated thicker streamers and a higher energy efficiency in the same reactor [175]. Positive pulsed arc electrohydraulic discharge was reported to have increased the energy efficiency of sulfadimethoxine when pulse duration was brought back from 100 μs to 20 μs [51]. Remarkably, also pulse frequency can influence reactor efficiency. In a bubble reactor with positive pulsed corona, the energy efficiency of 2,4-DCP degradation increased with increasing pulse frequency [176]. In contrast, the energy efficiency of indigo carmine reduction dropped for increasing pulse frequency in a positive pulsed streamer wetted-wall reactor [129], in a negative pulsed DBD spray reactor [144], and in a positive pulsed DBD hybrid film and spray reactor [154]. In a negative pulsed DBD falling water film reactor [177] and a positive pulsed corona spray reactor [143], no influence of the frequency was observed. In bipolar pulsed electrohydraulic reactors, breakdown voltage decreases with frequency [62, 96]. Reversing voltage polarity can also cause significant changes in plasma properties and thus treatment efficiency. In the study of Lee et al. [96], positive polarity of pulsed corona in electrohydraulic discharge greatly enhanced the energy efficiency of methyl orange decomposition in pin-to-plate electrode configuration in comparison with negative polarity. The authors explain this observation with the space charge effect, which causes positive corona streamers to be faster and longer, hence increasing radical production and plasma–water contact surface. In the study of Yasuoka et al. [178], higher efficiency was observed as well for positive polarity in a DC bubble discharge reactor. For phenol decomposition in gas phase discharge reactors with a 20% O_2 and 80% N_2 atmosphere, negative DC was found to be more efficient than positive one by Sano et al. [179], while according to Miyazaki et al. [108], no significant difference was seen for both polarities with pulsed power. In a pulsed DBD falling water film reactor, better results were obtained with negative polarity as compared to positive one [171]. In the bubble discharge reactor of Figure 3h with O_2, He, Ar, or Ne bubbles, positive DC voltage performed better than negative one for the decomposition of interfacial active agents [83]. For negative polarity, electrolysis occurred with formation of hydrogen and oxygen gases. For positive polarity, the interfacial anion agents were more concentrated at the bubble surface due to electrostatic attraction. Positive plasma species collided with the water surface, where a cathode drop is formed. These unknown species likely enhanced the decomposition of the agents.

The working gas determines many plasma features for a given input voltage waveform, such as breakdown voltage, electron density and temperature, plasma homogeneity and intensity, generated reactive species, etc. Air is the most frequently used working gas for water treatment plasma reactors due to its wide availability. Pure oxygen gas, however, is often found to give more efficient organic degradation, while nitrogen gas leads to lower efficiency [42, 44, 50]. This can be partly explained with formation of OH radicals and O_3 in oxygen. In N_2 containing gases, however, toxic aqueous nitric products are generated, which decrease solution pH and act as scavengers of oxidants such as OH radicals [50]. Noble gases like helium and argon are sometimes used, especially in bubble discharge reactors. Often, argon leads to the faster decomposition of phenols but performs worse than oxygen for other compounds [50]. In a wetted-wall reactor (Figure 6b) with argon, streamer discharge has been observed which had slightly better phenol decomposition energy efficiency than corona in oxygen [128]. Noble gases, however, are expected to be less economically feasible for use on larger scale due to their high price. Interestingly, treatment efficiency of electrohydraulic discharge reactors can also be altered by bubbling different gases through the solution under treatment, where bubbles are kept away from the discharge zone. Sahni and Locke observed a decrease in nitroform anion decomposition by pulsed underwater corona when the solution was oxygenated with O_2 gas or deoxygenated with N_2, He, or Ar gas as compared to discharge without prior bubbling, but the authors could not explain this effect [180]. Also, water content of the working gas should be taken into account, as it influences formation of important oxidants such as OH and H_2O_2.

Gas flow rate is another factor that needs to be taken into consideration. In three bubble discharge reactors, increasing bubble flow rate enhanced the decomposition process [158, 181, 182], while according to Reddy et al. [183] no significant effect was observed. In the hybrid reactor with bubbles from Figure 10e, oxidation rate first increased and then reached a stationary value with rising gas flow rate [153]. In the remote discharge reactor with bubbling from Chen et al. [184], increasing gas flow also enhanced decomposition. For gas phase discharge reactors, the influence of gas flow rate seems less pronounced. In a positive pulsed corona-like discharge over water, phenol decomposition was slower with increasing oxygen flow and slightly dropped with increasing airflow, while argon flow rate had no influence [185, 186]. In a pulsed DBD falling water film reactor, the effect of the oxygen flow rate on methylene blue degradation was not significant [171].

In reactors with moving solution, the water flow rate often influences the decomposition process. In coaxial falling water film reactors with corona discharge, phenol degradation rate was unchanged with faster water flow for positive pulsed voltage [128], while it dropped in case of negative DC, which was attributed to higher roughness of the water surface [127]. In a pulsed DBD falling water film reactor, the energy efficiency of methylene blue degradation decreased with increasing water flow rate [171]. In contrast, 4-chlorobenzoic acid was decomposed faster with increasing water flow for AC gliding arc discharge over falling water film [123]. Measurements in spray discharge reactors indicate the existence of an optimal water flow rate. In a positive pulsed streamer spray reactor, the energy efficiency of indigo carmine decomposition initially increased and then stabilized with increasing water flow rate due to saturation of aqueous ozone [140]. Moreover, the efficiency was independent of droplet size. In a similar corona reactor, the energy efficiency of oxalic acid decomposition first increased with rising water flow rate, reached a maximal value and subsequently dropped again [187]. The optimal water flow rate increased with applied pulse frequency. Rising water flow rate decreased breakdown voltage in a positive pulsed DBD spray reactor [145]. For the hybrid positive pulsed DBD spray and falling water film reactor of Figure 10g, the energy efficiency of rhodamine B decomposition enhanced with increasing water flow [154].

5. Summary and concluding remarks

Removal of hazardous micropollutants is often insufficient by means of modern conventional wastewater treatment plants. Preventive measures and optimization of conventional biological treatment are suggested as most cost-effective solutions. Nevertheless, preventive measures are limited by increasing demand, while the optimization of conventional techniques often has negligible effect on many persistent micropollutants. Therefore, advanced treatment techniques such as electrochemical separation, activated carbon, nanofiltration, and reverse osmosis have recently received more attention for their effective removal of micropollutants. These techniques, however, are associated with high costs and the additional problem of hazardous concentrate disposal. Advanced oxidation techniques are a promising alternative, as they are the most effective available methods to decompose biorecalcitrant organics. As a main drawback, their energy costs are high up to now, preventing their implementation on large scale. Alternatively, their application can be limited to the treatment of important micropollutant sources, such as hospital and industrial effluent.

Among the advanced oxidation techniques, water treatment by means of electrical discharge takes an interesting place since it is able to generate a wide spectrum of oxidative species, leading to a low selectivity of the degradation process. Further, the optimization of this technology is complex due to the wide variety in reactor design and materials, discharge types, and operational parameters. In this chapter, plasma reactors are comprehensively classified based on their design and materials, in contrast to other reviews where focus lies more on applied voltage and discharge type. Six main reactor types are distinguished. In electrohydraulic discharge reactors and bubble discharge reactors, plasma is generated directly in the liquid bulk, respectively, without and with external application of bubbles. In gas phase discharge reactors and spray discharge reactors, plasma is generated in het gas phase, respectively, over a water bulk or film and in contact with water drops or mist. Reactors that use a combination of these types simultaneously are classified as hybrid reactors. In the last type of reactors, referred to as remote discharge reactors, plasma is not generated in direct contact with the solution under treatment.

Most commonly, electrohydraulic discharge reactors use pulsed arc or positively pulsed corona discharge, where electrode material can have substantial influence on organic decomposition after plasma contact due to formation of erosion particles in water. For the case of arc discharge, energy efficiency is reported to be dependent on interelectrode distance. For pulsed corona, high-voltage pin curvature radius is an important parameter as well. The corona plasma volume can be enlarged by replacing the high-voltage pin with a multipin electrode or a high-voltage electrode covered with a thin porous ceramic layer. Diaphragm and capillary discharge are expected to have similar plasma features to corona discharge in pin-to-plate electrode configuration, which is in agreement with the similar energy efficiency. Contact glow discharge electrolysis is another common type of electrohydraulic discharge, where electrode material also plays an important role. In more exotic types of electrohydraulic discharge reactors, plasma formation is preceded by cavitation under application of RF or microwave power.

Adding external bubbles in the water bulk has the advantages of easier plasma onset, immediate mixing of the solution, minimizing electrode erosion and increasing radical density. Most commonly, bubbles are generated by pumping gas through a dielectric or an electrode, which is shaped as a nozzle, perforated plate, or porous ceramic. Choices in bubble gas, nozzle or perforation material, shape, dimensions, and orientation strongly influence plasma properties, which complicates reactor comparison. Often, the high-voltage electrode is positioned directly in contact with the bubble or in de gas phase in contact with the bubble. In that case, energy efficiency can be enhanced by increasing the number of nozzles or holes. Alternatively, bubbles can be positioned in between submerged electrodes. Some common bubble discharge reactors use a coaxial DBD configuration, where the gas is bubbled in axial direction. Energy efficiency in these systems can be increased by using a double barrier or adding glass beads in the water bulk. More exotic types of bubble discharge reactors have been reported in literature, with promising first results.

Corona and glow discharge in gas phase over grounded water bulk or film is mostly generated with pulsed power. Negative pulsed corona over flowing water film in a pilot system has been shown to have better or comparable energy efficiency than other advanced oxidation processes. Based on one report, energy efficiency of organic decomposition seems to be independent of the type of discharge, voltage amplitude, polarity of the applied voltage, and amount of pin electrodes. In contrast, DBD over water bulk or film has an energy efficiency that is reported to increase with increasing voltage amplitude and decreasing interelectrode distance. Interelectrode distance can be decreased substantially by adding a porous ceramic segment at the water surface. Movement of the water phase by stirring or by making it flow as a film along the discharge increases energy efficiency substantially, making water flow rate an important parameter for optimization. Based on this principle, coaxial reactors with falling water film are gaining more popularity. Such reactors can be further optimized by adjusting gas flow direction and electrode and barrier dimensions. A last common gas phase discharge reactor uses arc discharge over the water bulk or film. Larger arc-treated water surface and energy efficiency can be reached by using a couple of controlled electrodes or a second high-voltage electrode.

Spray discharge reactors have received more attention in recent years due to their high reported energy efficiencies. In the case of positive pulsed corona, treatment is more efficient for droplets near the inner reactor wall, while droplet size has no influence on energy efficiency. Such reactors can be further optimized by adding fluorocarbon wires along the inner reactor wall for larger droplet retention time. Other spray discharge reactors treat droplets with DBD, electrospray or arc discharge. In the case of gliding arc, the process can be optimized by addition of a couple of controlled electrodes or a second high-voltage electrode and by introducing the water perpendicular to the gas flow to improve contact with plasma. DC plasma torches can also be used for water treatment by injecting the solution into the torch as plasma forming gas.

One type of hybrid reactor is designed by placing a high-voltage electrode in the water phase and a ground electrode or second high-voltage electrode in the gas phase above the water surface, without or with the addition of bubbles. Their energy efficiency for organic decomposition is moderately higher than the one of hydraulic discharge reactors. Another hybrid reactor type naturally occurs by spraying water from a shower nozzle at an angle, causing a falling water film in combination with spray. In such case, organic decomposition in droplets is more efficient than in the water film. A more exotic hybrid reactor that deserves more attention consists of a spray discharge reactor where droplets are formed by electrospray with additional plasma treatment afterwards.

Remote discharge reactors can be encountered in many configurations. In the most standard design, plasma gas is remotely generated and sequentially bubbled through the solution, as in the well-known example of ozonation. Plasma gas bubbling can be combined with UV irradiation by generating plasma inside a quartz tube submerged in the solution. Research has indicated that such reactors may be competitive technology to systems with direct plasma treatment. Plasma can also be used solely for UV irradiation, as in plasma lamps, excimer lamps and microwave discharge electrodeless lamps. Recently, the generation and application of plasma-activated water for water disinfection is gaining popularity, while its effects on organic contamination are still largely unknown.

Additionally, the importance of voltage waveform, working gas and flow rates of gas, and water for further optimization are shortly reviewed in this chapter. Energy efficiency has different dependency for different reactor types on voltage parameters, such as voltage type, amplitude, polarity, sinusoidal frequency, pulse rise time, pulse duration, and pulse frequency. For example, positive polarity causes higher efficiency of electrohydraulic and bubble discharge reactors in all considered cases, explained with the space charge effect. On the other hand, negative polarity gives better performance in gas phase discharge reactors. The choice of working gas can significantly alter plasma chemistry and therefore treatment efficiency and by-product formation. While atmospheric air is often chosen due to its wide availability, oxygen generally enhances the process. Argon often performs better for phenol degradation but is expected to be less economically feasible for use on large scale. Increasing gas flow rate typically enhances decomposition in bubble discharge and remote discharge reactors, whereas its effect is less pronounced in gas phase discharge reactors. Also, adjustment of water flow can significantly increase energy efficiency. Measurements in spray discharge reactors, for instance, indicate the existence of an optimal water flow rate. While dependency of a few of these parameters on energy efficiency seems adequately established for a limited number of reactor types, a deeper literature study and more experimental investigation are required for additional confirmation and a better understanding in most of the cases. Due to the wide variety in plasma reactors and the distinct, unique features of every reactor, researchers are motivated to report new results in this field, including clear descriptions of reactor design and materials.

One needs to keep a few additional influences in mind when interpreting energy efficiencies reported in literature. First of all, solution parameters such as water temperature, pH, conductivity, and water matrix can have significant effect on plasma chemistry and reactions in the water bulk. Often, deionized water at room temperature is used as solvent in plasma reactors. In other cases, however, deviations in energy efficiency can be caused by a difference in solution parameters. Also, it should be remarked that micropollutant measurement is generally performed a certain time after plasma treatment with analytical chemistry methods, such as gas chromatography–mass spectrometry and liquid chromatography–mass spectrometry. During this time, postreactions with long-living oxidants can occur, decomposing the micropollutant to a greater extent. This effect is usually neglected by researchers, complicating accurate comparison between reported energy efficiencies. On the positive side, this aging effect can be beneficial in applications where sufficiently long hydraulic retention time is possible after plasma treatment. Last but not least, most of the reported electrical discharge reactors only treat small water volumes in batch mode without or with recirculation of water. As a result, the determined energy efficiency can be significantly different for the same reactor type in single-pass mode, where water is flowing through the system only once. The latter case is more representative for industrial application and thus deserves more attention.

In this chapter, only improvement of the plasma process in terms of reactor design, materials, and working parameters is discussed. Further optimization can be achieved by combining plasma discharge with other advanced treatment methods, such as adsorption, Fenton's reagent, photocatalysis, and ultrasonication. Such combinations have been reported before but require additional attention and further exploration. As should be noted, plasma technology can also be used for synthesis, pretreatment, regeneration, and posttreatment of materials and matter involved in water treatment processes, such as nanotubes, membranes, activated carbon, excess sludge, and organic concentrate from filtration. In-line application of these methods during the water treatment process needs to be considered as a possible alternative to direct water treatment with advanced oxidation processes since energy demand and overall costs can be pressed significantly this way. However, more experimental investigation and thorough cost analysis is necessary to confirm this claim.

Future application of plasma discharge for water treatment will largely depend on its effectiveness and energy efficiency as compared to other advanced oxidation processes and treatment methods in general, but additional criteria need to be taken into consideration as well. Sustainability, ease of operation, capital costs, and costs related to maintenance, gas input, and additional energy for pumps will also determine whether a system will be adopted on a large scale. Moreover, an extensive study of generated oxidation by-products and long-living oxidants in treated water is necessary to assure that overall toxicity is consistently and sufficiently decreased after plasma treatment. Up to now, reports on these topics are largely lacking in literature or limited to only a few specific cases.

Author details

Patrick Vanraes*, Anton Y. Nikiforov and Christophe Leys

*Address all correspondence to: Patrick.Vanraes@UGent.be

Department of Applied Physics, Research Unit Plasma Technology, Ghent University, Ghent, Belgium

References

[1] L. Rizzo, *et al.*, "Urban wastewater treatment plants as hotspots for antibiotic resistant bacteria and genes spread into the environment: a review," *Science of the Total Environment,* vol. 447, pp. 345–360, 2013.

[2] T. Hayes, *et al.*, "Herbicides: feminization of male frogs in the wild," *Nature,* vol. 419, pp. 895–896, 2002.

[3] S. Milla, *et al.*, "The effects of estrogenic and androgenic endocrine disruptors on the immune system of fish: a review," *Ecotoxicology,* vol. 20, pp. 305–319, 2011.

[4] I. R. Falconer, *et al.*, "Endocrine-disrupting compounds: a review of their challenge to sustainable and safe water supply and water reuse," *Environmental Toxicology,* vol. 21, pp. 181–191, 2006.

[5] R. McKinlay, *et al.*, "Endocrine disrupting pesticides: implications for risk assessment," *Environment International,* vol. 34, pp. 168–183, 2008.

[6] W. Mnif, *et al.*, "Effect of endocrine disruptor pesticides: a review," *International Journal of Environmental Research and Public Health,* vol. 8, pp. 2265–2303, 2011.

[7] K. Fent, *et al.*, "Ecotoxicology of human pharmaceuticals," *Aquatic Toxicology,* vol. 76, pp. 122–159, 2006.

[8] A. Pruden, *et al.*, "Antibiotic resistance genes as emerging contaminants: studies in northern Colorado," *Environmental Science & Technology,* vol. 40, pp. 7445–7450, 2006.

[9] A. L. Batt, *et al.*, "Comparison of the occurrence of antibiotics in four full-scale wastewater treatment plants with varying designs and operations," *Chemosphere,* vol. 68, pp. 428–435, 2007.

[10] X. Flores Alsina, "Conceptual design of wastewater treatment plants using multiple objectives," Doctor PhD, Universitat de Girona, Girona, 2008.

[11] Y.-P. Wang and R. Smith, "Design of distributed effluent treatment systems," *Chemical Engineering Science,* vol. 49, pp. 3127–3145, 1994.

[12] I. Michael, *et al.*, "Urban wastewater treatment plants as hotspots for the release of antibiotics in the environment: a review," *Water Research,* vol. 47, pp. 957–995, 2013.

[13] L. Benedetti, *et al.*, "Multi-criteria analysis of wastewater treatment plant design and control scenarios under uncertainty," *Environmental Modelling & Software,* vol. 25, pp. 616–621, 2010.

[14] Y. Luo, *et al.*, "A review on the occurrence of micropollutants in the aquatic environment and their fate and removal during wastewater treatment," *Science of the Total Environment,* vol. 473–474, pp. 619–641, 2014.

[15] O. A. Jones, *et al.*, "Questioning the excessive use of advanced treatment to remove organic micropollutants from wastewater," *Environmental Science & Technology,* vol. 41, pp. 5085–5089, 2007.

[16] Z.-h. Liu, *et al.*, "Removal mechanisms for endocrine disrupting compounds (EDCs) in wastewater treatment—physical means, biodegradation, and chemical advanced oxidation: a review," *Science of the Total Environment,* vol. 407, pp. 731–748, 2009.

[17] A. Joss, *et al.*, "Removal of pharmaceuticals and fragrances in biological wastewater treatment," *Water Research,* vol. 39, pp. 3139–3152, 2005.

[18] T. A. Ternes, *et al.*, "Peer reviewed: scrutinizing pharmaceuticals and personal care products in wastewater treatment," *Environmental Science & Technology,* vol. 38, pp. 392A-399A, 2004.

[19] O. H. Jones, *et al.*, "Human pharmaceuticals in wastewater treatment processes," *Critical Reviews in Environmental Science and Technology,* vol. 35, pp. 401–427, 2005.

[20] N. Tadkaew, *et al.*, "Removal of trace organics by MBR treatment: the role of molecular properties," *Water Research,* vol. 45, pp. 2439–2451, 2011.

[21] I. Sirés and E. Brillas, "Remediation of water pollution caused by pharmaceutical residues based on electrochemical separation and degradation technologies: a review," *Environment International,* vol. 40, pp. 212–229, 2012.

[22] J. M. Poyatos, *et al.*, "Advanced oxidation processes for wastewater treatment: state of the art," *Water, Air, and Soil Pollution,* vol. 205, pp. 187–204, 2010.

[23] O. Lefebvre, *et al.*, "Microbial fuel cells for energy self-sufficient domestic wastewater treatment—a review and discussion from energetic consideration," *Applied Microbiology and Biotechnology,* vol. 89, pp. 259–270, 2011.

[24] A. Helal, *et al.*, "Feasibility study for self-sustained wastewater treatment plants—using biogas CHP fuel cell, micro-turbine, PV and wind turbine systems," 2013.

[25] P. Verlicchi, *et al.*, "What have we learned from worldwide experiences on the management and treatment of hospital effluent?—an overview and a discussion on perspectives," *Science of the Total Environment,* vol. 514, pp. 467–491, 2015.

[26] P. Verlicchi, *et al.*, "Hospital effluents as a source of emerging pollutants: an overview of micropollutants and sustainable treatment options," *Journal of Hydrology,* vol. 389, pp. 416–428, 2010.

[27] C. Comninellis, *et al.*, "Advanced oxidation processes for water treatment: advances and trends for R&D," *Journal of Chemical Technology and Biotechnology,* vol. 83, pp. 769–776, 2008.

[28] V. Homem and L. Santos, "Degradation and removal methods of antibiotics from aqueous matrices—a review," *Journal of Environmental Management,* vol. 92, pp. 2304–2347, 2011.

[29] J. Rivera-Utrilla, *et al.*, "Pharmaceuticals as emerging contaminants and their removal from water. A review," *Chemosphere,* vol. 93, pp. 1268–1287, 2013.

[30] W. Yang, *et al.*, "Treatment of organic micropollutants in water and wastewater by UV-based processes: a literature review," *Critical Reviews in Environmental Science and Technology,* vol. 44, pp. 1443–1476, 2014.

[31] H. R. Ghatak, "Advanced oxidation processes for the treatment of biorecalcitrant organics in wastewater," *Critical Reviews in Environmental Science and Technology,* vol. 44, pp. 1167–1219, 2014.

[32] M. A. Oturan and J.-J. Aaron, "Advanced oxidation processes in water/wastewater treatment: principles and applications. A review," *Critical Reviews in Environmental Science and Technology,* vol. 44, pp. 2577–2641, 2014.

[33] M. Klavarioti, *et al.*, "Removal of residual pharmaceuticals from aqueous systems by advanced oxidation processes," *Environment International,* vol. 35, pp. 402–417, 2009.

[34] R. P. Joshi and S. M. Thagard, "Streamer-like electrical discharges in water: part II. Environmental applications," *Plasma Chemistry and Plasma Processing,* vol. 33, pp. 17–49, 2013.

[35] J.-L. Brisset and E. Hnatiuc, "Peroxynitrite: a re-examination of the chemical properties of non-thermal discharges burning in air over aqueous solutions," *Plasma Chemistry and Plasma Processing,* vol. 32, pp. 655–674, 2012.

[36] P. Lukes, *et al.*, "Aqueous-phase chemistry and bactericidal effects from an air discharge plasma in contact with water: evidence for the formation of peroxynitrite through a pseudo-second-order post-discharge reaction of $H2O_2$ and HNO_2," *Plasma Sources Science and Technology,* vol. 23, p. 015019, 2014.

[37] X. Jin, *et al.*, "Reaction kinetics of selected micropollutants in ozonation and advanced oxidation processes," *Water Research,* vol. 46, pp. 6519–6530, 2012.

[38] S. Sudhakaran and G. L. Amy, "QSAR models for oxidation of organic micropollutants in water based on ozone and hydroxyl radical rate constants and their chemical classification," *Water Research,* vol. 47, pp. 1111–1122, 2013.

[39] U. Von Gunten, "Ozonation of drinking water: Part I. Oxidation kinetics and product formation," *Water Research,* vol. 37, pp. 1443–1467, 2003.

[40] W. R. Haag and C. D. Yao, "Rate constants for reaction of hydroxyl radicals with several drinking water contaminants," *Environmental Science & Technology,* vol. 26, pp. 1005–1013, 1992.

[41] M. E. Sillanpää, *et al.*, "Degradation of chelating agents in aqueous solution using advanced oxidation process (AOP)," *Chemosphere,* vol. 83, pp. 1443–1460, 2011.

[42] M. A. Malik, "Water purification by plasmas: which reactors are most energy efficient?" *Plasma Chemistry and Plasma Processing,* vol. 30, pp. 21–31, 2010.

[43] P. J. Bruggeman and B. R. Locke, "Assessment of potential applications of plasma with liquid water," in *Low Temperature Plasma Technology—Methods and Applications*, P. K. Chu and L. XinPei, Eds., Boca Raton, FL: CRC Press, 2014, pp. 367–399.

[44] B. Jiang, *et al.*, "Review on electrical discharge plasma technology for wastewater remediation," *Chemical Engineering Journal*, vol. 236, pp. 348–368, 2014.

[45] P. Bruggeman and C. Leys, "Non-thermal plasmas in and in contact with liquids," *Journal of Physics D: Applied Physics*, vol. 42, p. 053001, 2009.

[46] B. Locke, *et al.*, "Electrohydraulic discharge and nonthermal plasma for water treatment," *Industrial & Engineering Chemistry Research*, vol. 45, pp. 882–905, 2006.

[47] E.-J. Lee, *et al.*, "Application of underwater dielectric barrier discharge as a washing system to inactivate *Salmonella typhimurium* on perillar leaves," in *Plasma Sciences (ICOPS), 2015 IEEE International Conference on*, 2015, pp. 1–1.

[48] P. Vanraes, *et al.*, "Removal of atrazine in water by combination of activated carbon and dielectric barrier discharge," *Journal of Hazardous Materials*, vol. 299, pp. 647–655, 2015.

[49] B. R. Locke and K.-Y. Shih, "Review of the methods to form hydrogen peroxide in electrical discharge plasma with liquid water," *Plasma Sources Science and Technology*, vol. 20, p. 034006, 2011.

[50] M. Hijosa-Valsero, *et al.*, "Decontamination of waterborne chemical pollutants by using atmospheric pressure nonthermal plasma: a review," *Environmental Technology Reviews*, vol. 3, pp. 71–91, 2014.

[51] N. Parkansky, *et al.*, "Submerged arc breakdown of sulfadimethoxine (SDM) in aqueous solutions," *Plasma Chemistry and Plasma Processing*, vol. 28, pp. 583–592, 2008.

[52] N. Parkansky, *et al.*, "Submerged arc breakdown of methylene blue in aqueous solutions," *Plasma Chemistry and Plasma Processing*, vol. 32, pp. 933–947, 2012.

[53] Q.-X. Lin, *et al.*, "The research on pulsed arc electrohydraulic discharge with discharge electrode and its application to removal of bacteria," *IEEE Transactions on Plasma Science*, vol. 43, pp. 1029–1039, 2015.

[54] L. Zhu, *et al.*, "The research on the pulsed arc electrohydraulic discharge and its application in treatment of the ballast water," *Journal of Electrostatics*, vol. 71, pp. 728–733, 2013.

[55] N. Parkansky, *et al.*, "Decomposition of dissolved methylene blue in water using a submerged arc between titanium electrodes," *Plasma Chemistry and Plasma Processing*, vol. 33, pp. 907–919, 2013.

[56] P. Lang, *et al.*, "Oxidative degradation of 2,4,6-trinitrotoluene by ozone in an electrohydraulic discharge reactor," *Environmental Science & Technology*, vol. 32, pp. 3142–3148, 1998.

[57] D. M. Angeloni, *et al.*, "Removal of methyl-tert-butyl ether from water by a pulsed arc electrohydraulic discharge system," *Japanese Journal of Applied Physics*, vol. 45, pp. 8290–8293, 2006.

[58] N. Karpel Vel Leitner, *et al.*, "Generation of active entities by the pulsed arc electrohydraulic discharge system and application to removal of atrazine," *Water Research*, vol. 39, pp. 4705–4714, 2005.

[59] L. V. Hoang and B. Legube, "Degradation of 4-chlorophenol by pulsed arc discharge in water-estimation of the energy consumption," *IEEE Transactions on Dielectrics and Electrical Insulation*, vol. 16, pp. 1604–1608, 2009.

[60] P. Šunka, *et al.*, "Potential applications of pulse electrical discharges in water," *Acta Physica Slovaca*, vol. 54, pp. 135–145, 2004.

[61] Š. Potocký, *et al.*, "Needle electrode erosion in water plasma discharge," *Thin Solid Films*, vol. 518, pp. 918–923, 2009.

[62] P. Baroch, *et al.*, "Generation of plasmas in water: utilization of a high-frequency, low-voltage bipolar pulse power supply with impedance control," *Plasma Sources Science and Technology*, vol. 20, p. 034017, 2011.

[63] T. Takeda, *et al.*, "Morphology of high-frequency electrohydraulic discharge for liquid-solution plasmas," *IEEE Transactions on Plasma Science*, vol. 36, pp. 1158–1159, 2008.

[64] Q. Lu, *et al.*, "Degradation of 2,4-dichlorophenol by using glow discharge electrolysis," *Journal of Hazardous Materials*, vol. 136, pp. 526–531, 2006.

[65] Q. Lu, *et al.*, "Glow discharge induced hydroxyl radical degradation of 2-naphthylamine," *Plasma Science and Technology*, vol. 7, p. 2856, 2005.

[66] S. Mededovic, *et al.*, "Aqueous-phase mineralization of S-triazine using pulsed electrical discharge," *IJPEST*, vol. 1, pp. 82–90, 2007.

[67] S. Mededovic and B. R. Locke, "Side-chain degradation of atrazine by pulsed electrical discharge in water," *Industrial & Engineering Chemistry Research*, vol. 46, pp. 2702–2709, 2007.

[68] P. Sunka, *et al.*, "Generation of chemically active species by electrical discharges in water," *Plasma Sources Science and Technology*, vol. 8, p. 258, 1999.

[69] B. Sun, *et al.*, "Optical study of active species produced by a pulsed streamer corona discharge in water," *Journal of Electrostatics*, vol. 39, pp. 189–202, 1997.

[70] P. Lukes, *et al.*, "The catalytic role of tungsten electrode material in the plasmachemical activity of a pulsed corona discharge in water," *Plasma Sources Science and Technology*, vol. 20, p. 034011, 2011.

[71] H. Zheng, *et al.*, "p-Nitrophenol Enhanced Degradation in High-Voltage Pulsed Co-

rona Discharges Combined with Ozone System," *Plasma Chemistry and Plasma Processing,* vol. 33, pp. 1053–1062, 2013.

[72] C. W. Chen, *et al.*, "Ultrasound-assisted plasma: a novel technique for inactivation of aquatic microorganisms," *Environmental Science & Technology,* vol. 43, pp. 4493–4497, 2009.

[73] M. M. Sein, *et al.*, "Studies on a non-thermal pulsed corona plasma between two parallel-plate electrodes in water," *Journal of Physics D: Applied Physics,* vol. 45, p. 225203, 2012.

[74] P. Lukes, *et al.*, "Effect of ceramic composition on pulse discharge induced processes in water using ceramic-coated wire to cylinder electrode system," *Czechoslovak Journal of Physics,* vol. 52, pp. D800-D806, 2002.

[75] P. Lukes, *et al.*, "The role of surface chemistry at ceramic/electrolyte interfaces in the generation of pulsed corona discharges in water using porous ceramic-coated rod electrodes," *Plasma Processes and Polymers,* vol. 6, pp. 719–728, 2009.

[76] Y. J. Liu and X. Z. Jiang, "Phenol degradation by a nonpulsed diaphragm glow discharge in an aqueous solution," *Environmental Science & Technology,* vol. 39, pp. 8512–8517, 2005.

[77] L. Wang, "4-Chlorophenol degradation and hydrogen peroxide formation induced by DC diaphragm glow discharge in an aqueous solution," *Plasma Chemistry and Plasma Processing,* vol. 29, pp. 241–250, 2009.

[78] V. Goncharuk, *et al.*, "The diaphragm discharge and its use for water treatment," *Journal of Water Chemistry and Technology,* vol. 30, pp. 261–268, 2008.

[79] X. Jin, *et al.*, "An improved multi-anode contact glow discharge electrolysis reactor for dye discoloration," *Electrochimica Acta,* vol. 59, pp. 474–478, 2012.

[80] T. Maehara, *et al.*, "Influence of conductivity on the generation of a radio frequency plasma surrounded by bubbles in water," *Plasma Sources Science and Technology,* vol. 20, p. 034016, 2011.

[81] T. Ishijima, *et al.*, "Efficient production of microwave bubble plasma in water for plasma processing in liquid," *Plasma Sources Science and Technology,* vol. 19, p. 015010, 2010.

[82] N. Remya and J.-G. Lin, "Current status of microwave application in wastewater treatment—a review," *Chemical Engineering Journal,* vol. 166, pp. 797–813, 2011.

[83] K. Yasuoka, *et al.*, "An energy-efficient process for decomposing perfluorooctanoic and perfluorooctane sulfonic acids using dc plasmas generated within gas bubbles," *Plasma Sources Science and Technology,* vol. 20, p. 034009, 2011.

[84] H. Aoki, *et al.*, "Plasma generation inside externally supplied Ar bubbles in water,"

Plasma Sources Science and Technology, vol. 17, p. 025006, 2008.

[85] D. C. Johnson, *et al.,* "Treatment of methyl tert-butyl ether contaminated water using a dense medium plasma reactor: a mechanistic and kinetic investigation," *Environmental Science & Technology,* vol. 37, pp. 4804–4810, 2003.

[86] W. Bian, *et al.,* "Enhanced degradation of p-chlorophenol in a novel pulsed high voltage discharge reactor," *Journal of Hazardous Materials,* vol. 162, pp. 906–912, 2009.

[87] Y. Zhang, *et al.,* "Methyl orange degradation by pulsed discharge in the presence of activated carbon fibers," *Chemical Engineering Journal,* vol. 159, pp. 47–52, 2010.

[88] N. Lu, *et al.,* "Electrical characteristics of pulsed corona discharge plasmas in chitosan solution," *Plasma Science and Technology,* vol. 16, pp. 128–133, 2014.

[89] M. Kurahashi, *et al.,* "Radical formation due to discharge inside bubble in liquid," *Journal of Electrostatics,* vol. 42, pp. 93–105, 1997.

[90] K. Sato, *et al.,* "Water treatment with pulsed discharges generated inside bubbles," *Electrical Engineering in Japan,* vol. 170, pp. 1–7, 2010.

[91] A. Yamatake, *et al.,* "Water purification by atmospheric DC/pulsed plasmas inside bubbles in water," *International Journal of Environmental Science and Technology,* vol. 1, pp. 91–95, 2007.

[92] J. Li, *et al.,* "Degradation of phenol in water using a gas–liquid phase pulsed discharge plasma reactor," *Thin Solid Films,* vol. 515, pp. 4283–4288, 2007.

[93] N. Lu, *et al.,* "Treatment of dye wastewater by using a hybrid gas/liquid pulsed discharge plasma reactor," *Plasma Science and Technology,* vol. 14, pp. 162–166, 2012.

[94] A. Y. Nikiforov, "An application of AC glow discharge stabilized by fast air flow for water treatment," *IEEE Transactions on Plasma Science,* vol. 37, pp. 872–876, 2009.

[95] H. Wang, *et al.,* "Enhanced generation of oxidative species and phenol degradation in a discharge plasma system coupled with TiO_2 photocatalysis," *Applied Catalysis B: Environmental,* vol. 83, pp. 72–77, 2008.

[96] H. Lee, *et al.,* "The effect of liquid phase plasma for photocatalytic degradation of bromothymol blue," *Science of Advanced Materials,* vol. 6, pp. 1627–1631, 2014.

[97] T. Miichi, *et al.,* "Generation of radicals using discharge inside bubbles in water for water treatment," *Ozone: Science & Engineering,* vol. 24, pp. 471–477, 2002.

[98] P. Vanraes, *et al.,* "Electrical and spectroscopic characterization of underwater plasma discharge inside rising gas bubbles," *Journal of Physics D: Applied Physics,* vol. 45, p. 245206, 2012.

[99] R. Zhang, *et al.,* "Plasma induced degradation of indigo carmine by bipolar pulsed dielectric barrier discharge (DBD) in the water–air mixture," *Journal of Environmental Sciences (China),* vol. 16, pp. 808–812, 2004.

[100] R.-b. Zhang, *et al.*, "Enhancement of the plasma chemistry process in a three-phase discharge reactor," *Plasma Sources Science and Technology*, vol. 14, pp. 308–313, 2005.

[101] A. Anpilov, *et al.*, "Electric discharge in water as a source of UV radiation, ozone and hydrogen peroxide," *Journal of Physics D: Applied Physics*, vol. 34, p. 993, 2001.

[102] E. M. Barkhudarov, *et al.*, "Multispark discharge in water as a method of environmental sustainability problems solution," *Journal of Atomic and Molecular Physics*, vol. 2013, pp. 1–12, 2013.

[103] S. Horikoshi, *et al.*, "A novel liquid plasma AOP device integrating microwaves and ultrasounds and its evaluation in defluorinating perfluorooctanoic acid in aqueous media," *Ultrasonics Sonochemistry*, vol. 18, pp. 938–942, 2011.

[104] C. Yamabe, *et al.*, "Water treatment using discharge on the surface of a bubble in water," *Plasma Processes and Polymers*, vol. 2, pp. 246–251, 2005.

[105] P. Bruggeman, *et al.*, "Water surface deformation in strong electrical fields and its influence on electrical breakdown in a metal pin–water electrode system," *Journal of Physics D: Applied Physics*, vol. 40, p. 4779, 2007.

[106] P. Bruggeman, *et al.*, "Influence of the water surface on the glow-to-spark transition in a metal-pin-to-water electrode system," *Plasma Sources Science and Technology*, vol. 17, p. 045014, 2008.

[107] L. O. d. B. Benetoli, *et al.*, "Effect of temperature on methylene blue decolorization in aqueous medium in electrical discharge plasma reactor," *Journal of the Brazilian Chemical Society*, vol. 22, pp. 1669–1678, 2011.

[108] Y. Miyazaki, *et al.*, "Pulsed discharge purification of water containing nondegradable hazardous substances," *Electrical Engineering in Japan*, vol. 174, pp. 1–8, 2011.

[109] M. Magureanu, *et al.*, "Pulsed corona discharge for degradation of methylene blue in water," *Plasma Chemistry and Plasma Processing*, vol. 33, pp. 51–64, 2013.

[110] M. Dors, *et al.*, "Phenol oxidation in aqueous solution by gas phase corona discharge," *Journal of Advanced Oxidation Technologies*, vol. 9, pp. 139–143, 2006.

[111] A. Sharma, *et al.*, "Destruction of pentachlorophenol using glow discharge plasma process," *Environmental Science & Technology*, vol. 34, pp. 2267–2272, 2000.

[112] P. Jamróz, *et al.*, "Atmospheric pressure glow discharges generated in contact with flowing liquid cathode: production of active species and application in wastewater purification processes," *Plasma Chemistry and Plasma Processing*, vol. 34, pp. 25–37, 2014.

[113] I. Even-Ezra, *et al.*, "Application of a novel plasma-based advanced oxidation process for efficient and cost-effective destruction of refractory organics in tertiary effluents and contaminated groundwater," *Desalination and Water Treatment*, vol. 11, pp. 236–244, 2009.

[114] D. Gerrity, *et al.*, "An evaluation of a pilot-scale nonthermal plasma advanced oxidation process for trace organic compound degradation," *Water Research,* vol. 44, pp. 493–504, 2010.

[115] A. Mizrahi and I. M. Litaor, "The kinetics of the removal of organic pollutants from drinking water by a novel plasma-based advanced oxidation technology," *Desalination and Water Treatment,* vol. 52, pp. 5264–5275, 2013.

[116] Y. Hu, *et al.*, "Application of dielectric barrier discharge plasma for degradation and pathways of dimethoate in aqueous solution," *Separation and Purification Technology,* vol. 120, pp. 191–197, 2013.

[117] S. P. Li, *et al.*, "Degradation of nitenpyram pesticide in aqueous solution by low-temperature plasma," *Environmental Technology,* vol. 34, pp. 1609–1616, 2013.

[118] J. Feng, *et al.*, "Degradation of aqueous 3,4-dichloroaniline by a novel dielectric barrier discharge plasma reactor," *Environmental Science and Pollution Research,* vol. 22, pp. 4447–4459, 2015.

[119] M. Hijosa-Valsero, *et al.*, "Removal of priority pollutants from water by means of dielectric barrier discharge atmospheric plasma," *Journal of Hazardous Materials,* vol. 262, pp. 664–673, 2013.

[120] G. Fridman, *et al.*, "Floating electrode dielectric barrier discharge plasma in air promoting apoptotic behavior in melanoma skin cancer cell lines," *Plasma Chemistry and Plasma Processing,* vol. 27, pp. 163–176, 2007.

[121] E. Marotta, *et al.*, "Advanced oxidation process for degradation of aqueous phenol in a dielectric barrier discharge reactor," *Plasma Processes and Polymers,* vol. 8, pp. 867–875, 2011.

[122] O. Lesage, *et al.*, "Degradation of 4-chlorobenzoïc acid in a thin falling film dielectric barrier discharge reactor," *Industrial & Engineering Chemistry Research,* vol. 53, pp. 10387–10396, 2014.

[123] O. Lesage, *et al.*, "Treatment of 4-chlorobenzoic acid by plasma-based advanced oxidation processes," *Chemical Engineering and Processing: Process Intensification,* vol. 72, pp. 82–89, 2013.

[124] S. Li, *et al.*, "Degradation of thiamethoxam in water by the synergy effect between the plasma discharge and the TiO_2 photocatalysis," *Desalination and Water Treatment,* vol. 53, pp. 3018–3025, 2014.

[125] P. Baroch, *et al.*, "Special type of plasma dielectric barrier discharge reactor for direct ozonization of water and degradation of organic pollution," *Journal of Physics D: Applied Physics,* vol. 41, p. 085207, 2008.

[126] S. Rong and Y. Sun, "Wetted-wall corona discharge induced degradation of sulfadiazine antibiotics in aqueous solution," *Journal of Chemical Technology & Biotechnology,* vol. 89, pp. 1351–1359, 2013.

[127] N. Sano, *et al.*, "Decomposition of phenol in water by a cylindrical wetted-wall reactor using direct contact of gas corona discharge," *Industrial & Engineering Chemistry Research,* vol. 42, pp. 5423–5428, 2003.

[128] M. Sato, *et al.*, "Decomposition of phenol in water using water surface plasma in wetted-wall reactor," *International Journal of Plasma Environmental Science & Technology,* vol. 1, pp. 71–75, 2007.

[129] T. Yano, *et al.*, "Decolorization of indigo carmine solution using nanosecond pulsed power," *Dielectrics and Electrical Insulation, IEEE Transactions on,* vol. 16, pp. 1081–1087, 2009.

[130] M. Morimoto, *et al.*, "Indigo carmine solution treatment by nanosecond pulsed power with a dielectric barrier electrode," *Dielectrics and Electrical Insulation, IEEE Transactions on,* vol. 22, pp. 1872–1878, 2015.

[131] K. Faungnawakij, *et al.*, "Removal of acetaldehyde in air using a wetted-wall corona discharge reactor," *Chemical Engineering Journal,* vol. 103, pp. 115–122, 2004.

[132] Y. Chen, *et al.*, "A discharge reactor with water-gas mixing for methyl orange removal," *International Journal of Plasma Environmental Science and Technology,* 2008.

[133] S. Ognier, *et al.*, "Analysis of mechanisms at the plasma–liquid interface in a gas–liquid discharge reactor used for treatment of polluted water," *Plasma Chemistry and Plasma Processing,* vol. 29, pp. 261–273, 2009.

[134] A. G. Bubnov, *et al.*, "Plasma-catalytic decomposition of phenols in atmospheric pressure dielectric barrier discharge," *Plasma Chemistry and Plasma Processing,* vol. 26, pp. 19–30, 2006.

[135] S.-P. Rong, *et al.*, "Degradation of sulfadiazine antibiotics by water falling film dielectric barrier discharge," *Chinese Chemical Letters,* vol. 25, pp. 187–192, 2014.

[136] J.-L. Brisset, *et al.*, "Chemical reactivity of discharges and temporal post-discharges in plasma treatment of aqueous media: examples of gliding discharge treated solutions," *Industrial & Engineering Chemistry Research,* vol. 47, pp. 5761–5781, 2008.

[137] R. Burlica, *et al.*, "Organic dye removal from aqueous solution by glidarc discharges," *Journal of Electrostatics,* vol. 62, pp. 309–321, 2004.

[138] J. Janca, *et al.*, "Investigation of the chemical action of the gliding and "point" arcs between the metallic electrode and aqueous solution," *Plasma Chemistry and Plasma Processing,* vol. 19, pp. 53–67, 1999.

[139] T. Kobayashi, *et al.*, "The effect of spraying of water droplets and location of water droplets on the water treatment by pulsed discharge in air," *IEEE Transactions on Plasma Science,* vol. 38, pp. 2675–2680, 2010.

[140] T. Sugai, *et al.*, "The effect of flow rate and size of water droplets on the water treat-

ment by pulsed discharge in air," *IEEE Transactions on Plasma Science,* vol. PP, p. 1, 2015.

[141] T. Sugai, *et al.*, "Improvement of efficiency for decomposition of organic compound in water using pulsed streamer discharge in air with water droplets by increasing of residence time," in *Pulsed Power Conference, 2009. PPC'09. IEEE,* 2009, pp. 1056–1060.

[142] G. An, *et al.*, "Degradation of phenol in mists by a non-thermal plasma reactor," *Chemosphere,* vol. 84, pp. 1296–1300, 2011.

[143] S. Preis, *et al.*, "Formation of nitrates in aqueous solutions treated with pulsed corona discharge: the impact of organic pollutants," *Ozone: Science & Engineering,* vol. 36, pp. 94–99, 2014.

[144] S. Jiang, *et al.*, "Investigation of pulsed dielectric barrier discharge system on water treatment by liquid droplets in air," *Dielectrics and Electrical Insulation, IEEE Transactions on,* vol. 22, pp. 1866–1871, 2015.

[145] J. Kornev, *et al.*, "Generation of active oxidant species by pulsed dielectric barrier discharge in water-air mixtures," *Ozone: Science & Engineering,* vol. 28, pp. 207–215, 2006.

[146] Z. Wang, *et al.*, "Plasma decoloration of dye using dielectric barrier discharges with earthed spraying water electrodes," *Journal of Electrostatics,* vol. 66, pp. 476–481, 2008.

[147] M. Elsawah, *et al.*, "Corona discharge with electrospraying system for phenol removal from water," *IEEE Transactions on Plasma Science,* vol. 40, pp. 29–34, 2012.

[148] C. M. Du, *et al.*, "Degradation of 4-chlorophenol using a gas–liquid gliding arc discharge plasma reactor," *Plasma Chemistry and Plasma Processing,* vol. 27, pp. 635–646, 2007.

[149] D. Moussa, *et al.*, "Plasma-chemical destruction of trilaurylamine issued from nuclear laboratories of reprocessing plants," *Industrial & Engineering Chemistry Research,* vol. 45, pp. 30–33, 2006.

[150] D. R. Grymonpré, *et al.*, "Hybrid gas–liquid electrical discharge reactors for organic compound degradation," *Industrial & Engineering Chemistry Research,* vol. 43, pp. 1975–1989, 2004.

[151] P. Lukes, *et al.*, "Hydrogen peroxide and ozone formation in hybrid gas–liquid electrical discharge reactors," *Industry Applications, IEEE Transactions on,* vol. 40, pp. 60–67, 2004.

[152] P. Lukes, *et al.*, "Generation of ozone by pulsed corona discharge over water surface in hybrid gas–liquid electrical discharge reactor," *Journal of Physics D: Applied Physics,* vol. 38, pp. 409–416, 2005.

[153] H. Ren, *et al.*, "Oxidation of ammonium sulfite by a multi-needle-to-plate gas phase

pulsed corona discharge reactor," *Journal of Physics: Conference Series*, vol. 418, p. 012128, 2013.

[154] Y. Nakagawa, *et al.*, "Decolorization of rhodamine B in water by pulsed high-voltage gas discharge," *Japanese Journal of Applied Physics*, vol. 42, pp. 1422–1428, 2003.

[155] E. Njatawidjaja, *et al.*, "Decoloration of electrostatically atomized organic dye by the pulsed streamer corona discharge," *Journal of Electrostatics*, vol. 63, pp. 353–359, 2005.

[156] Q. Tang, *et al.*, "Gas phase dielectric barrier discharge induced reactive species degradation of 2,4-dinitrophenol," *Chemical Engineering Journal*, vol. 153, pp. 94–100, 2009.

[157] J. Wohlers, *et al.*, "Application of an air ionization device using an atmospheric pressure corona discharge process for water purification," *Water, Air, and Soil Pollution*, vol. 196, pp. 101–113, 2008.

[158] A. Yamatake, *et al.*, "Water treatment by fast oxygen radical flow with DC-driven microhollow cathode discharge," *IEEE Transactions on Plasma Science*, vol. 34, pp. 1375–1381, 2006.

[159] D. Dobrynin, *et al.*, "Cold plasma inactivation of Bacillus cereus and Bacillus anthracis (anthrax) spores," *IEEE Transactions on Plasma Science*, vol. 38, pp. 1878–1884, 2010.

[160] Z. Zhang, *et al.*, "Killing of red tide organisms in sea enclosure using hydroxyl radical-based micro-gap discharge," *IEEE Transactions on Plasma Science*, vol. 34, pp. 2618–2623, 2006.

[161] H. S. Kim, *et al.*, "Use of plasma gliding arc discharges on the inactivation of *E. coli* in water," *Separation and Purification Technology*, vol. 120, pp. 423–428, 2013.

[162] L. Jie, *et al.*, "Degradation of organic compounds by active species sprayed in a dielectric barrier corona discharge system," *Plasma Science and Technology*, vol. 11, p. 211, 2009.

[163] T. C. Wang, *et al.*, "Multi-tube parallel surface discharge plasma reactor for wastewater treatment," *Separation and Purification Technology*, vol. 100, pp. 9–14, 2012.

[164] D. Zhu, *et al.*, "Wire–cylinder dielectric barrier discharge induced degradation of aqueous atrazine," *Chemosphere*, vol. 117, pp. 506–514, 2014.

[165] M. Tichonovas, *et al.*, "Degradation of various textile dyes as wastewater pollutants under dielectric barrier discharge plasma treatment," *Chemical Engineering Journal*, vol. 229, pp. 9–19, 2013.

[166] Y. S. Mok, *et al.*, "Application of dielectric barrier discharge reactor immersed in wastewater to the oxidative degradation of organic contaminant," *Plasma Chemistry and Plasma Processing*, vol. 27, pp. 51–64, 2007.

[167] J. E. Foster, *et al.*, "A comparative study of the time-resolved decomposition of meth-

ylene blue dye under the action of a nanosecond repetitively pulsed DBD plasma jet using liquid chromatography and spectrophotometry," *IEEE Transactions on Plasma Science,* vol. 41, pp. 503–512, 2013.

[168] R. Ma, *et al.*, "Non-thermal plasma-activated water inactivation of food-borne pathogen on fresh produce," *Journal of Hazardous Materials,* vol. 300, pp. 643–651, 2015.

[169] P. Olszewski, *et al.*, "Optimizing the electrical excitation of an atmospheric pressure plasma advanced oxidation process," *Journal of Hazardous Materials,* vol. 279, pp. 60–66, 2014.

[170] S. Horikoshi, *et al.*, "Microwave discharge electrodeless lamps (MDEL). III. A novel tungsten-triggered MDEL device emitting VUV and UVC radiation for use in wastewater treatment," *Photochemical & Photobiological Sciences,* vol. 7, p. 303, 2008.

[171] M. Magureanu, *et al.*, "Decomposition of methylene blue in water using a dielectric barrier discharge: optimization of the operating parameters," *Journal of Applied Physics,* vol. 104, p. 103306, 2008.

[172] B. Jaramillo-Sierra, *et al.*, "Phenol degradation in aqueous solution by a gas–liquid phase DBD reactor," *European Physical Journal Applied Physics,* vol. 56, p. 24026, 2011.

[173] R. Ono, *et al.*, "Effect of pulse width on the production of radicals and excited species in a pulsed positive corona discharge," *Journal of Physics D: Applied Physics,* vol. 44, p. 485201, 2011.

[174] T. Sugai, *et al.*, "Investigation of optimum pulse width of applied voltage for water treatment by pulsed streamer discharge in air spraying water droplets," in *Power Modulator and High Voltage Conference (IPMHVC), 2010 IEEE International,* 2010, pp. 213–216.

[175] T. Sugai and Y. Minamitani, "Influence of rise rate of applied voltage for water treatment by pulsed streamer discharge in air-sprayed droplets," *IEEE Transactions on Plasma Science,* vol. 41, pp. 2327–2334, 2013.

[176] J. Zhang, *et al.*, "Degradation of 2,4-dichorophenol by pulsed high voltage discharge in water," *Desalination,* vol. 304, pp. 49–56, 2012.

[177] M. Magureanu, *et al.*, "Degradation of pharmaceutical compound pentoxifylline in water by non-thermal plasma treatment," *Water Research,* vol. 44, pp. 3445–3453, 2010.

[178] K. Yasuoka, *et al.*, "Degradation of perfluoro compounds and F-recovery in water using discharge plasmas generated within gas bubbles," *International Journal of Environmental Science and Technology,* vol. 4, pp. 113–117, 2010.

[179] N. Sano, *et al.*, "Decomposition of organic compounds in water by direct contact of gas corona discharge: influence of discharge conditions," *Industrial & Engineering Chemistry Research,* vol. 41, pp. 5906–5911, 2002.

[180] M. Sahni and B. R. Locke, "Quantification of reductive species produced by high voltage electrical discharges in water," *Plasma Processes and Polymers,* vol. 3, pp. 342–354, 2006.

[181] Y. Jin, *et al.*, "Optimizing decolorization of methylene blue and methyl orange dye by pulsed discharged plasma in water using response surface methodology," *Journal of the Taiwan Institute of Chemical Engineers,* vol. 45, pp. 589–595, 2014.

[182] R. Zhang, *et al.*, "Kinetics of decolorization of azo dye by bipolar pulsed barrier discharge in a three-phase discharge plasma reactor," *Journal of Hazardous Materials,* vol. 142, pp. 105–110, 2007.

[183] P. M. K. Reddy, *et al.*, "Degradation and mineralization of methylene blue by dielectric barrier discharge non-thermal plasma reactor," *Chemical Engineering Journal,* vol. 217, pp. 41–47, 2013.

[184] X. Chen, *et al.*, "Degradation of 4-chlorophenol in a dielectric barrier discharge system," *Separation and Purification Technology,* vol. 120, pp. 102–109, 2013.

[185] M. Sato, *et al.*, "Aqueous phenol decomposition by pulsed discharges on the water surface," *Industry Applications, IEEE Transactions on,* vol. 44, pp. 1397–1402, 2008.

[186] Z. He, *et al.*, "The formation, role and removal of No3-N during corona discharge in air for phenol removal," *Environmental Technology,* vol. 26, pp. 285–292, 2005.

[187] P. Ajo, *et al.*, "Pulsed corona discharge in water treatment: the effect of hydrodynamic conditions on oxidation energy efficiency," *Industrial & Engineering Chemistry Research,* vol. 54, pp. 7452–7458, 2015.

5

Industrial Applications of Laser-Induced Breakdown Spectroscopy

Yoshihiro Deguchi and Zhenzhen Wang

Abstract

Laser-induced breakdown spectroscopy (LIBS) is an analytical detection technique based on atomic emission spectroscopy to measure elemental composition. With the development of lasers and detection systems, applications of LIBS encompass a broad range, including physics, engineering, space missions, environment, etc. due to the unique features of little or no sample preparation, noncontact, fast response, and multielemental analysis. The fundamental and application have been extensively studied to improve LIBS technique. This chapter largely targets the engineering fields, especially practical applications. Laser-induced breakdown spectroscopy will be discussed in this chapter including its fundamentals, industrial applications, and challenges.

Keywords: Laser-induced breakdown spectroscopy (LIBS), Industrial applications, Challenge

1. Introduction

Laser-induced plasma was introduced as a spectroscopic emission source only two years after the invention of laser. LIBS is the acronym of "laser-induced breakdown spectroscopy", and it is also called LIPS (laser-induced plasma spectroscopy), laser spark spectroscopy (LSS), and other related names [1]. LIBS is an analytical detection technique based on atomic emission spectroscopy to measure elemental composition. A laser pulse focuses in or on a sample, which can be gas, liquid, aerosol, or solid, to form the micro-plasma. The spectra emitted are used to determine the elemental constituents of measured samples.

1.1. Theory

The initiation, formation, and decay of the plasma are complex physical and chemical processes. In the LIBS process, a laser beam is focused on a small area of the sample. When the laser energy exceeds the breakdown threshold, the plasma with high temperature and high density is produced in the portion. The core of plasma is firstly produced by the absorption of the incident laser energy, such as multiphoton ionization. The creation of the plasma core induces the rapid growth of plasma through the absorption of the laser light by electrons and the electron impact ionization process in it. After the termination of the laser pulse, the plasma continues expanding because of its high temperature and pressure gradients compared with the ambient conditions. At the same time, the recombination of electrons and ions proceeds due to the collision process and the temperature decreases gradually compared with that in the plasma generation process. Emission signals arise in the plasma cooling period [2]. In the plasma, the ions, atoms, and molecules distributed in the different levels transit from the high energy level to the low energy level, emitting the strong emission spectra. The emission intensity from the atomized species provides the elemental compositions of the materials. The light corresponding to a unique wavelength of each element is emitted from excited atoms in plasma, as shown in Figure 1.

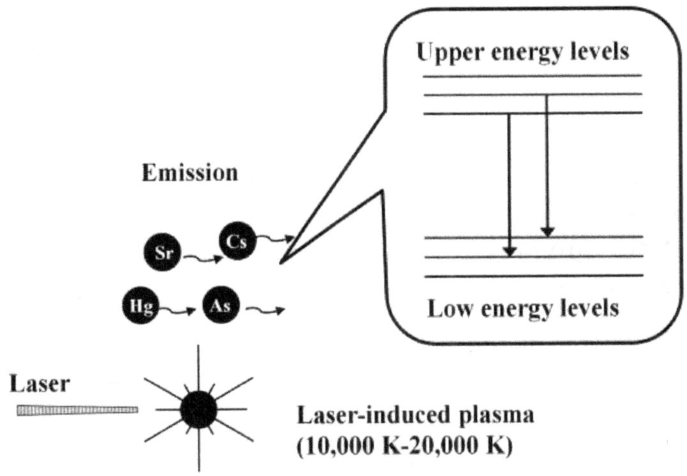

Figure 1. Laser-induced plasma processes

A calibration of the LIBS signal is necessary for quantitative analysis. Despite the fact that the LIBS processes involved are complex, the emission intensity from the atomized species can be described by the following equation with the assumption of uniform plasma temperature [2]:

$$I_{\mathrm{i}} = n_{\mathrm{i}} K_{\mathrm{i,j}} g_{\mathrm{i,j}} \exp\left(-\frac{E_{\mathrm{i,j}}}{kT}\right) \tag{1}$$

where I_{i} is the emission intensity of species i, n_{i} is the concentration of species i, $K_{\mathrm{i,j}}$ is a variable that includes the Einstein A coefficient from the upper energy level j, $g_{\mathrm{i,j}}$ is the statistical weight of species i at the upper energy level j, $E_{i,j}$ is the upper-level energy of species i, k is the Boltzmann constant, and T is the plasma temperature. Equation (1) is applicable under the conditions of local thermodynamic equilibrium (LTE). In Eq. (1), there are several factors that affect the emission intensity I_i, including plasma temperature, plasma nonuniformity, matrix effects, etc. It is very difficult to solve the LIBS process theoretically because it contains laser–material interactions, rapid temperature changes over 10,000 K in a nanosecond or picosecond timescale, and plasma cooling phenomena including the recombination process of ions, electrons, and neutrals. Therefore, the appropriate correction factors must be contained in $K_{i,j}$ to obtain the quantitative results.

The factors that affect the target signal in LIBS process generally include the background noise, stability of plasma, nonuniformity of plasma, matrix effects, dirt on measurement windows, etc. The main background noise of LIBS is the continuum emission from plasma itself. Atomic emissions appear after a certain time delay, which means that LIBS signals appear during the plasma cooling process. Therefore, it is important to choose the appropriate delay time and gate width to reduce the effect of the background noise. The fluctuation of plasma signal is an intrinsic characteristic in LIBS. Use of the intensity ratio and plasma temperature correction can reduce the fluctuation to some extent. The laser-induced plasma has its structure in it. In the plasma generation process, the plasma structure depends largely on the laser density pattern and measured material conditions. The LIBS signal depends on the measurement area across the plasma. In this sense, the correction method also includes the effects of plasma nonuniformity. Matrix effects are the combined effects of all components other than the measured species. The changes of these components may cause the alteration of LIBS signal intensity even if the number density of the measured species is the same in the measurement period. Matrix effects are usually corrected by the experimental calibrations. The attenuation of LIBS signals by the contamination of measurement windows is automatically neglected because the signal intensity ratio is usually used for the elemental composition analyses. Considering the measurement stability and soundness of windows, however, the cleanliness of measurement windows should be maintained, especially in the practical applications.

1.2. Geometric arrangement and measurement species

The LIBS apparatus fundamentally consists of laser, measured materials, lens, spectrometer, ICCD camera, and related devices. A typical geometric arrangement of LIBS is shown in Figure 2. Lasers such as a pulsed Nd:YAG laser are used as a light source. The output laser beam is focused onto the measurement area using the focal lens to induce the plasma. The plasma emission is focused onto the optical fiber. Emission signals are finally detected by the combination of a spectrometer, an ICCD camera, and auxiliary equipment. According to the measured materials of solid, liquid, and gas phases, different measurement chamber or platform can be employed. It is worth noting that the reflection of a laser light from the windows must be considered carefully. Its reflection often results in the damages of optics due to the high-energy laser light. The reflection from plasma is sometimes tricky and malicious for LIBS systems. Damages of optics by the reflection from plasma cause troubles in some cases, especially in the analyses of liquids.

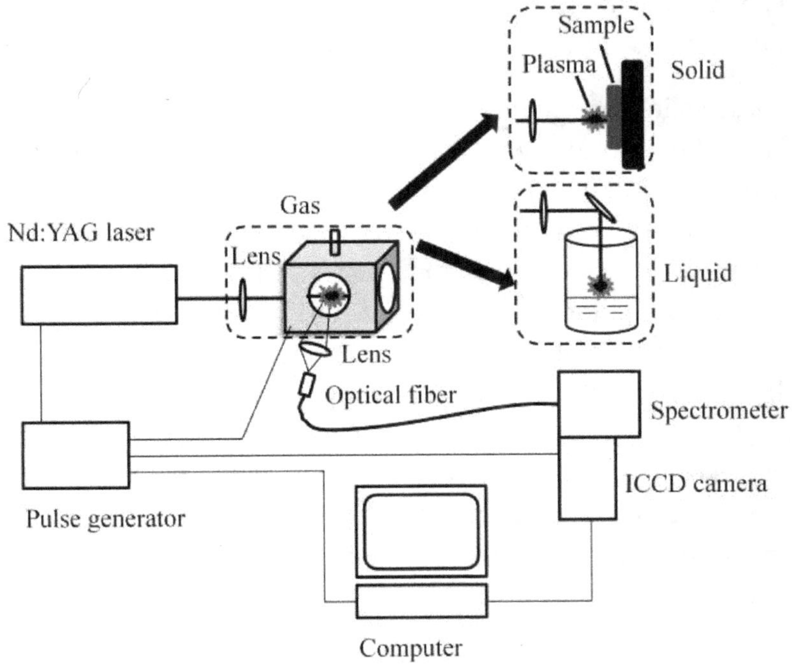

Figure 2. Typical geometric arrangement of LIBS

The basic components of LIBS system are similar but the component specifications are tailored to the particular applications. These specifications include the equipment types, physical parameters, and the technical specifications belonging to operational performance.

2. Fundamentals

Since it is the basis of successful application of LIBS technology, the fundamental study is of great importance. With the development of LIBS, the mechanism of laser–material interaction, plasma generation, plasma–environment interaction, self-absorption effect, signal enhancement, and some other fundamental research have been studied extensively to promote LIBS technique [3-7]. Various quantitative analytical methods have also been studied and improved, such as traditional calibration method, internal calibration method, calibration-free method, partial least squares (PLS) method, etc.

2.1. Plasma and its models

Plasma is a local assembly of atoms, ions, and free electrons, overall electrically neutral. Plasma is characterized by a variety of parameters. The degree of ionization is the most basic parameter. The ratio of electrons to other species is less than 10% in the plasma, called weakly ionized plasma. On the other hand, highly ionized plasma may have atoms stripped of many of their electrons, resulting in very high electron to atom or ion ratios. The plasma produced in LIBS typically belongs to the category of weakly ionized plasma. The goal of LIBS technique is to create the optically thin plasma, which is in local thermodynamic equilibrium and whose elemental composition is the same as that of the sample.

The LIBS plasma features inhomogeneities that can lead to spatial differentiation. This fact is important in choosing the temporal window in order to accumulate spectroscopic data. The spatial and temporal evolution of LIBS plasma from a steel target was monitored using time of flight and shadowgraph techniques [8]. Two regions in the plume were observed, one characterized by air and continuum emissions produced by shock wave ionization, and the other one by emissions from ablated material. The sufficiently high laser fluence and acquisition delay time are necessary to assure the homogeneity for the analytical applications. The homogeneity of LIBS plasma was investigated using the curve of growth method employing five Fe(I) lines [9]. In that formalism, the line shapes as a function of temperature and concentration were modeled. The agreement between modeled and experimental line shapes implied that the Stark effect was the dominant broadening mechanism in the plasma. The temperatures obtained from neutral and ion spectral lines were studied [10]. The different temperatures studied can be obtained from Boltzmann and Saha plots. The difference was explained by the spatial variation of the plasma temperature and densities leading to a difference in spatial locus for populations in the upper levels of transitions for neutrals and ions.

Plasma models are becoming more comprehensive and detailed. A radiative model of LIBS plasma expanding into a vacuum was validated by the experiments [11]. The inverse problem was specifically addressed, which means finding the initial conditions by the comparison of the calculated synthetic spectra and the experimentally measured ones. The composition of the material was effectively deduced from the calculated spectra. The plasma was considered to be characterized by a single temperature and electron density. The combination of the original modeling work on laser-evaporated plasma plume expansion into a vacuum and ablation leading to vaporization and particle formation was studied [12]. The interaction of a nanosecond pulse with a copper target was modeled in vacuum. Some of the parameters were studied including the melting and evaporation of the target, the plume expansion and plasma formation, the ionization degree and density profiles of neutral; once-ionized; and doubly ionized copper and electrons, and the resultant plasma shielding.

2.2. LIBS detection ability

Most fundamental studies focused on signal enhancement to improve the detection limit. The measured results showed that the spatial confinement and fast discharge would be able to enhance the signal from several times to dozens of times, while dual pulse is able to enhance signal 100–1000 times [13-15]. Besides signal enhancement, there were also some other studies worth mentioning. The self-absorption in laser-induced plasma was studied. The results suggested that the self-absorption effect could be alleviated by the selection of suitable atomic line, operating at higher pulse energy and detecting with longer delay [16]. The pressure effect on the plasma emission from fundamental 0.1 to 40 MPa in bulk seawater was investigated [17]. The time-resolved LIBS emission results demonstrated that plasma emission is weakly dependent on the ambient pressure during the early stage of plasma and the pressure has a significant influence on the plasma form during plasma evaluation at a later stage of plasma.

The detection ability of trace species using LIBS has been improved with the development of LIBS devices. The utilization of short pulse laser for plasma generation has been extensively studied [18,19]. Short pulse irradiation allowed for a specificity of excitation that could yield LIBS signals more tightly correlated to particular chemical species and showed significantly lower background emission. A new method to control the LIBS plasma generation process is necessary for the enhancement of detection limit, i.e., low pressure and short pulse LIBS [20-22]. Because of the pressure, volatility, and quenching effects of liquid, the plasma lifetime of liquid sample is shorter compared with that of solid and gas phases. Meanwhile, sputtering of liquid sample by LIBS plasma often raises the problem of the measurement windows. The sensitivity, stability, and repeatability of LIBS signal are much lower, leading to the increasing difficulty of its analyses. Numerous papers have reported LIBS measurement of different forms of liquid phase materials including the solidification, liquid bulk, liquid surface, and others [23-28], which shows different detection features and detection limit.

2.3. Quantitative analysis

The ultimate goal of LIBS technique is to provide a quantitative analysis with high precision and accuracy. Usually, a quantitative analysis begins with determining the response of a system for a given concentration or mass of the analyte of interest, which usually takes the form of a calibration curve. The calibration is usually strongly dependent on the analysis conditions, such as the stability of the laser pulse energy, the sample and sampling procedure, the physical and chemical properties of the sample, etc. The dependence of elemental signals of LIBS on the plasma temperature attributes to a very complex process in plasma. Several studies have reported the LTE condition of plasma in several types of plasmas [29]. The plasma temperature is a very important factor for the quantification of the LIBS measurement. There are several calibration methods to analyze the measured species quantitatively, including the traditional calibration method, internal calibration method, calibration-free method, etc. [30,31].

As for the simple samples, the emission intensity of the measured species is linear with the species content under the ideal condition. The traditional calibration model is relatively simple and convenient. However, the influences of matrix effect and element interference are not considered in the model. The accuracy becomes worse when the complex samples are measured or the experimental parameters fluctuate. The internal calibration method is a commonly used spectral analysis model with strict conditions. The elements with the features of high content, low detection limit, and good stability are mainly selected as the internal calibration elements. Usually, the compositions of the calibration sample and measured sample are not entirely consistent. When the measured samples contain various elements, the accuracy will be affected due to the matrix effect.

A new procedure is proposed for calibration-free quantitative elemental analysis of materials using LIBS technique. The method based on an algorithm developed and patented overcomes the matrix effects. The precise and accurate quantitative results on elemental composition of materials can be acquired without the use of calibration curves. Some applications of the method have been illustrated, e.g., the quantitative analysis of the composition of metallic alloys [32]. This model of CF-LIBS is applicable under the conditions of LTE and optically thin, as well as the assumed conditions without the element interference and self-absorption. Research recently focused on the correction for self-absorption. Multivariate analysis (MVA) is an effective mathematical and statistical approach for LIBS data analysis, since it can utilize much quantitative information from the complex LIBS spectra. Partial least squares (PLS) is such an MVA method and has shown great potential for LIBS quantitative measurement. The model utilizes the multiline spectral information of the measured element and characterizes the signal fluctuations due to the variation of plasma characteristic parameters, such as plasma temperature, electron number density, and total number density, for signal uncertainty reduction [33,34]. LIBS can be used to provide the quantitative analysis of a variety of samples in the laboratory and in the field. However, each application has some unique characteristics that must be dealt with in order to optimize performance. In the real applications of LIBS, the procedures for obtaining quantitative results reproducibly will be developed.

A much deeper understanding of LIBS fundamental physics is the key to overcome the bottlenecks for wide applications of LIBS, such as the relatively low measurement repeatability due to the plasma property and morphology fluctuations, the relatively low accuracy suffered from matrix effects, etc. The plasma generation and evolution processes are complicated processes. Much more work is still required to improve the qualitative and quantitative analyses, as well as the applications of LIBS technique.

3. Applications

LIBS has attracted great attention in various industries as a qualitative and quantitative analytical detection technique due to its noncontact, fast-response, and multidimensional features. With the development of laser and detection systems, LIBS has been applied in various fields, including combustion, metallurgy, food, human, the Mars project, and so on. Especially, the advantages of this method are more significant in the areas of combustion, metallurgy, and harsh environments. Many applications have successfully demonstrated the monitoring of plant control factors using LIBS. LIBS has been actively applied to commercial plants such as iron and steel making, thermal power, waste disposal, and so on. Environmental monitoring and safety applications are also the growing fields for LIBS. Applications of LIBS have covered all industry fields, including analyses from food, plant to space missions, which will be discussed in detail in the next section. In these cases, ruggedness and reliability become important requirements.

3.1. Applications for plants

LIBS, with the features of excellent temporal and spatial resolutions, appears to be a very promising analysis method in steel industry where element distribution measurements of materials at all stages of production provide the information of material quality and production process. By the continuous monitoring of element distribution, the raw materials with narrow composition tolerances can be available ahead of further processing. LIBS measurement of geological materials on the conveyor belts was studied and discussed preliminarily [35,36]. A multispectral line calibration method was proposed for the quantitative analysis of elemental compositions. Its feasibility and superiority over a single-wavelength determination have been confirmed by comparison with the traditional chemical analysis of the copper content in the ore. Two iron ore samples were employed to complete the mineralogical classification using a combination of LIBS and principal components analysis (PCA)/principal components regression (PCR) [37,38]. The combined method of LIBS and PCR was applied to determine the elemental compositions of a series of run-of-mine iron ore samples, which exhibited the potential for in situ determination of ore composition. The calibration models of LIBS have also been studied and discussed in the measurement of ores. The different data-driven multivariate statistical predictive algorithms, such as Principal Components Regression (PCR), Partial Least Squares Regression (PLSR), Multi-Block Partial Least Squares (MB-PLS), and Serial Partial Least Squares Regression (S-PLSR), were compared for the quantitative analysis in iron ore measured using LIBS to improve the performance of the quantitative measurements [39,40]. The on-line measurement system of LIBS has been discussed for the real applications. Figure 3 shows a LIBS system for on-line measurement using extractive sampling. For example, an analytical instrument based on LIBS technique was developed to operate on-line in the harsh environment of iron-ore pelletizing plants. The detection system was successful for the measurements of Si, Ca, Mg, Al, and graphitic C contents in different iron ore slurries prior to filtration and pelletizing [41]. A method for automated quantitative analysis of ores was developed using a commercial LIBS instrument fitted with a developed computer-controlled auto-sampler [42]. The preparation and analysis time for each sample was less than 5 min. The similar method was suitable for a range of ores and minerals.

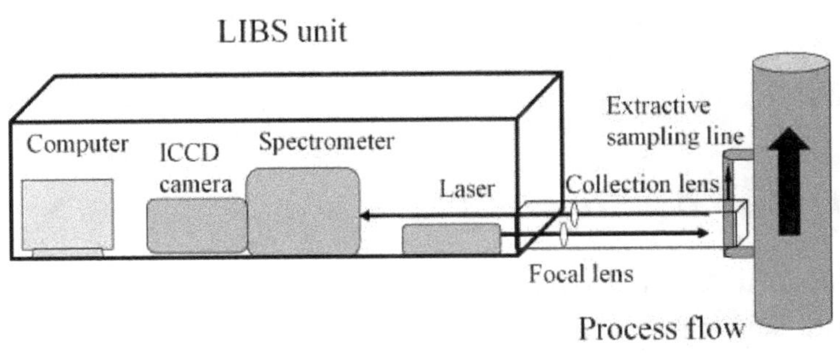

Figure 3. A LIBS system for on-line measurement using extractive sampling

Operating characteristics of coal-fired boilers are heavily influenced by factors such as the differences in fuel properties and combustion conditions. In order to achieve the optimal operation of multiple coal-fired boilers, it is necessary to accurately understand the coal quality and the state of combustion, and to adjust the control parameters. LIBS technique has been widely applied to analyze the compositional characterization of coal [43-45]. A nonlinearized multivariate dominant-factor-based partial least-squares (PLS) model was applied to coal elemental concentration measurement using LIBS [46]. Unburned carbon in fly ash is an important factor to estimate the combustion efficiency of boiler. Fly ash and bottom ash resulting from the coal combustion in a coal-fired power plant were analyzed using binders. Once the experimental conditions and features are optimized, application of LIBS may be a promising technique for combustion process control even in on-line mode [47,48]. Software-controlled LIBS systems including LIBS apparatus and sampling equipment have been designed for possible application to power plants for on-line quality analysis of pulverized coal and unburned carbon in fly ash [49,50], which shows the capability of reliable and real-time measurement using LIBS. LIBS has been applied for detection of unburned carbon in fly ash, char, and pulverized coal without any sample preparation. Figure 4 presents the examples of LIBS spectral lines obtained from fly ash. The calibration difficulty of aerosol sample was surpassed by using the correction factors for quantitative measurement. This automated LIBS apparatus was applied in a boiler-control system of a power plant with the objective of achieving optimal and stable combustion [51,52], which enabled real-time measurement of unburned carbon in fly ash, as shown in Figure 5. The boiler control in the real power plant was demonstrated to achieve an optimized operation without time consumption.

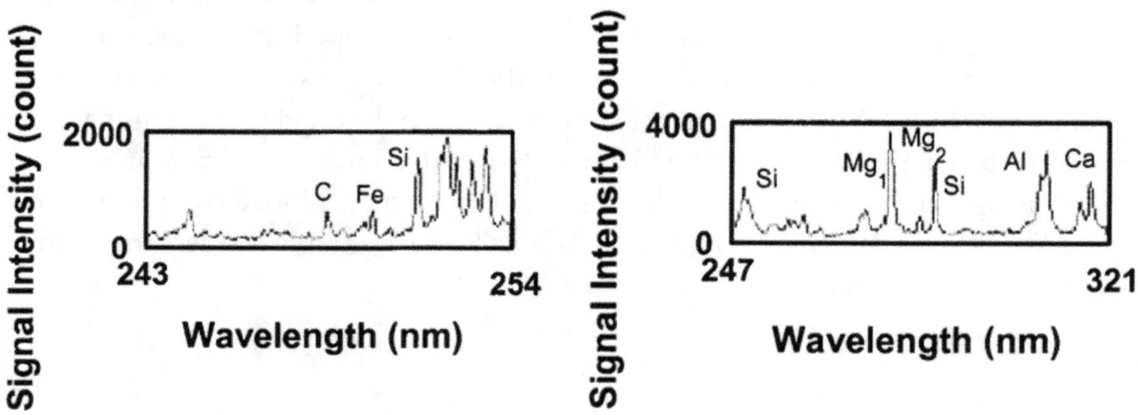

(a) High-resolution spectrum (b) Wide-range spectrum

Figure 4. Fly ash LIBS spectra [52]

The safe and rational utilization is very important for nuclear power application. The radioactive contamination is a serious problem for the environment and human health. The radioactive materials released from the nuclear power plant are one of the main sources. Simultaneously, nuclear weapons testing fallout, some industry waste discharge, and radioactive substances employed for research also contribute to the issue [53,54]. The atmosphere, water and soil are polluted by these released radioactive materials. There are several serious pollutions to environment and human not only in the surrounding area of the nuclear power plant but also in the outlying regions [55-57]. LIBS has been investigated as a potential analytical tool to improve operations and safeguards for electrorefiners such as those used in processing spent nuclear fuel [58]. Detection of uranium and other nuclear materials is very important for nuclear safeguards and security. The spatial and temporal evolutions of uranium species in laser-produced plasmas were investigated. A set of optimal operating conditions was determined based on the experimental results, which is important for obtaining the

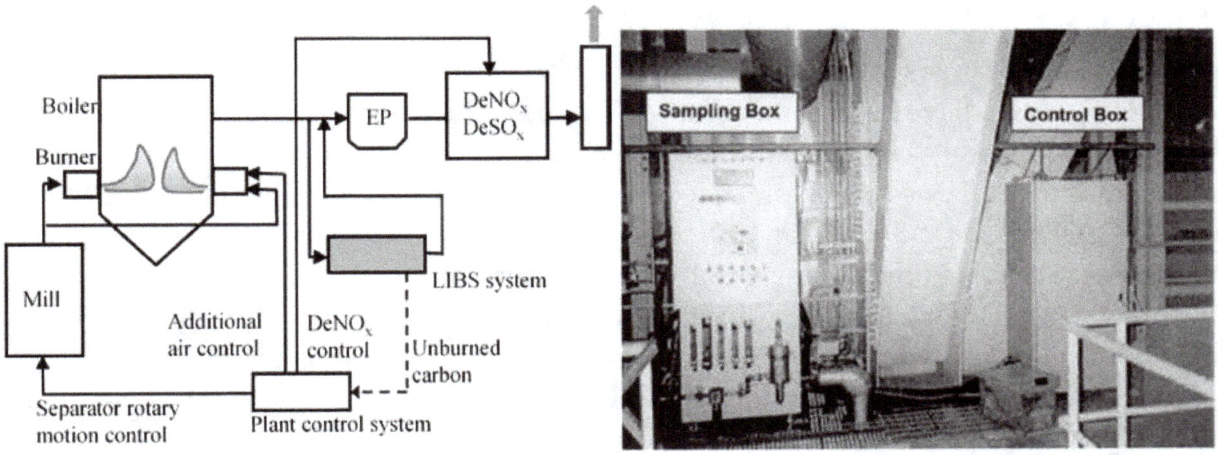

(a) Optimal boiler control system (b) Photograph of the measurement apparatus

Figure 5. Unburned-carbon measurement in thermal power plant [52]

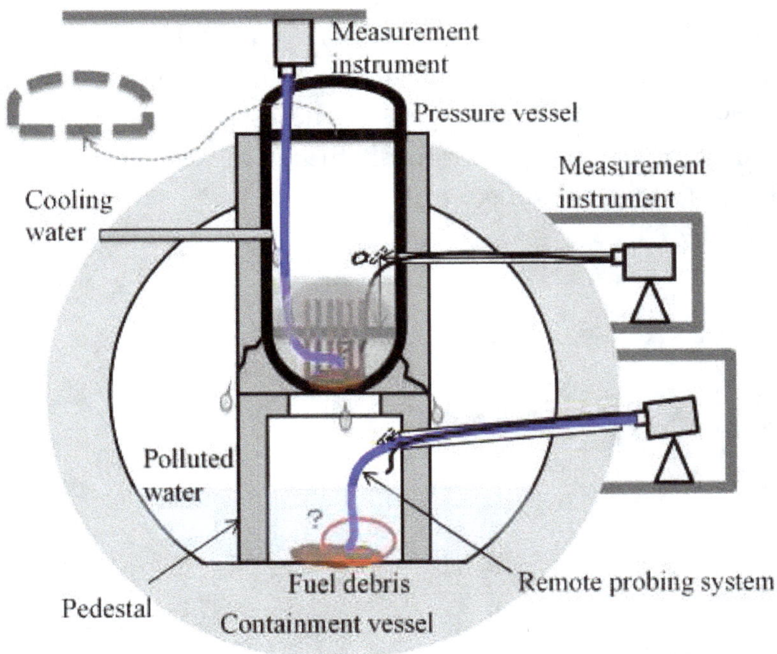

Figure 6. Image of the inside of the post-accident Fukushima Daiichi nuclear power plant and the inspection technique using optical fiber [60]

optimal spectral intensity from samples containing very small amounts of uranium [59]. LIBS can be applied to monitor radioactive elements, which is of utmost importance in case of leakage of radioactive materials from a nuclear power plant [60,61]. Figure 6 shows a schematic of the imaging observation using the imaging fiber, which was equipped for additional electric delivery. A transportable fiber coupled LIBS instrument was developed, which is feasible for the material analysis of underwater debris under a high-radiation field.

3.2. Applications for food, humans, and archaeology

As for LIBS applications to food, composition and contamination measurements of flours of wheat, barley, etc., have been demonstrated. The feasibility of quantifying trace elements in powdered food samples by spatially resolved LIBS has been demonstrated under a reduced argon atmosphere. The selection of the location in the plasma is crucial for obtaining the best signal-to-background ratio of analytical signal and to avoid background continuum and line broadening. The operating parameters affected the plasma property were optimized and used for further analysis of trace elements in starch-based food samples. Spatially resolved LIBS has been shown to be an accurate technique for determining trace elements of ppm level in starch-based food samples directly with an acceptable precision without any tedious digestion and dilution procedure [62]. A procedure for the analysis of K, P, Mg, and Ca in crop plant samples using a commercially available LIBS spectrometer was also developed. Real plant samples employed as the calibration standards were analyzed by ICP-OES or AAS after microwave digestion. A satisfactory agreement between LIBS and AAS/ICP-OES results was achieved [63]. The trace and ultra-trace element detection and qualitative analysis in fresh vegetables have been demonstrated using LIBS technique [64]. For a typical root vegetable such as potato, spectral analysis of the plasma emission reveals more than 400 lines emitted by 27 elements and 2 molecules, C_2 and CN. Many elements such as Mg, Al, Cu, Cr, K, Mn, Rb, Cd, and Pb have been measured by LIBS, as shown in Figure 7. These results demonstrate the potential of an interesting tool for botanical and agricultural studies as well for food quality/safety and environment pollution assessment and control.

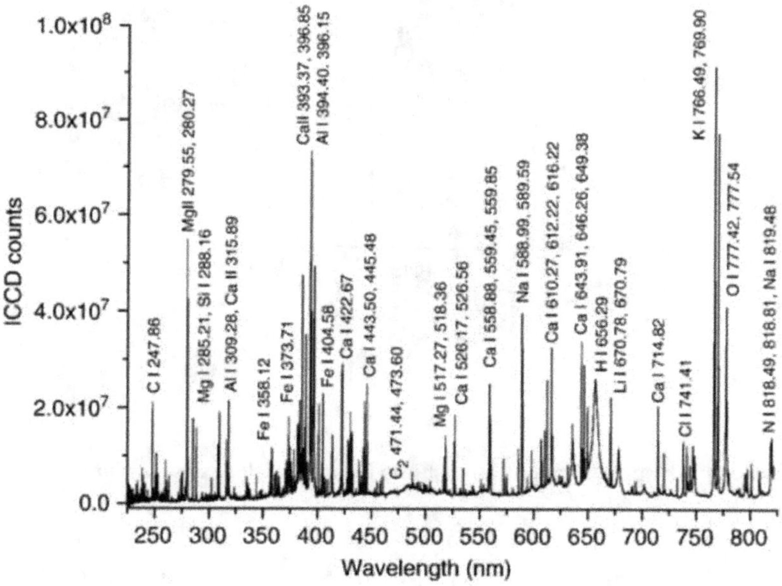

Figure 7. Typical LIBS spectrum of a fresh potato [64]

The application of LIBS to the analysis of important minerals and the accumulation of potentially toxic elements in calcified tissue has been reported [65], which exemplified for quantitative detection and mapping of Al, Pb, and Sr in representative samples, including teeth

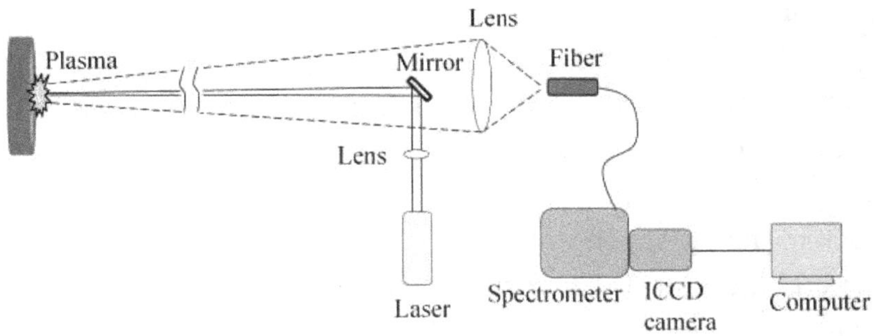

Figure 8. LIBS schematic diagram for measurements from a distance

and bones. In order to identify and quantify the major and trace elements in the tissues, one- and two-dimensional profiles and maps were generated. The state of the tooth has also been diagnosed using prominent constituent transitions in laser-excited tooth [66]. The spectroscopic observations in conjunction with discriminate analysis showed that calcium attached to the hydroxyapatite structure of the tooth was affected severely at the infected part of the tooth. It is possible to distinguish the healthy and caries infected tooth using emission spectroscopy and ICCD imaging of the expanding plasma.

Advancement in LIBS technique has led to its increased use in the fields of conservation, art history, and archaeology [67,68]. A prototype LIBS system was used to determine the elemental composition of multilayer structures in a metal jug from the mid-twentieth century [69]. The piece was highly deteriorated due to environmental damage. The LIBS technique was used as part of a historical investigation that required the determination of the material employed. The jug was selectively sampled at different points on the surface using the stereoscope. By sampling at different points, the surface composition was determined. Furthermore, the presence of two layers of Pb and Cu and their thicknesses were determined through in-depth analysis.

3.3. Applications for space and underwater explorations

One of the more exotic and exciting applications of LIBS instrumentation is for space missions to planet surfaces. Current technological developments of lasers, spectrometers and detectors have made the use of LIBS for space exploration feasible. Figure 8 shows the LIBS schematic diagram for measurements from a distance. LIBS technique greatly increases the scientific return from new missions by providing extensive data relating to planetary geology, which is one of the main goals of space exploration. Meanwhile, the geologic analysis can provide some information of a planet's history, e.g., whether earlier conditions were favorable for life. Several studies have addressed the feasibility of LIBS for space exploration [70-72]. The feasibility of LIBS for stand-off analysis of geological samples under Martian atmospheric conditions has been demonstrated. Under Martian conditions, the analyzed signals appear to be somewhat enhanced compared to that recorded at atmospheric pressure due to the increased ablation.

It is a big challenge to apply LIBS to ocean in situ detection, which has been studied with the development of LIBS. In order to apply LIBS to in situ elemental analysis in the deep ocean, the multielemental analysis of high-pressure aqueous solutions has been studied. The affected factors, such as pressure, laser energy, and so on, have been discussed [73,74]. The potential of LIBS for the underwater chemical characterization of archaeological materials has been also demonstrated [75,76], which involves the delivery of a focused laser pulse toward the distant target through the aqueous media and then the transmission of the light emitted by the laser-induced plasma back to the detection system. Figure 9 shows the LIBS spectra corresponding to different submersed materials obtained in laboratory. The performance of the remote LIBS system was evaluated in a measurement campaign in the Mediterranean Sea. The pictures taken during the on-site trials using LIBS are illustrated in Figure 10. The seashell as the biomineralization product records the growth development and the ocean ecosystem evolution. Therefore, the seashell has been studied as a representative for marine research [77]. LIBS-Raman combination was introduced to obtain the compositional distribution of scallop shell on the surface and also in the shallow layers, which suggested that the micro-chemical diagnostics of LIBS-Raman was a potential way to construct a 3D analysis for the shell research. There are also other techniques that combine LIBS and other methods for the 3D surface analysis.

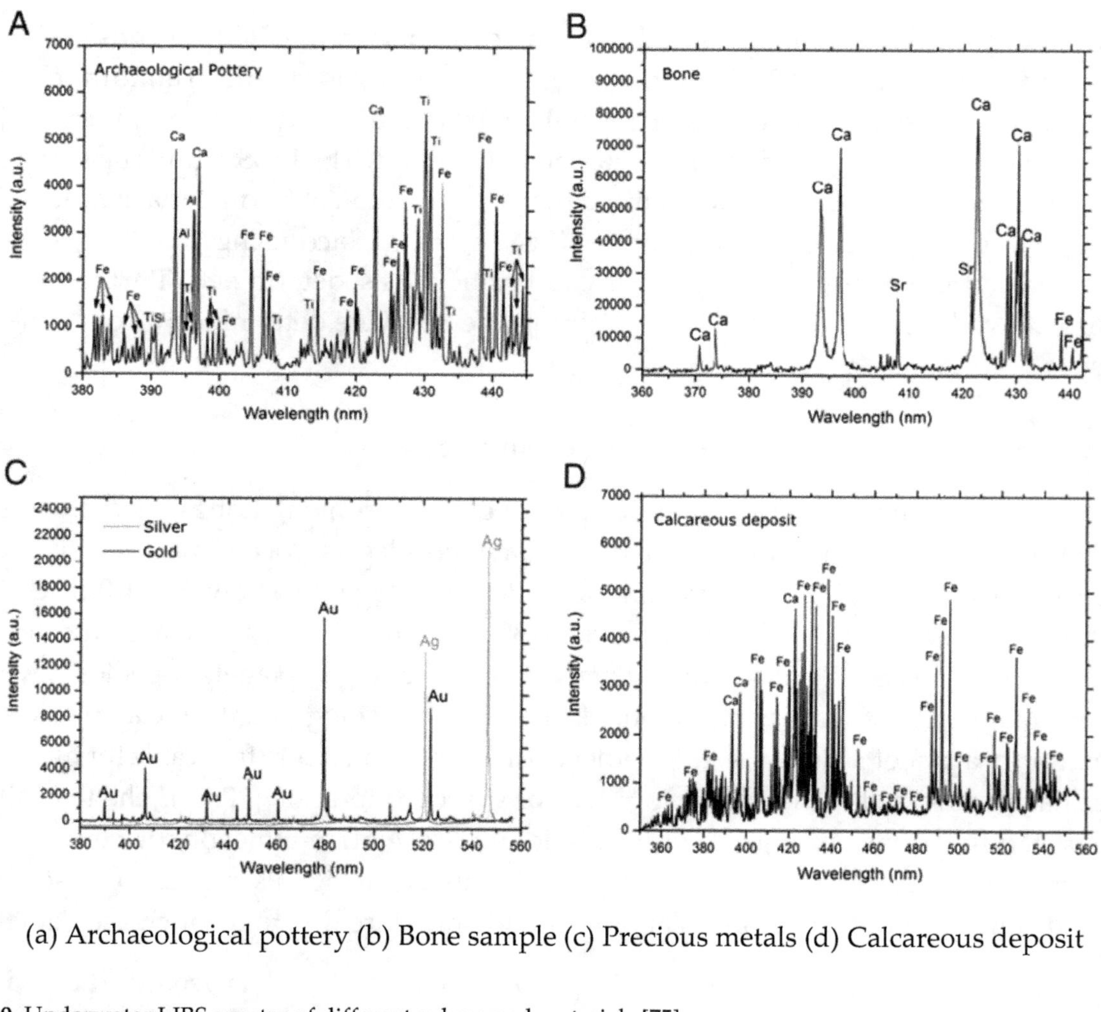

(a) Archaeological pottery (b) Bone sample (c) Precious metals (d) Calcareous deposit

Figure 9. Underwater LIBS spectra of different submersed materials [75]

(a) Remote LIBS instrument on the research vessel

(b) Diver working at a 30 m depth

Figure 10. Pictures taken during the on-site trials on the Mediterranean Sea [75]

3.4. Other applications

Applications of LIBS have covered all industry fields, including analyses from food, plant, to space missions. Apart from the applications mentioned above, there are a variety of other applications of LIBS. In engine applications, LIBS has been used to measure the fuel–air ratios in combustion. If the fuel composition does not change, the fuel–air ratio can be inferred from the elemental analysis of unburned and burned gases. It is useful to know that the equivalence ratio can be inferred from burned gas measurement because the elemental composition does not change during reactions [78]. LIBS can be also used for the elemental analysis of particles, such as soot, which contain not only carbon but also metallic elements [79,80]. Tighter environmental regulations recently have focused on global limit of harmful substances released from industry, traffic, and domestic waste. Due to the sensitive and fast analysis features, LIBS has the capability to be used as a continuous-emission monitor for toxic metals, such as Be, Cd, Cr, Hg, and Pb. The sampling methodology and signal processing have been improved [81-83]. The utilization of LIBS technique has been extensively studied in different phase conditions, i.e., solid, liquid, and gas materials, which show different laser-induced plasma processes. The wide pressure dependence and various atmospheric compositions have been studied to understand the LIBS phenomena [84]. One of the challenging targets of LIBS is the enhancement of detection limit of gas phase materials. A new method to control the LIBS plasma generation process has been proposed for the enhancement of detection limit, i.e., low pressure and short pulse LIBS.

LIBS is a promising technique for in situ elemental analysis. The advancement of portable LIBS systems becomes a key technique. There have been a growing number of applications using LIBS in life science, medical fields, and so on. A new mobile instrument for LIBS analysis was developed, which is based on double-pulse LIBS and a calibration-free LIBS technique. Some applications have been presented including the analysis of cultural heritage, environmental diagnostics, and metallurgy [85]. Laser-induced breakdown spectroscopy and Raman spectroscopy have several features that make a combined instrument for remote analysis. These two techniques are very useful and feasible as the combination of elemental compositions from LIBS and molecular vibrational information from Raman spectroscopy strongly complement each other. Remote LIBS and Raman spectroscopy spectra were taken together on a number of mineral samples [86,87]. Figure 11 shows the Raman spectra, the combined Raman and LIBS spectra, and the LIBS spectra of calcite ($CaCO_3$) in air in the 534–699 nm wavelength range. The Raman lines in Figure 11(b) are marked with the letter "R." On the other hand, an approach to further increase the sensitivity of LIBS for the determination of traces is the combination of LIBS and laser-induced fluorescence, which has been studied. The combination of LIBS and LIF allows linking the multispecies capability of LIBS for a broad range of analytes with the high sensitivity of LIF for individual selected species [88].

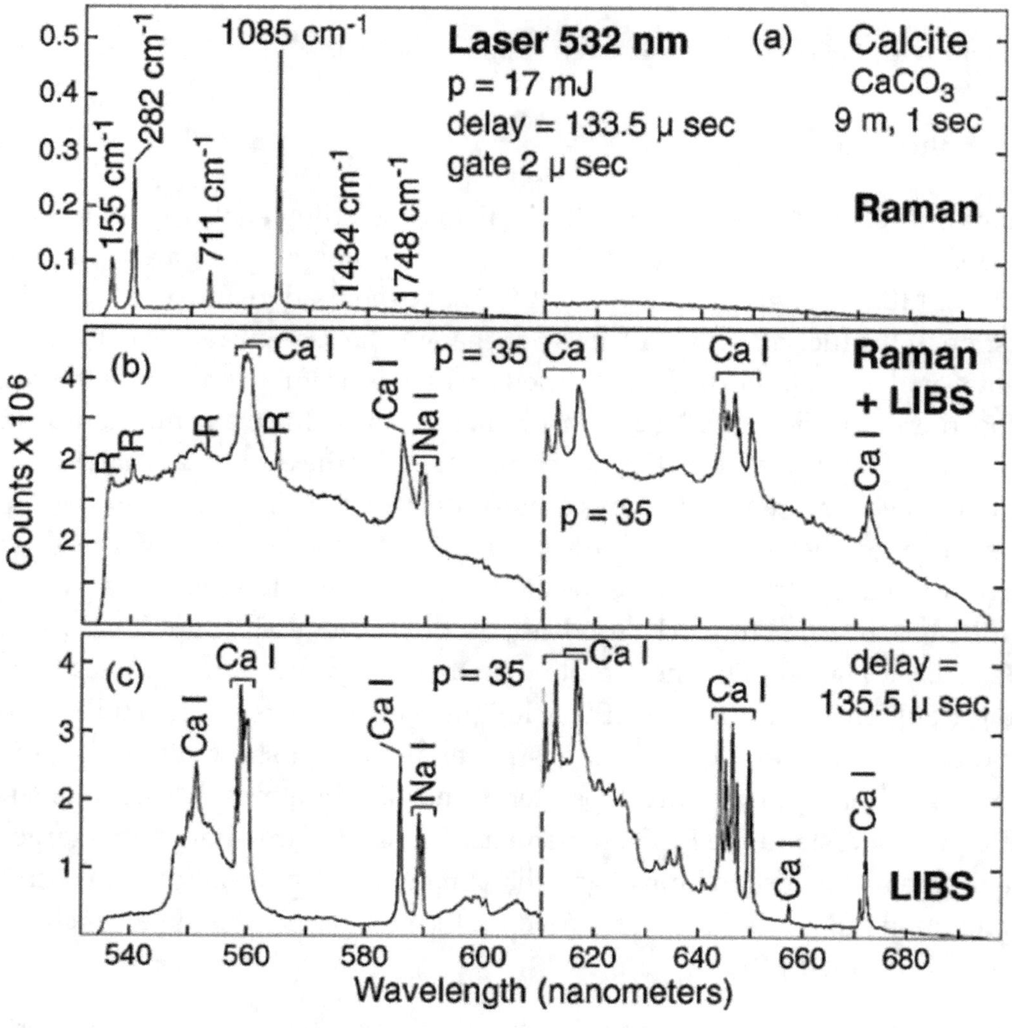

(a) Raman spectra (b) Combined Raman and LIBS spectra (c) LIBS spectra

Figure 11. Combined remote Raman and LIBS spectra of calcite in the 534–699 nm range [86]

4. Challenges

LIBS features various merits of the noncontact, fast response, and multidimensional detection and has been widely studied and applied in different fields as the qualitative and quantitative analytical detection technique. However, one of the major drawbacks of LIBS is the difficulty of quantitative analysis. There are numerous correction methods for LIBS to achieve the quantitative information, which are usually application-dependent. On the other hand, it has become increasingly important to monitor factors in plant conditions in order to improve the operation of industrial plants. Furthermore the long-term continuous use of LIBS system is a considerable factor for industrial applications. As a consequence, the improvement of measurement accuracy, quantitative analysis, and real-time measurement is very necessary for the operation of the overall plants using LIBS system.

4.1. Accuracy and durability

The laser-induced plasma processes are different from the phase samples in various applications. The measurement methods and parameters should be determined according to the specific conditions. There are several important factors that need to be considered when obtaining quantitative information using LIBS. Choosing the appropriate experimental parameters, therefore, is important to make the theoretical treatment applicable for quantitative measurements. On the other hand, data processing and modeling play an important role in LIBS for the analytical results of the measured spectra. An ideal data processing method should be based on a deep understanding of plasma physics and should be capable of minimizing the noise effects, compensating for the signal fluctuations, and reducing the matrix effects. There have been several calibration methods such as the Boltzmann plot method using many emission lines to increase the correction precision. The calibration methods should be developed to realize the quantitative analyses with the precision and accuracy of a measurement. As for the on-line application in the industry, the system simplicity and real-time measurement capability are also significant factors to be considered. The methods for quantitative analyses should be workable and satisfactory for practical applications.

The real advantage of LIBS technique is that the results are delivered continuously and in real time compared with periodic sampling and standard analytical methods with the time consumption. Consequently, LIBS gives a more representative reading of the state of the process, particularly when rapid perturbations occur, and allows process optimization and quality improvement. Current research aims to develop the commercial equipment for continuous industrial applications. However, in these applications the long-term stability and durability of LIBS devices, especially lasers, is one of the challenges. LIBS employs pulsed lasers and their lifetime often limits the plant applications, especially the long-term continuous use for plant monitoring and control. In a harsh environment, actually, all the devices should be paid attention to, including lasers.

4.2. Instrument development

The applications of LIBS technique have recently been proposed in the fields of materials science, industrial process control, environmental protection, cultural heritage conservation, etc. All of these applications would benefit from a mobile instrument. Therefore, the availability of affordable commercial instrumentation, the standardization of measurement procedures, and the calibration standards are required for reproducible and reliable quantitative LIBS analysis in situ.

The focus of a laser beam of LIBS, for instance, is one of the most important factors to be considered when applying LIBS to industrial processes with the change of a target profile. 3D profile information of the object is required for the positioning of a focused laser beam. The noncontact-type profile measurement systems, in general, can be divided into three categories, including a measurement machine integrated with a triangulation laser probe [89], a measuring machine integrated with a laser line projector and one/two CCD cameras [90-92], and a measurement machine integrated with a structured fringes projector and two CCD cameras [93]. To digitize small complex objects with dimensions smaller than about 30 mm, using a measurement machine integrated with a triangulation laser probe is a good strategy due to its small spot size. In general, phase shifting algorithm is applied to calculate the phase map and the 3D profile of an object using the structured fringes projection system [94]. If a 3D profile measurement system can be integrated with a LIBS system, the measured 3D profile information of the object can be used for the real-time positioning of a focused laser beam in a LIBS system.

It has become increasingly important to monitor factors in plant conditions in order to improve the operation of industrial plants. As a consequence, improved on-line monitoring techniques for plant control factors are necessary to enhance the capability of maintaining the overall plant operation. The associated monitoring and control techniques are necessary for the continued operational improvement. Emphasis is placed mainly on instrument development for applications as well as fundamental scientific investigations.

Author details

Yoshihiro Deguchi[1*] and Zhenzhen Wang[2]

*Address all correspondence to: ydeguchi@tokushima-u.ac.jp

1 Graduate School of Advanced Technology and Science, The University of Tokushima, Tokushima, Japan

2 State Key Laboratory of Multiphase Flow in Power Engineering, Xi’an Jiaotong University, Xi’an, China

References

[1] Miziolek AW, Palleschi V, Schechter I. *Laser Induced Breakdown Spectroscopy*. Cambridge University Press: Cambridge, United Kingdom; 2008. p. 1-4.

[2] Deguchi Y. *Industrial Applications of Laser Diagnostics*. CRS Press, Taylor & Francis: New York, USA; 2011. p. 107-113.

[3] Guo LB, Li CM, Hu W, Zhou YS, Zhang BY, Cai ZX, Zeng XY, Lu YF. Plasma confinement by hemispherical cavity in laser-induced breakdown spectroscopy. *Appl Physics Lett*. 2011; 98: 131501-1-131501-3. doi:10.1063/1.3573807

[4] Lui SL, Cheung NH. Resonance-enhanced laser induced plasma spectroscopy for sensitive elemental analysis: Elucidation of enhancement mechanisms. *Appl Physics Lett*. 2002; 81(27): 5114-5116. DOI: 10.1063/1.1532774

[5] Lui SL, Cheung NH. Resonance-enhanced laser induced plasma spectroscopy: Ambient gas effects. *Spectrochim Acta Part B-Atomic Spectroscopy*. 2003; 58: 1613-1623. DOI: 10.1016/S0584-8547(03)00139-3

[6] Hou ZY, Wang Z, Liu JM, Ni WD, Li Z. Signal quality improvement using cylindrical confinement for laser induced breakdown spectroscopy. *Optics Express*. 2013; 21(13): 15974-15979. DOI:10.1364/OE.21.015974

[7] Li XF, Zhou WD, Cui ZF. Temperature and electron density of soil plasma generated by LA-FPDPS. *Front Physics*. 2012; 7(6): 721-727. DOI 10.1007/s11467-012-0254-z

[8] Corsi M, Cristoforetti G, Hildalgo M, Iriarte D, Legnaioli S, Palleschi V, Salvetti A, Tognoni E. Temporal and spatial evolution of a laser-induced plasma from a steel target. *Appl Spectro*. 2003; 57(6): 715-721. DOI: 10.1366/000370203322005436

[9] Aguilera JA, Bengoechea J, Aragón C. Curves of growth of spectral lines emitted by a laser-induced plasma: influence of the temporal evolution and spatial inhomogeneity of the plasma. *Spectrochim Acta Part B-Atomic Spectroscopy*. 2003; 58: 221-237. DOI: 10.1016/S0584-8547(02)00258-6

[10] Aguilera JA, Aragón C. Characterization of a laser-induced plasma by spatially resolved spectroscopy of neutral atom and ion emissions. Comparison of local and spatially integrated measurements. *Spectrochim Acta Part B-Atomic Spectroscopy*. 2004; 59: 1861-1876. DOI:10.1016/j.sab.2004.08.003

[11] Gornushkin IB, Kazakov AY, Omenetto N, Smith BW, Winefordner JD. Experimental verification of a radiative model of laser-induced plasma expanding into vacuum. *Spectrochim Acta Part B-Atomic Spectroscopy*. 2005; 60: 215-230. DOI:10.1016/j.sab.2004.11.009

[12] Bogaerts A, Chen Z, Gijbels R, Vertes A. Laser ablation for analytical sampling: what can we learn from modeling? *Spectrochim Acta Part B-Atomic Spectroscopy*. 2003; 58: 1867-1893. DOI:10.1016/j.sab.2003.08.004

[13] Sun DX, Su MG, Dong CZ, Wen GH. A comparative study of the laser induced breakdown spectroscopy in single- and collinear double-pulse laser geometry. *Plasma Sci Technol.* 2014; 16(4): 374-379. DOI: 10.1088/1009-0630/16/4/13

[14] Zhou WD, Su XJ, Qian HG, Li KX, Li XF, Yu YL, Ren ZJ. Discharge character and optical emission in a laser ablation nanosecond discharge enhanced silicon plasma. *J Analy Atomic Spectrometry.* 2013; 28: 702-710. DOI: 10.1039/c3ja30355a

[15] Li KX, Zhou WD, Shen QM, Ren ZJ, Peng BJ. Laser ablation assisted spark induced breakdown spectroscopy on soil sample. *J Analy Atomic Spectrometry.* 2010; 25: 1475-1481.DOI: 10.1039/b922187e

[16] Sun LX, Yu HB. Correction of self-absorption effect in calibration-free laser-induced breakdown spectroscopy by an internal reference method. *Talanta.* 2009; 79: 388-395. DOI:10.1016/j.talanta.2009.03.066

[17] Hou HM, Tian Y, Li Y, Zheng RE. Study of pressure effects on laser induced plasma in bulk seawater. *J Analy Atomic Spectrometry.* 2014; 29: 169-175. DOI: 10.1039/c3ja50244a

[18] Megan RLH, Joseph M, Robert O, Jennifer B, Yamac D, Caroline M, James BS. Ultrafast laser-based spectroscopy and sensing: applications in LIBS, CARS, and THz spectroscopy. *Sensors.* 2010; 10: 4342-4372. DOI: 10.3390/s100504342

[19] Eland KL, Stratis DN, Lai T, Berg MA, Goode SR, Angel SM. Some comparisons of LIBS measurement using nanosecond and picosecond laser pulses. *Appl Spectroscopy.* 2001; 55(3): 279-285. DOI: 10.1366/0003702011951894

[20] Wang ZZ, Deguchi Y, Kuwahara M, Yan JJ, Liu JP. Enhancement of laser-induced breakdown spectroscopy (LIBS) detection limit using a low-pressure and short-pulse laser-induced plasma process. *Appl Spectroscopy.* 2013; 67(11): 1242-1251. DOI: 10.1366/13-07131

[21] Wang ZZ, Deguchi Y, Kuwahara M, Zhang XB, Yan JJ, Liu JP. Sensitive measurement of trace mercury using low pressure laser-induced plasma. *Jpn J Appl Physics.* 2013; 52: 11NC05-1-11NCO5-6. DOI: 10.7567/JJAP.52.11NC05

[22] Zhang XB, Deguchi Y, Wang ZZ, Yan JJ, Liu JP. Detection characteristics of iodine by low pressure and short pulse laser-induced breakdown spectroscopy (LIBS). *J Analy Atomic Spectrometry.* 2014; 29: 1082-1089. DOI: 10.1039/c4ja00044g

[23] Yaroshchyk P, Morrison RJS, Body D, Chadwick BL. Quantitative determination of wear metals in engine oils using LIBS: The use of paper substrates and a comparison between single- and double-pulse LIBS. *Spectrochim Acta Part B-Atomic Spectroscopy.* 2005; 60: 1482-1485. DOI: 10.1016/j.sab.2005.09.002

[24] Chen ZJ, Li HK, Liu M, Li RH. Fast and sensitive trace metal analysis in aqueous solutions by laser-induced breakdown spectroscopy using wood slice substrates. *Spectrochim Acta Part B-Atomic Spectroscopy.* 2008; 63: 64-68. DOI: 10.1016/j.sab.2007.11.010

[25] Sakka T, Yamagata H, Oguchi H, Fukami K, Ogata YH. Emission spectroscopy of laser ablation plume: Composition analysis of a target in water. *Appl Surf Sci*. 2009; 255: 9576-9580. DOI: 10.1016/j.apsusc.2009.04.086

[26] Sakka T, Masai S, Fukami K, Ogata YH. Spectral profile of atomic emission lines and effects of pulse duration on laser ablation in liquid. *Spectrochim Acta Part B-Atomic Spectroscopy*. 2009; 64: 981-985. DOI: 10.1016/j.sab.2009.07.018

[27] St-Onge L, Kwong E, Sabsabi M, Vadas EB. Rapid analysis of liquid formulations containing sodium chloride using laser-induced breakdown spectroscopy. *J Pharmaceut Biomed Analy*. 2004; 36: 277-284. DOI: 10.1016/j.jpba.2004.06.004

[28] Kumar A, Yueh FY, Miller T, Singh JP. Detection of trace elements in liquids by laser-induced breakdown spectroscopy with a Meinhard nebulizer. *Appl Optics*. 2003; 42(30): 6040-6046. DOI: 10.1364/AO.42.006040

[29] Zhao YH, Horlick G. A spectral study of charge transfer and Penning processes for Cu, Zn, Ag, and Cd in a glow discharge. *Spectrochim Acta Part B-Atomic Spectroscopy*. 2006; 61: 660-673. DOI: 10.1016/j.sab.2006.05.010

[30] Noll R. *Laser-Induced Breakdown Spectroscopy: Fundamentals and Applications*. Springer Berlin Heidelberg, 2012. p. 206-220. DOI: 10.1007/978-3-642-20668-9

[31] Cremers DA, Radziemski LJ. *Handbook of Laser-Induced Breakdown Spectroscopy*. John Wiley & Sons Ltd, England. 2006. p. 201-203.

[32] Ciucci A, Corsi M, Palleschi V, Rastelli S, Salvetti A, Tognoni E. New procedure for quantitative elemental analysis by laser-induced plasma spectroscopy. *Appl Spectroscopy*. 1999; 53(8): 960-964(5). DOI: 10.1366/0003702991947612

[33] Li XW, Wang Z, Lui SL, Fu YT, Li Z, Liu JM, Ni WD. A partial least squares based spectrum normalization method for uncertainty reduction for laser-induced breakdown spectroscopy measurements. *Spectrochim Acta Part B-Atomic Spectroscopy*. 2013; 88: 180-185. DOI: 10.1016/j.sab.2013.07.005

[34] Wang Z, Feng J, Li LZ, Ni WD, Li Z. A non-linearized PLS model based on multivariate dominant factor for laser-induced breakdown spectroscopy measurements. *J Analy Atomic Spectrometry*. 2011; 26: 2175-2182. DOI: 10.1039/c1ja10113g

[35] Lorenzen CJ, Carlhoff C, Hahn U, Jogwich M. Applications of laser-induced emission spectral analysis for industrial process and quality control. *J Analy Atomic Spectrometry*. 1992; 7: 1029-1035. DOI: 10.1039/ja9920701029

[36] Wan X, Wang P. Remote quantitative analysis of minerals based on multispectral line-calibrated laser-induced breakdown spectroscopy (LIBS). *Appl Spectroscopy*. 2014; 68(10): 1132-1136. DOI: 10.1366/13-07203

[37] Death DL, Cunningham AP, Pollard LJ. Multi-element and mineralogical analysis of mineral ores using laser induced breakdown spectroscopy and chemometric analy-

sis. *Spectrochim Acta Part B-Atomic Spectroscopy*. 2009; 64: 1048-1058. DOI: 10.1016/j.sab.2009.07.017

[38] Death DL, Cunningham AP, Pollard LJ. Multi-element analysis of iron ore pellets by laser-induced breakdown spectroscopy and principal components regression. *Spectrochim Acta Part B-Atomic Spectroscopy*. 2008; 63: 763-769. DOI: 10.1016/j.sab.2008.04.014

[39] Yaroshchyk P, Death DL, Spencer SJ. Quantitative measurements of loss on ignition in iron ore using laser-induced breakdown spectroscopy and partial least squares regression analysis. *Appl Spectroscopy*. 2010; 64(12): 1335-1341. DOI: 10.1366/000370210793561600

[40] Yaroshchyk P, Death DL, Spencer SJ. Comparison of principal components regression, partial least squares regression, multi-block partial least squares regression, and serial partial least squares regression algorithms for the analysis of Fe in iron ore using LIBS. *J Analy Atomic Spectrometry*. 2012; 27: 92-98. DOI: 10.1039/c1ja10164a

[41] Barrette L, Turmel S. On-line iron-ore slurry monitoring for real-time process control of pellet making processes using laser-induced breakdown spectroscopy: graphitic vs. total carbon detection. *Spectrochim Acta Part B-Atomic Spectroscopy*. 2001; 56: 715-723. DOI: 10.1016/S0584-8547(01)00227-0

[42] Rosenwasser S, Asimellis G, Bromley B, Hazlett R, Martin J, Pearce T, Zigler A. Development of a method for automated quantitative analysis of ores using LIBS. *Spectrochim Acta Part B-Atomic Spectroscopy*. 2001; 56: 707-714. DOI: 10.1016/S0584-8547(01)00191-4

[43] Mateo MP, Nicolas G, Yañez A. Characterization of inorganic species in coal by laser-induced breakdown spectroscopy using UV and IR radiations. *Appl Surf Sci*. 2007; 254: 868-872. DOI: 10.1016/j.apsusc.2007.08.043

[44] Yuan TB, Wang Z, Li LZ, Hou ZY, Li Z, Ni WD. Quantitative carbon measurement in anthracite using laser-induced breakdown spectroscopy with binder. *Appl Optics*. 2012; 51(7): B22-B29. DOI: 10.1364/AO.51.000B22

[45] Ctvrtnickova T, Mateo MP, Yañez A, Nicolas G. Characterization of coal fly ash components by laser-induced breakdown spectroscopy. *Spectrochim Acta Part B-Atomic Spectroscopy*. 2009; 64: 1093-1097. DOI: 10.1016/j.sab.2009.07.032

[46] Feng J, Wang Z, Li L, Li Z, Ni WD. A nonlinearized multivariate dominant factor-based partial least squares (PLS) model for coal analysis by using laser-induced breakdown spectroscopy. *Appl Spectroscopy*. 2013; 67(3): 291-300. DOI: 10.1366/11-06393

[47] Ctvrtnickova T, Mateo MP, Yañez A, Nicolas G. Laser induced breakdown spectroscopy application for ash characterisation for a coal fired power plant. *Spectrochim Acta Part B-Atomic Spectroscopy*. 2010; 65: 734-737. DOI: 10.1016/j.sab.2010.04.020

[48] Blevins LG, Shaddix CR, Sickafoose SM, Walsh PM. Laser-induced breakdown spectroscopy at high temperatures in industrial boilers and furnaces. *Appl Optics*. 2003; 42(30): 6107-6118. DOI: 10.1364/AO.42.006107

[49] Yin W, Zhang L, Dong L, Ma WG, Jia ST. Design of a laser-induced breakdown spectroscopy system for on-line quality analysis of pulverized coal in power plants. *Appl Spectroscopy*. 2009; 63(8): 865-872. DOI: 10.1366/000370209788964458

[50] Zhang L, Ma W, Dong L, Yang XJ, Hu ZY, Li ZX, Zhang YZ, Wang L, Yin WB, Jia ST. Development of an apparatus for on-line analysis of unburned carbon in fly ash using laser-induced breakdown spectroscopy (LIBS). *Appl Spectroscopy*. 2011; 65(7): 790-796. DOI: 10.1366/10-06213

[51] Noda M, Deguchi Y, Iwasaki S, Yoshikawa N. Detection of carbon content in a high-temperature and high-pressure environment using laser-induced breakdown spectroscopy. *Spectrochim Acta Part B-Atomic Spectroscopy*. 2002; 57: 701-709. DOI: 10.1016/S0584-8547(01)00403-7

[52] Kurihara M, Ikeda K, Izawa Y, Deguchi Y, Tarui H. Optimal boiler control through real-time monitoring of unburned carbon in fly ash by laser-induced breakdown spectroscopy. *Appl Optics*. 2003; 42(30): 6159-6165. DOI: 10.1364/AO.42.006159

[53] Papp Z, Dezső Z, Daróczy S. Significant radioactive contamination of soil around a coal-fired thermal power plant. *J Environ Radioactivity*. 2009; 59: 191-205. DOI: 10.1016/S0265-931X(01)00071-6

[54] Brim H, Venkateswaran A, Kostandarithes HM, Fredrickson JK, Daly MJ. Engineering deinococcus geothermalis for bioremediation of high-temperature radioactive waste environments. *Appl Environ Microbiol*. 2003; 69(8): 4575-4582. DOI: 10.1128/AEM.69.8.4575-4582.2003

[55] Sato I, Kudo H, Tsuda S. Removal efficiency of water purifier and adsorbent for iodine, cesium, strontium, barium and zirconium in drinking water. *J Toxicologic Sci*. 2011; 36(6): 859-834.

[56] Kim CK, Byun JI, Chae JS, Choi HY, Choi SW, Kim DJ, Kim YJ, Lee DM, Park WJ, Yim SA, Yun JY. Radiological impact in Korea following the Fukushima nuclear accident. *J Environ Radioactivity*. 2012; 111: 70-82.DOI 10.1016/j.jenvrad.2011.10.018

[57] Manolopoulou M, Vagena E, Stoulos S, Ioannidou A, Papastefanou C. Radioiodine and radiocesium in Thessaloniki, Northern Greece due to the Fukushima nuclear accident. *J Environ Radioactivity*. 2011; 102: 796-797. DOI: 10.1016/j.jenvrad.2011.04.010

[58] Hanson C, Phongikaroon S, Scott JR. Temperature effect on laser-induced breakdown spectroscopy spectra of molten and solid salts. *Spectrochim Acta Part B-Atomic Spectroscopy*. 2014; 97: 79-85. DOI: 10.1016/j.sab.2014.04.012

[59] LaHaye NL, Harilal SS, Diwakar PK, Hassanein A. Persistence of uranium emission

in laser-produced plasmas. *J Appl Physics*. 2014; 115: 163301-1-163301-8. DOI: 10.1063/1.4873120

[60] Saeki M, Iwanade A, Ito C, Wakaida I, Thornton B, Sakka T, Ohba H. Development of a fiber-coupled laser-induced breakdown spectroscopy instrument for analysis of underwater debris in a nuclear reactor core. *J Nuclear Sci Technol*. 2014; 51(7-8): 930-938.DOI: 10.1080/00223131.2014.917996

[61] Gong YD, Choi DW, Han BY, Yoo JH, Han SH, Lee YH. Remote quantitative analysis of cerium through a shielding window by stand-off laser-induced breakdown spectroscopy. *J Nuclear Mater*. 2014; 453: 8-15. DOI: 10.1016/j.jnucmat.2014.06.022

[62] Cho HH, Kim YJ, Jo YS, Kitagawa K, Arai N, Lee YI. Application of laser-induced breakdown spectrometry for direct determination of trace elements in starch-based flours. *J Analy Atomic Spectrometry*. 2001; 16: 622-627. DOI: 10.1039/b100754h

[63] Pouzar M, Černohorský T, Průšová M, Prokopčáková P, Krejčová A. LIBS analysis of crop plants. *J Analy Atomic Spectrometry*. 2009; 24: 953-957. DOI: 10.1039/b903593a

[64] Juvé V, Portelli R, Boueri M, Baudelet M, Yu J. Space-resolved analysis of trace elements in fresh vegetables using ultraviolet nanosecond laser-induced breakdown spectroscopy. *Spectrochim Acta Part B-Atomic Spectroscopy*. 2008; 63: 1047-1053. DOI: 10.1016/j.sab.2008.08.009

[65] Samek O, Beddows DCS, Telle HH, Kaiser J, Liška M, Cáceres JO, Gonzáles Ureña A. Quantitative laser-induced breakdown spectroscopy analysis of calcified tissue samples. *Spectrochim Acta Part B-Atomic Spectroscopy*. 2001; 56, 865-875. DOI: 10.1016/S0584-8547(01)00198-7

[66] Thareja RK, Sharma AK, Shukla S. Spectroscopic investigations of carious tooth decay. *Med Engin Physics*. 2008; 30, 1143-1148. DOI: 10.1016/j.medengphy.2008.02.005

[67] Nevin A, Spoto G, Anglos D. Laser spectroscopies for elemental and molecular analysis in art and archaeology. *Appl Physics A: Mater Sci Process*. 2012; 106: 339-361. DOI: 10.1007/s00339-011-6699-z

[68] Giakoumaki A, Melessanaki K, Anglos D. Laser-induced breakdown spectroscopy (LIBS) in archaeological science—applications and prospects. *Analy Bioanaly Chem*. 2007; 387: 749-760. DOI : 10.1007/s00216-006-0908-1

[69] Osorio LM, Cabrera LVP, García MAA, Reyes TF, Ravelo I. Portable LIBS system for determining the composition of multilayer structures on objects of cultural value. *J Physics: Conference Series*. 2011; 274: 1-6. DOI: 10.1088/1742-6596/274/1/012093

[70] McCanta MC, Dobosh PA, Dyar MD, Newsom HE. Testing the veracity of LIBS analyses on Mars using the LIBSSIM program. *Planetary Space Sci*. 2013; 81: 48-54. DOI: 10.1016/j.pss.2013.03.004

[71] Cho Y, Sugita S, Kameda S, Miura YN, Ishibashi K, Ohno S, Kamata S, Arai T, Morota T, Namiki N, Matsui T. High-precision potassium measurements using laser-induced breakdown spectroscopy under high vacuum conditions for in situ K-Ar dating of planetary surfaces. *Spectrochim Acta Part B-Atomic Spectroscopy*. 2015; 106: 28-35. DOI: 10.1016/j.sab.2015.02.002

[72] Knight AK, Scherbarth NL, Cremers DA, Ferris MJ. Characterization of laser-induced breakdown spectroscopy (LIBS) for application to space exploration. *Appl Spectroscopy*. 2000; 54(3): 331-340. DOI: 10.1366/0003702001949591

[73] Hou HM, Tian T, Li Y, Zheng RE. Study of pressure effects on laser induced plasma in bulk seawater. *J Analy Atomic Spectrometry*. 2014; 29: 169-175. DOI: 10.1039/c3ja50244a

[74] Marion Lawrence-Snyder, Jonathan P. Scaffidi, William F. Pearman, Christopher M. Gordon, S. Michael Angel. Issues in deep ocean collinear double-pulse laser induced breakdown spectroscopy: Dependence of emission intensity and inter-pulse delay on solution pressure. *Spectrochim Acta Part B-Atomic Spectroscopy*. 2014; 99: 172-178. DOI: 10.1016/j.sab.2014.06.008

[75] Guirado S, Fortes FJ, Lazic V, Laserna JJ. Chemical analysis of archeological materials in submarine environments using laser-induced breakdown spectroscopy. On-site trials in the Mediterranean Sea. *Spectrochim Acta Part B-Atomic Spectroscopy*. 2012; 74-75: 137-143. DOI: 10.1016/j.sab.2012.06.032

[76] Fortes FJ, Guirado S, Metzinger A, Laserna JJ. A study of underwater stand-offlaser-induced breakdown spectroscopy for chemical analysis of objects in the deep ocean. *J Analy Atomic Spectrometry*. 2015; 30: 1050-1056. DOI: 10.1039/c4ja00489b

[77] Lu Y, Li YD, Li Y, Wang YF, Wang S, Bao ZM, Zheng RE. Micro spatial analysis of seashell surface using laser-induced breakdown spectroscopy and Raman spectroscopy. *Spectrochim Acta Part B-Atomic Spectroscopy*. 2015; 110: 63-69. DOI: 10.1016/j.sab.2015.05.012

[78] Ferioli F, Buckley SG, Puzinauskas PV. Real-time measurement of equivalence ratio using laser-induced breakdown spectroscopy. *Int J Engine Res*. 2006; 7: 447-457. DOI: 10.1243/14680874JER01206

[79] Lombaert K, Morel S, Le Moyne L, Adam P, Tardieu de Maleissye J, Amouroux J. Nondestructive analysis of metallic elements in diesel soot collected on filter: Benefits of laser induced breakdown spectroscopy. *Plasma Chem Plasma Process*. 2004; 24(1): 41-56. DOI: 10.1023/B:PCPP.0000004881.17458.0d

[80] Strauss N, Fricke-Begemann C, Noll R. Size-resolved analysis of fine and ultrafine particulate matter by laser-induced breakdown spectroscopy. *J Analy Atomic Spectrometry*. 2010; 25: 867-874. DOI: 10.1039/b927493f

[81] Zhang HS, Yueh FY, Singh JP. Laser-induced breakdown spectrometry as a multime-

tal continuous-emission monitor. *Appl Optics*. 1999; 38(9): 1459-1466. DOI: 10.1364/AO.38.001459

[82] Buckley SG, Johnsen HA, Hencken KR, Hahn DW. Implementation of laser-induced breakdown spectroscopy as a continuous emissions monitor for toxic metals. *Waste Manag*. 2000; 20: 455-462. DOI: 10.1016/S0956-053X(00)00011-8

[83] Vander Wal RL, Ticich TM, West JR, Householder PA. Trace metal detection by laser-induced breakdown spectroscopy. *Appl Spectroscopy*. 1999; 53(10): 1226-1236.

[84] Effenberger AJ, Scott JR. Effect of atmospheric conditions on LIBS spectra. *Sensors*. 2010; 10: 4907-4925. DOI: 10.3390/s100504907

[85] Bertolini A, Carelli G, Francesconi F, Francesconi M, Marchesini L, Marsili P, Sorrentino F, Cristoforetti G, Legnaioli S, Palleschi V, Pardini L, Salvetti A. Modì: a new mobile instrument for in situ double-pulse LIBS analysis. *Analy Bioanaly Chem*. 2006; 385: 240-247. DOI: 10.1007/s00216-006-0413-6

[86] Sharma SK, Misra AK, Lucey PG, Lentz RCF. A combined remote Raman and LIBS instrument for characterizing minerals with 532 nm laser excitation. *Spectrochim Acta Part A-Mol Biomol Spectroscopy*. 2009; 73: 468-476. DOI: 10.1016/j.saa.2008.08.005

[87] Wiens RC, Sharma SK, Thompson J, Misra A, Lucey PG. Joint analyses by laser-induced breakdown spectroscopy (LIBS) and Raman spectroscopy at stand-off distances. *Spectrochim Acta Part A-Mol Biomol Spectroscopy*. 2005; 61: 2324-2334. DOI: 10.1016/j.saa.2005.02.031

[88] Hilbk-Kortenbruck F, Noll R, Wintjens P, Falk H, Becker C. Analysis of heavy metals in soils using laser-induced breakdown spectrometry combined with laser-induced fluorescence. *Spectrochim Acta Part B-Atomic Spectroscopy*. 2001; 56: 933-945. DOI: 10.1016/S0584-8547(01)00213-0

[89] Shiou FJ, Chen HH, Tsai CH. Development of a non-contact complete 3D surface measurement system for small complex objects using a triangulation laser probe. *J Chinese Instit Engin*. 2010; 33(2): 141-151. DOI: 10.1080/02533839.2010.9671606

[90] Fan KC, Tsai TH. Optimal shape error analysis of the matching image for a freeform surface. *Robotics Compu Integr Manufact*. 2001; 17: 215-222. DOI: 10.1016/S0736-5845(00)00029-6

[91] Nguyen HC, Lee BR. Laser-vision-based quality inspection system for small-bead laser welding. *Int J Precision Engin Manufact*. 2014; 15(3): 415-423.DOI: 10.1007/s12541-014-0352-7

[92] Shiou FJ, Deng YW. Simultaneous flatness and surface roughness measurement of a plastic sheet using a fan-shaped laser beam scanning system. *Key Engin Mater*. 2008; 381-382: 233-236. DOI: 10.4028/www.scientific.net/KEM.381-382.233

[93] GOM GmbH. User's manual of ATOS 4.7, GOM GmbH, Braunschweig, Germany; 2001.

[94] Chang M, Lin, KH. Non-contact scanning measurement utilizing a spacing mapping method. *Optics and Lasers in Engin.* 1998; 30(6): 503-512. DOI: 10.1016/S0143-8166(98)00053-0

6

Optically Thick Laser-Induced Plasmas in Spectroscopic Analysis

Fatemeh Rezaei

Abstract

Studies on the plasma physics has been grown over the past few decades as a major research field. The plasma can be produced by different sources such as acr, spark, electric discharge, laser and so on. The spectral radiation of the plasma which acts as its fingerprint, contains valuable information about plasma features. Characterization of plasmas by spectroscopic measurement is a powerful tool for increasing the knowledge and applications of these kinds of radiation sources. Therefore, the spectral diagnostics methods are proposed which are based on measurement of spectral lines intensity, estimation of continuous and absorption radiation, and as well as determination of shifts and halfwiths of the spectrum [1]. The fundamental characteristic parameters of the plasma, i.e., the number densities of plasma species, electron temperature, and as well as particle transport property at each plasma space can be determined by optical emission spectroscopy and utilizing appropriate methods [2]. For accurate evaluation of plasma parameters, its thickness must be thoroughly considered. Generally, the plasmas can be separated into two categories of thin and thick groups. In thin plasmas, the re-absorption of radiation is negligible. Consequently, in spectroscopic analysis, the non-self-absorbed spectral radiation is evaluated by considering the summation of all spectral emissions along the line of sight. In optically thick plasmas, the radiation trapping happens which leads to the self-absorption phenomenon in spectroscopic analysis that is explained with details in below section.

Keywords: Thick plasma, self absorption

1. Introduction

Studies on plasma physics have been grown over the past few decades as a major research field. Plasma can be produced by different sources such as arc, spark, electric discharge, laser, and so on. The spectral radiation of the plasma, which acts as its fingerprint, contains valuable information about plasma features. Characterization of plasmas by spectroscopic measurement is a powerful tool for increasing the knowledge and applications of these kinds of radiation sources. Therefore, spectral diagnostics methods are proposed which are based on measurements of spectral line intensity, estimation of continuous and absorption radiation, as well as the determination of shifts and half-widths of the spectrum [1]. The fundamental characteristic parameters of the plasma, i.e., the number densities of plasma species, electron temperature, and the particle transport property at each plasma space can be determined by optical emission spectroscopy and by utilizing appropriate methods [2]. For accurate evaluation of plasma parameters, its thickness must be thoroughly considered. Generally, the plasmas can be separated into two categories of thin and thick groups. In thin plasmas, the reabsorption of radiation is negligible. Consequently, in spectroscopic analysis, the non-self-absorbed spectral radiation is evaluated by considering the summation of all spectral emissions along the line of sight. In optically thick plasmas, radiation trapping happens, which leads to the self-absorption phenomenon in spectroscopic analysis that is explained in details in the following section.

2. Self-absorption effect

In thick plasmas, when light is emitted from the interior hot parts of plasma and travels to the outside cold regions, the light may be absorbed by the same sort of emitting atoms and molecules. Consequently, the resultant spectrum in a spectrograph will be weakened so that the plasma itself absorbs its emission. This particular kind of absorption of a light source is called self-absorption. The main error that happens in the evaluation of the plasma parameters is the erosion of the spectral intensity due to self-absorption. This phenomenon results in peak height reduction and growth of spectral line widths. In some cases, absorption in the center of the spectral line is more severe than its sides, so that the self-absorption appears as a self-reversal [3]. Self-reversal happens especially in strong resonance lines and in inhomogeneous thick plasmas. In this case, a central dip is observed in spectral lines due to the cold absorbing atoms from the outer parts of the plasma plume. In most cases, the self-absorption is shown as a height reduction, which will not be well recognized from the shape of the spectrum. Self-absorption is mainly severe for atomic lines with low excitation energies of upper levels or spectral lines with high transition probabilities. Furthermore, resonant lines are particularly influenced by the self-absorption effect.

For the spectroscopic purposes in a reabsorbed plasma, the spectral intensity has a complicated relation with plasma parameters, as well as with emission coefficients. Here, the self-absorption effect is investigated for the laser-induced plasmas in local thermal equilibrium (LTE) condition. The details of optically thick plasma calculations in collisional radiative models is indicated in ref. [4].

Ref. [3] completely investigated the effects of different spectral distributions such as Doppler, resonance, and natural broadening on the magnitude of self-absorption. They focused mainly on self-absorption treatment in arcs and sparks sources. Moreover, they investigated the influences of uniformly excited sources and non-homogenous sources on the amount of line self-absorption. In this study, most of the attention is concentrated on laser as a plasma sources. In the laser-induced breakdown spectroscopy (LIBS) technique, it is observed that in nearly all strong lines of a spectrum and for concentrations more than approximately 3% in the sample, the plasma can behave as a thick medium [5].

It should be noted that self-absorption can be comprehensively studied for different spectroscopic techniques in all intervals of electromagnetism emission, including gamma and X-ray spectroscopy to radiofrequency region, and it comprises the relativistic synchrotron emission as well. In this chapter, the attention is focused on spectral intervals from UV to IR region and on plasma produced by laser radiation.

Self-absorption coefficient of a particular line is usually defined as the ratio of spectral peak height in the presence of self-absorption to its peak magnitude in the condition without self-absorption as:

$$SA = \frac{I(\lambda_j)}{I_0(\lambda_j)}. \tag{1}$$

or it can be expressed as the ratio of line width as below:

$$\frac{\Delta\lambda}{\Delta\lambda_0} = (SA)^{-0.5}. \tag{2}$$

or as the ratio of integrated intensities by

$$\frac{\overline{I}}{\overline{I_0}} \approx \frac{I(\lambda_0)\Delta\lambda}{I_0(\lambda_0)\Delta\lambda_0} = (SA)^{0.5}. \tag{3}$$

Several research groups proposed different methods of line ratios [6, 7], duplicating mirror [8], curve of growth (COG) [9–12], and calculation models [13–17] for the identification and evaluation of the self-absorption of the considered spectrum. Then, after diagnostic stages, appropriate corrective methods corrected the self-absorbed spectral line intensities before utilizing them for analytical goals. Finally, they calculated the plasma parameters after suitable correction methods, but some groups used these self-absorbed lines straightly by applying appropriate theoretical models without any correction, in spite of its complicated calculations [4, 18–21].

3. Homogenous plasmas

In homogenous plasmas, it is assumed that the plasma parameters are uniform, i.e. they have the same magnitudes in the entire plasma volume. The main investigated parameters are electron and species temperatures T, and number densities of all different elements in the plasma n.

The intensity of a particular spectral line along the line profile related to transition between two ionic or atomic levels l and u (lower and upper levels), can be evaluated by radiation transfer equation. This equation describes the radiation intensity changes after passing a distance dl of a plasma by taking into account the contribution of the emission within this distance and emission reduction because of absorption along dl [4]:

$$dI_\lambda = \left(\varepsilon_\lambda - k(\lambda) I_\lambda\right) dl, \tag{4}$$

where, ε_λ is the emission coefficient in thin plasmas, and k(λ) is the absorption coefficient. In this equation, the source function is expressed as:

$$S_\lambda = \frac{\varepsilon_\lambda}{k(\lambda)}. \tag{5}$$

The optical depth or optical thickness is defined as multiplication of absorption coefficient by the geometrical thickness of plasma as τ=k(λ)dl.

In a two-level system, by neglecting the stimulated emission, the spectral emission can be calculated by considering the spontaneous emission coefficient (in SI units) as follows:

$$\varepsilon_{spec}(\lambda) = N_u A_{ul} h\nu \, V(\lambda). \tag{6}$$

In the above equation, N_u indicates the number of atoms in the upper level and by assumption of holding the local thermal equilibrium (LTE) condition, it is calculated by Boltzmann distribution function. h is the Plank constant, and ν is the spectral line frequency. Here, for more simplification, it is assumed that the line profile distribution for the emission coefficient and the absorption coefficient is similar as Voigt profile V(λ).

The absorption coefficient of the two-level system, in SI units, is explained by taking into account the absorption and induced emission between low and high levels of u and l as:

$$k(\lambda) = \frac{h}{\lambda}(B_{ul} N_l - B_{lu} N_u) V(\lambda). \tag{7}$$

B_{ul} and B_{lu} are Einstein coefficients due to absorption and induced emission, respectively, and they are dependent to spontaneous probability coefficient, A_{ul} by:

$$A_{ul} = \frac{g_l}{g_u} B_{lu} \frac{8\pi h\nu^3}{c^3}, g_u B_{ul} = g_l B_{lu}. \tag{8}$$

where, g_l and g_u are the degeneracy of the lower and upper levels, respectively. c is the light velocity in the vacuum. N_l is atomic density in lower level. Therefore, by substituting the above equations in Eq. (7), the absorption coefficient of the mentioned spectral line can be expressed as:

$$k(\lambda) = \frac{g_u A_{ul} N_{tot} \lambda^2}{8\pi Z} e^{-\frac{E_l}{k_B T}} (1 - e^{-\frac{h\nu}{k_B T}}) V(\lambda). \tag{9}$$

here, E_l is the energy of lower level, N_{tot} is the total number densities of species, T is the plasma temperature, Z is the partition function, and K_B is the Boltzman constant.

It should be mentioned that in the collisional-radiative plasma, the calculation procedure is the same as in LTE model, except that the estimation of number densities are performed by rate equation as mentioned in ref. [4].

In this chapter, at first, some corrective methods will be explained for evaluation of self-absorption coefficient in laser-induced homogeneous plasmas by utilizing suitable experimental and numerical methods. Then, the effect of self-absorption on inhomogoneous plasmas will be discussed. Afterward, the impressive parameters on spectral lines charactristics affected by self-absorption will be expressed.

4. Different corrective methods in homogeneous plasmas

One of the methods for observation of self-absorption in the experiment is bending of calibration curve (or curve of growth) constructed by standard samples at high concentrations. Consequently, the self-absorbed spectral lines must be corrected to reach the condition of thin plasmas for prediction of the accurate magnitude of sample concentration without any reduction in intensity. Therefore, in the following sections, some of the corrective methods proposed by different research groups are mentioned.

4.1. Ratio of two spectral line features (width, peak, and surface)

Amamou et al. [6, 7] calculated the self-absorption for both Gaussian and Lorentzian line profiles with a Simplex algorithm program fitting method [22] in a homogenous laser-produced plasma. They fitted the experimental results with theoretical calculation and then, their expressions for Lorentzian profiles were used for quantification of the transition probabilities, ratios, and the ratios of optical thicknesses as well. They introduced different correction factors by considering the ratio of the peaks, line widths, and surfaces of two spectral lines for both of the considered line profiles. The correction factor for the line height can be evaluated by considering the ratio of peak intensities of spectral lines in case of non-self-absorbed atomic line to the case of the self-absorbed line as follows:

$$F_{\max}^{L} = \frac{I_0^{Thin}}{I_0^{Thick}} = \frac{\tau_0}{1-\exp(-\tau_0)}. \tag{10}$$

As well as by taking into account the FWHM ratios, the correction factor for the line width can be calculated by the following equation:

$$F_{\Delta\lambda}^{L} = \frac{\Delta\lambda_L^{Thin}}{\Delta\lambda_L^{Thick}} = \left(-1 - \frac{\tau_0}{\log([1+\exp(-\tau_0)]/2)}\right)^{-1/2}. \tag{11}$$

For an optical thickness of less than 4, it is assumed that the line surface is proportional to the multiplication of line width by its height. Hence, the correction factor for the line surface can be estimated as:

$$F_S^{L} = \frac{S_L^{Thin}}{S_L^{Thick}} = \frac{\tau_0}{1-\exp(-\tau_0)}\left(-1 - \frac{\tau_0}{\log([1+\exp(-\tau_0)]/2)}\right)^{-1/2}. \tag{12}$$

here, τ_0 is the optical thickness at central line wavelength (or maximum of optical thickness). The results of these calculations are obtained from the plasma created by laser irradiation on a silicate solid sample placed in a xenon and hydrogen atmosphere for the various multiples of Si II lines. Figure 1 illustrates the evolution of self-absorption correction factors for the above-mentioned parameters in a Lorentzian distribution.

4.2. Simple theoretical equation

El Sherbini et al. [15] presented a simple relation for correcting the self-absorption effect in a homogenous plasma. This model is applicable when the Stark broadening parameter of selected spectral lines is known, as well as when the plasma electron density is available from experiments.

In this work, the intensity of a spectral line in thick condition (erg.s^{-1}.cm^{-3}) along the line profile due to the transition between two levels (*j* and *i*) is expressed by

$$I(\lambda) = \frac{8\pi hc^2}{\lambda_0^5}\frac{n_j}{n_i}\frac{g_i}{g_j}(1-e^{-k(\lambda)l}). \tag{13}$$

Generally, the absorption coefficient k(λ) is described by a Voigt profile, which is convolution of a Lorentzian and a Gaussian distribution. In laser sources plasmas, Lorentzian width is associated with the Stark effect and the Gaussian line width is dominated by the Doppler broadening. In a typical LIBS experiment, the Gaussian contribution of a spectral line width is negligible compared to the Lorentzian component, hence, the optical depth k(λ)l can be calculated by

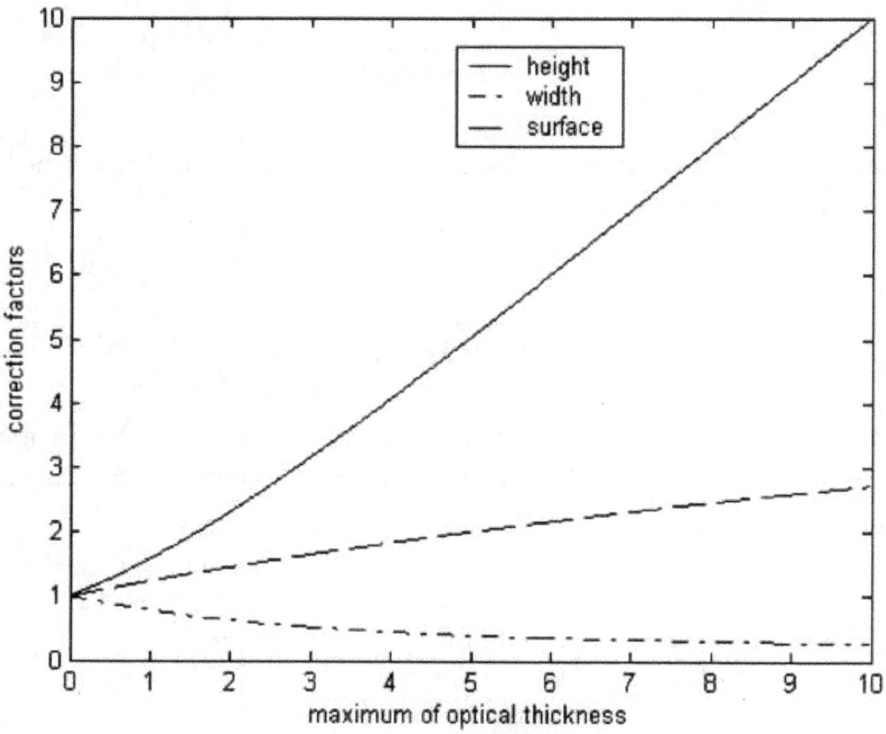

Figure 1. Self-absorption correction factors for: height "solid line", width "alternate dashed line", and surface "long dashed line", in Lorentzian line profile.

$$k(\lambda)l \approx K\frac{\Delta\lambda_0}{4(\lambda-\lambda_0)^2+\Delta\lambda_0^2}, \tag{14}$$

Here, $K = 2\frac{e^2}{mc^2}n_i f \lambda_0^2 l$ and the line width is $\Delta\lambda_0 = \Delta\lambda_L$. In the condition of thin plasmas with $k(\lambda)l \ll 1$, the above equation can be approximated as

$$I_0(\lambda) \approx \frac{8\pi hc^2}{\lambda_0^5}\frac{n_j}{n_i}\frac{g_i}{g_j}k(\lambda)l. \tag{15}$$

Consequently, $I_0(\lambda)$ illustrates the line intensity in case of negligible self-absorption. According to Eqs. (13) and (15), it is clearly seen that, in the case of self-absorption, the line intensity at its peak (i.e. for $\lambda = \lambda_0$) has less value compared to that in the case of thin plasmas. Then, the self-absorption coefficient, SA, can be expressed as

$$\frac{I(\lambda_0)}{I_0(\lambda_0)} = \frac{\left(1-e^{-k(\lambda_0)l}\right)}{k(\lambda_0)l} = \Delta\lambda_0\frac{\left(1-e^{\frac{-K}{\Delta\lambda_0}}\right)}{K} = SA. \tag{16}$$

SA is a coefficient between 0 and 1. SA equals to 1 in the condition of thin plasma and 0 in highly absorbing plasma. Moreover, the line width magnitude is influenced by the self-absorption effect. For obtaining an equation relating FWHM of an optically thick line $\Delta\lambda$ to thin spectral line width $\Delta\lambda_0$ and SA, the following equation can be used:

$$\frac{I(\lambda)}{I(\lambda_0)} = \left(\frac{1 - e^{-K \frac{\Delta\lambda_0}{4(\lambda-\lambda_0)^2 + \Delta\lambda_0^2}}}{1 - e^{-\frac{K}{\Delta\lambda_0}}} \right). \tag{17}$$

By numerically solving the above equation for $\Delta\lambda$ and taking into account the definition of FWHM as $\lambda = \lambda_0 \pm \Delta\lambda/2$, the intensity of $I(\lambda)$ equals to $I(\lambda_0)/2$. Hence, the exact equation between the measured spectral width ($\Delta\lambda$) and related non-self-absorbed line width ($\Delta\lambda_0$) can be evaluated. Then, after appropriate calculation, provided that $\Delta\lambda$ and n_e are measured from the experiment and w_s magnitude is inserted from the literature, the SA coefficient can be obtained as:

$$SA = \left(\frac{\Delta\lambda}{2\omega}\frac{1}{n_e}\right)^{1/\alpha}. \tag{18}$$

In the above equation, n_e can be measured from the non-self-absorbed spectral line of hydrogen H_α at 656.27 nm. For evaluation of the mentioned method, the experiment is performed on several Al spectral lines radiated from pure aluminum (99.9%) samples. The experiments are done with different equipment, one, at the Physics Department of Cairo University (Egypt) and another, at the Applied Laser Spectroscopy Laboratory in Pisa (Italy). At Cairo University, the experiment is performed with using a single pulse Nd:YAG laser with 160 mJ laser energy, 6 ns pulse duration, and 1064 nm laser wavelength. At Pisa Laboratory, the measurement is done utilizing a mobile double-pulse laser with 8 ns FWHM and laser energy of 80 + 80 mJ with a 2 μs delay between the pulses in collinear configuration.

In Figure 2, the temporal evolutions of the self-absorption coefficients SA for three spectral lines of Al I at 394.4 nm, Al II line at 281.6 nm, and Al II line at 466.3 nm are shown. In this figure, it is seen that the Al I spectral line at 394.4 nm exposes to a higher self-absorption (in the spectra taken at Cairo) compared to the other two cases. This is probably because of the higher electron density (produced by the higher laser energy) so that, based on the Saha equation, it provides a larger amount of neutral atoms in the plasma. Moreover, it is clearly observed that the ionized aluminum lines illustrate a low to moderate self-absorption at later delay times, but they are approximately optically thin for delay times lower than 3 μs. Furthermore, the increase of plasma optical thickness at longer delay times is proved for the Al I spectral line at 394.4 nm and is likely because of the plasma plume cooling which induces a growth in the population of the atomic and ionic lower energy levels.

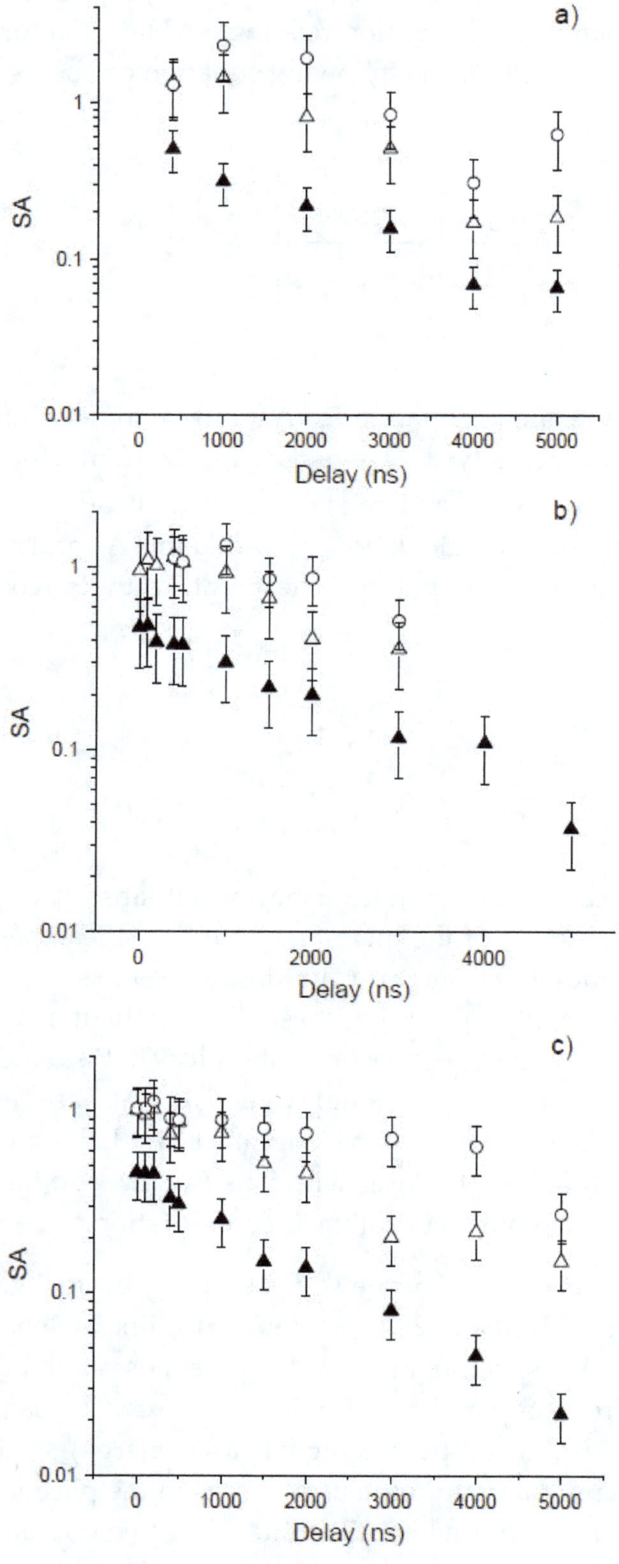

Figure 2. Temporal evolution of self-absorption coefficient SA for Al I spectral line at 394.4 nm (solid triangles), Al II line at 281.6 nm (open triangles), and Al II line at 466.3 nm (open circles). Measurements performed at Cairo University by impact of a single pulse with laser energy of 760 mJ.

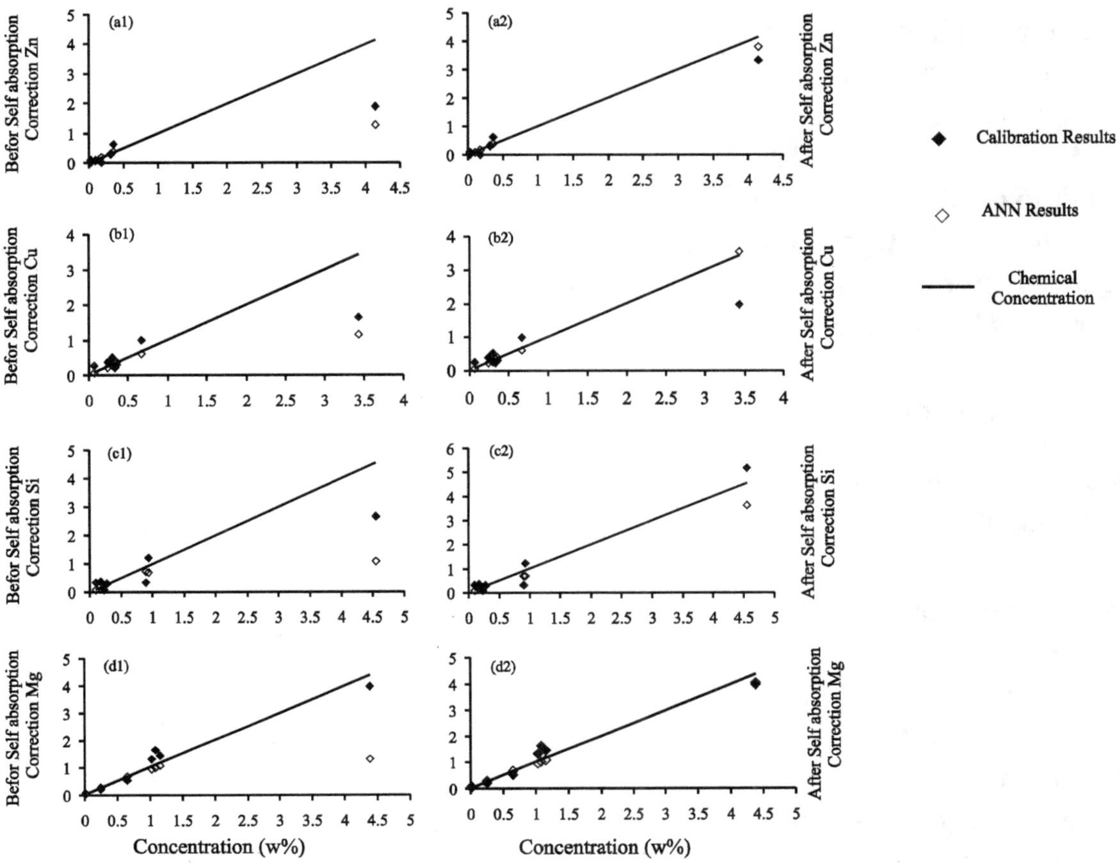

Figure 3. Comparison between the results of ANN and calibration methods with the chemical concentrations (1) without self-absorption correction and (2) with self-absorption correction for the elements of (a) Zn 334.40 nm, (b) Cu 324.40 nm, (c) Si 288.16 nm, and (d) Mg 285.21 nm.

By utilizing the above simple equation, Rezaei et al. [23] corrected the aluminum intensities and then, they predicted the known concentrations in the standard samples with two calibration curves and artificial neural network (ANN) to compare the accuracies of these methods. They used laser-induced breakdown spectroscopy (LIBS) technique for concentration predictions of six elements: Mn, Si, Cu, Fe, Zn, and Mg in seven Al samples. Then, the calibration curve and ANN techniques acquired by six samples are applied for prediction of the elements concentration of the seventh standard sample. In this experiment, a Q-switched Nd:YAG laser at 1,064 nm with a repetition rate of 10 Hz, diagonal output beam of 2.3 mm, and pulse width of 8 ns is focoused on samples. Laser pulse operating at TEM00 mode is adjusted for 50 mJ. The spectra are recorded with an ICCD with exposure time of 1 s and gate width of 5 μs. A comparison between two prediction methods of ANN and calibration curve with their real concentrations in standard samples for four elements of Zn, Cu, Mg, and Si is shown in Figure 3. As it can clearly be seen, at high concentrations, a considerable deviation from real data appeared in Figures 3(a1–d1) in the cases before correcting the self-absorption effect, while when taking into account the self-absorption effect, ANN prediction improves very much in Figures 3(a2–d2). As it is expected, the predictions of the farthest right points on Figure 3, i.e. before considering that self-absorption is not very reliable. Since, these data are a bit greater than the concentration ranges of the training analysis. Results indicate that after self-absorption correction, and at high concentrations, except for Si, the ANN method illustrates more accurate results with a lower relative error compared to the calibration curve method. These results express that the ANN approach is better than the traditional calibration technique for concentration prediction after self-absorption correction, so that the predicted intensities with ANN are nearer to the real emission spectra. The reason for this fact is that ANN obeys a nonlinear behavior at training stage. Moreover, the used constraint on ANN at the training step causes considerable improvement in its accuracy, while the calibration curve follows a linear function which induces higher errors in its predictions.

4.3. Curve of growth

Curve of growth (COG) method makes a relation between the emission intensity and the optical depth. First, this technique was applied for light sources of resonance vapor lamps [24] and flames [25, 26]. Then, Gornushkin et al. [27] applied a COG method for laser-induced plasma spectroscopy. Recently, Aragon and Aguilera represented several effects of different parameters such as variations of optical depth [11], plasma inhomogeneity [10], and delay time [12] on evolutions of COG curves. They fitted the theoretical COG equations to the experimental results and then, extracted plasma parameters, such as number density of neutral emitting atoms and damping constant. Moreover, they utilized the COG curves for estimation of the magnitude of self-absorption parameter and for the evaluation of the concentration at which transition from thin to thick plasma happens. They proposed that the integrated intensity of a spectral line ($W.m^{-2}.sr^{-1}$) in an optically thick plasma can be calculated from [11]:

$$I = I_P(\nu_0)\int_0^{\infty}(1-e^{-\tau(\nu)})d\nu, \tag{19}$$

where, ν_0 is the central frequency (Hz) of the spectral line, $I_P(\nu_0)$ is the Planck blackbody distribution ($W.m^{-2}.Sr^{-1}.Hz^{-1}$), and $\tau(\nu)$ is the optical depth which in a homogeneous plasma in LTE condition can be expressed as:

$$\tau(\nu) = k'(\nu)l = \frac{e^2}{4\varepsilon_0 mc} N f_{ij} l \frac{g_i e^{-\frac{E_i}{K_B T}}}{Z(T)} \times \left(1 - e^{-\frac{E_j - E_i}{K_B T}}\right) L(\nu), \tag{20}$$

$k'(\nu)$ is the effective absorption coefficient (m^{-1}), which includes the contribution of induced emission and absorption, l is the plasma length (m), f_{ij} is the transition oscillator strength (dimensionless), and $L(\nu)$ is the normalized Voigt line profile (Hz^{-1}) as follows:

$$L(\lambda) = \frac{2(\ln 2/\pi)^{1/2}}{\Delta\nu_D}\frac{a}{\pi}\int_{-\infty}^{+\infty}\frac{e^{-t^2}dt}{(y-t)^2+a^2}. \tag{21}$$

In Eq. (21), the dimensionless parameters of y and a are defined as

$$a = \frac{(\Delta\nu_N + \Delta\nu_L)}{\Delta\nu_D}\sqrt{\ln 2} \approx \frac{\Delta\nu_L}{\Delta\nu_D}\sqrt{\ln 2}. \quad (22)$$

$$y = \frac{2(\nu - \nu_0)}{\Delta\nu_D}\sqrt{\ln 2}, \quad (23)$$

where, $\Delta\nu_D$ $\Delta\nu_N$ and $\Delta\nu_L$ are are the Doppler, natural and Lorentzian line widths (Hz), respectively. It should be noted that the optical depth of a spectral line depends on the multiplication of $Nf_{ij}l$. The relation between the line intensity I and $Nf_{ij}l$ is expressed by Eq. (19) and is called a curve of growth equation. The main problem in utilizing the LIBS technique for analysis is the complex relation between number densities of emitting species N and the concentration in the sample x. In this study, it is assumed that the matrix effects are negligible, so that N is proportional to x by the following equation:

$$N = N' \frac{x}{100}. \quad (24)$$

In the above equation, N indicates the number densities of emitting elements (m^{-3}) in the plasma for the sample which contains 100% concentration. By inserting N in Eq. (24) to Eq. (20), the optical depth can be calculated versus wavelength λ(m) as:

$$\tau(\lambda) = 10^{-2} k_t N' x l L(\lambda). \quad (25)$$

The coefficient k_t, which is dependent to the transition parameters, can be calculated by determination of the plasma temperature as

$$k_t = \frac{e^2 \lambda_0^2}{4\varepsilon_0 m c^2} f_{ij} \frac{g_i e^{-\frac{E_i}{K_B T}}}{Z(T)} \left(1 - e^{-\frac{E_j - E_i}{K_B T}}\right). \quad (26)$$

In the above equation, λ_0 is the central wavelength (m), c is the light speed in vacuum ($m.s^{-1}$), e is the electron charge (C), g_i is degeneracy of lower level, m is electron mass (kg), f_{ij} is the transition oscillator strength (dimensionless), K_B is Boltzmann constant ($J.K^{-1}$), Z(T) is partition function (dimensionless), ε_0 is is the permittivity of free space ($F.m^{-1}$), E_i and E_j are energies of lower and higher levels, respectively. By applying Eq. (26) for the optical depth, the spectral line intensity can be obtained versus the concentration of the emitting element in the sample as below:

$$I(x) = I_p(\lambda_0)\int_0^{\infty}\left(1 - e^{-AxL(\lambda)}\right)d\lambda. \quad (27)$$

A parameter in Eq. (27) is the plasma perpendicular radiating area (m^2), which is defined by

$$A = 10^{-2} k_t N' l. \quad (28)$$

The asymptotic behavior of the COG in low and high concentrations can be obtained by

$$I_{low} \cong I_p Ax. \quad (29)$$

and

$$I_{high} \cong I_p\left[2A(\Delta\lambda_L)\right]^{1/2} x^{1/2}. \quad (30)$$

In double logarithmic scale, by considering the entire concentration range, a linear equation can be depicted. The intersection point of the asymptotes shows the following abscissa:

$$x_{\text{int}} = \frac{2(\Delta\lambda_L)}{A}. \quad (31)$$

This point determines the transition limit from thin to thick plasma and starting of the saturation of the spectral line. For instance, the measured spectral intensities of two neutral Fe lines are depicted versus concentration for the Fe-Ni samples. Here, the experiment is performed by focusing of a Q-switched Nd:YAG laser with 1064 nm wavelength and 4.5 ns line width at 20 Hz repetition rate on Fe-Ni alloy samples in atmospheric air. The pulse energy is adjusted for 100 mJ by utilizing of an optical attenuator. Moreover, the plasma emission is collected with temporal resolution, using a delay time of 5 μs from the laser pulse and a gate width of 1 μs. For comparison, the theoretical results are shown in this figure, too. These curves are illustrated in linear and double-logarithmic scales.

As it is seen in Figure 4 the atomic line at 375.82 nm, which contains a higher value of the k_t coefficient rather than the spectral line at 379.50 nm, exposes to a more intense self-absorption in the LIP experiment, which appears as a nonlinear COG curve at lower concentrations.Then, this group developed COG curves by introducing the CSigma graph [2], which comprises different lines of various elements in similar ionization states for LIBS technique. The method is based on the Saha, Boltzmann, and radiative transfer equations for plasmas in LTE condition. Cσ graphs rely on the evaluation of cross-section of a line, i.e. σ_l for each of the experimental

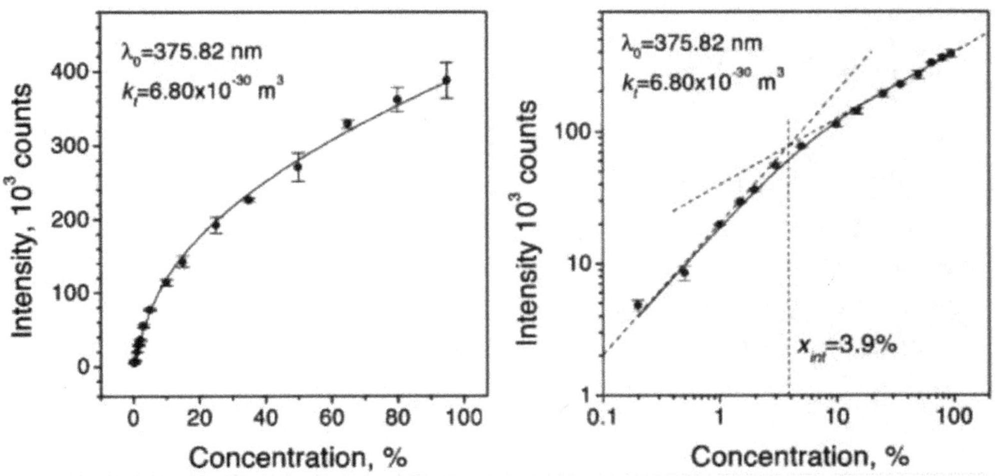

Figure 4. Theoretical and experimental COG for Fe I spectral line on a linear form in left and a double-logarithmic scale on right side. The magnitudes of k_t and x_{int} are indicated in this figure.

results, by knowing the electron density, temperature, and the line atomic data. Then, they fitted the experimental Cσ graphs to calculated curves and four parameters of βA (β is the system instrumental factor equal to the multiplication of the spectral efficiency by the solid angle of detection and A is the transverse area of the plasma region in which the emission is detected), Nl (columnar density), T, N_e could be determined for characterization of LIBS plasma for different ionic and neutral species. The details of the mathematical calculations are expressed in ref. [2].

4.4. Calibration free

Bulajic et al. [14] devised an algorithm for self-absorption correction which was first utilized for three different certified steel NIST samples and for three ternary alloys (Au, Ag, Cu) with known concentrations. Then, it was suggested as a tool for automatic correction of different standardless materials by laser-induced breakdown spectroscopy by using a calibration-free algorithm. The results illustrated that the self-absorption corrected calibration-free method presents reliable conclusions and improves the accuracy by nearly one order of magnitude. The main advantages of applying calibration curve is minimizing the matrix effect, which induces errors in precise evaluation of plasma parameter.

In this work, for each value of SA, 30 different samples are generated. Each synthetical line is appropriately fitted with the analytical software, which yields an estimation value for the parameters $\Delta\lambda_L$ and $\Delta\lambda_G$. In Figure 5, statistical results of $\Delta\lambda_L$ for different values of SA are reported, which explains how, for self-absorbed spectral lines, the experimental Lorentzian width deviates from the 'real' magnitude. Furthermore, by begining from the measured Lorentzian width, it is not feasible to obtain the true value because of the distortion of Voigt profile and the dispersion of the profile of the calculated line widths. It is found more reliable methods for acquiring the true spectral Lorentzian width, by starting from the total line width (i.e. Gaussian plus Lorentzian), which is proved to depend on the SA parameter based on the following expression (see Demtroder [28]):

$$\Delta\lambda_{obs} = \frac{\Delta\lambda_{true}}{\sqrt{SA}} \approx \frac{\sqrt{\Delta\lambda_L^2 + \Delta\lambda_G^2}}{\sqrt{SA}}. \quad (32)$$

Actually, a very good agreement is seen between the values of the total broadening quantities calculated according to the fit of the simulated self-absorbed profiles and those calculated by Eq. (32), when the magnitude of $\Delta\lambda_{true}$ is known. Hence, by assumption of knowing the self-absorption coefficient SA, the total true width can easily be found by utilizing Eq. (32), and then, the contribution of Doppler broadening $\Delta\lambda_G$ and instrumental broadening $\Delta\lambda_L$ will be obtained.

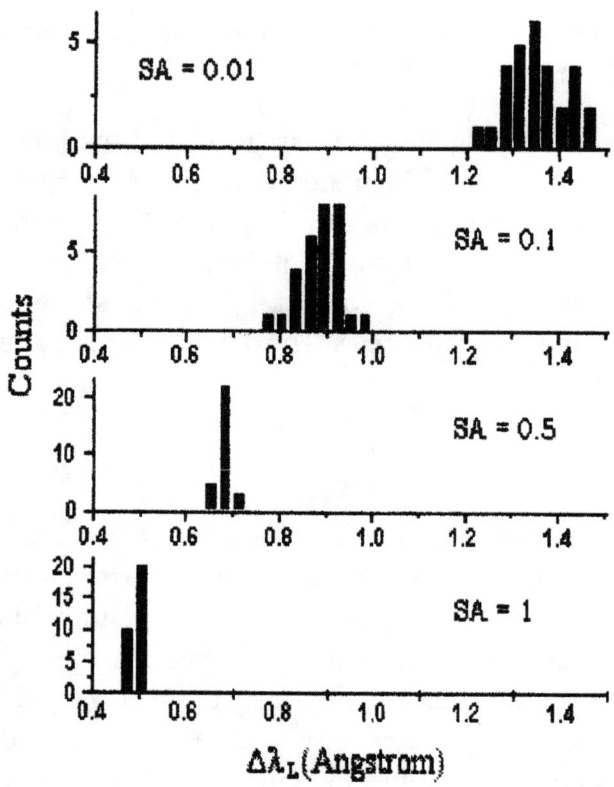

Figure 5. Statistical analysis of Lorentzian width magnitudes attained by fitting with simulated lines with $\Delta\lambda_L$ and $\Delta\lambda_G$ fixed to 0.5 and 0.1 A, respectively, for spectral line of Cu at 324.7 nm, and SA are 1, 0.5, 0.1, and 0.01 from the bottom.

Moreover, Figure 6 illustrates the simulation result of the copper line profile at 324.7 nm at different values of self-absorption for comparing with precious alloy sample containing 40% of Cu. Here, the simulation is performed by curve of growth method as mentioned before. In this figure, the 'flat-top' shape, which is a representation of the self-absorption effect, is clearly seen. Since the peak height reduction can't be well recognized as self-absorption effect, an algorithm is developed for automatic evaluation of self-absorption phenomenon. The performance of self-absorption correction is presented into the CF-LIBS procedure with a recursive algorithm, which is shown by a diagram in mentioned reference.

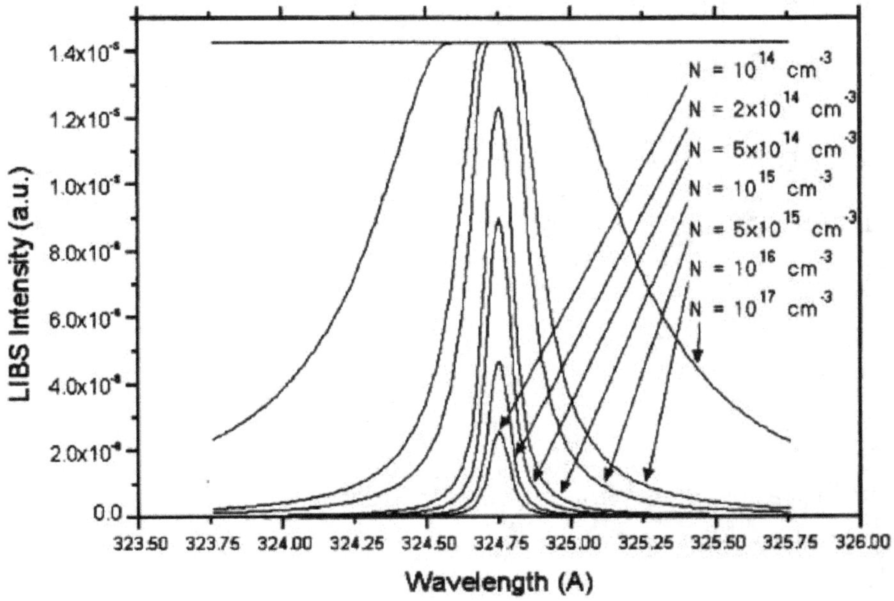

Figure 6. Simulation of Cu spectral line profile at 324.7 nm at different degrees of self-absorption.

4.5. Internal reference method

Sun and Yu [17] introduced a simplified procedure for correction of self-absorption by calibration-free laser-induced breakdown spectroscopy technique (CF-IBS). They utilized an internal reference line for each species. Then, they made a comparison between this spectral line with the other line intensity from the same species of the reference line to evaluate the self-absorption magnitude of the other spectral lines.

They started their method, i.e. internal standard self-absorption correction (IRSAC) by assumption of the plasma being in LTE condition, they estimated the measured integral line intensity as below:

$$I_\lambda^{ij} = f_\lambda^b F C_s A_{ij} \frac{g_i}{Z_s(T)} e^{-E_i / k_B T}, \tag{33}$$

where, A_{ij}, K_B, C_s, g_i, and λ are the transition probability, Boltzmann constant, the total densities of the emitting species s, the degeneracy of upper level i and wavelength of the transition, respectively. f_λ^b is self-absorption coefficient (the same SA) at wavelength λ, which has a magnitude between 0 and 1. F is a constant which includes the optical efficiency of the collection system, the total density of plasma and its volume. $Z_s(T)$ is the partition function of the analyzed spectral line. Here, the transition parameters of g_i, A_{ij} and E_i are inserted from spectral databases, and the magnitudes of F, C_s and T are extracted from the experimental data. According to the calibration-free method, the concentration of all the elements in the sample can be calculated by

$$C_s = \frac{1}{F} U_s(T) e^{q_s}. \tag{34}$$

where, $q_s = \ln \frac{C_s F}{U_s(T)}$. The self-absorption coefficient can be estimated by considering the ratio of the other emission intensities to an internal reference line for each species as

$$\frac{f_\lambda^b}{f_{\lambda_R}^b} = \frac{I_\lambda^{ij}}{I_{\lambda_R}^{mn}} \frac{A_{mn} g_m}{A_{ij} g_i} e^{-E_m - E_i / k_B T}, \tag{35}$$

here, $I_{\lambda_R}^{mn}$, λ_R, and $f_{\lambda_R}^b$ are spectral line intensity, wavelength, and self-absorption coefficient, respectively, of the mentioned reference line. A_{mn}, E_m, and g_m are the transition parameters related to the atomic levels *m* and *n*. It is assumed that the internal reference line has negligible self-absorption so that $f_{\lambda_R}^b \approx 1$. Hence, the self-absorption coefficient of other spectral lines are calculated by the following equation:

$$f_\lambda^b = \frac{I_\lambda^{ij} A_{mn} g_m}{I_{\lambda_R}^{mn} A_{ij} g_i} e^{-E_m - E_i / k_B T}. \tag{36}$$

Finally, the corrected line intensities without any self-absorption can be evaluated from the signal ratio of the measured spectral line intensity to the self-absorption coefficient as

$$\hat{I}_\lambda^{ij} = \frac{I_\lambda^{ij}}{f_\lambda^b} = \frac{I_{\lambda_R}^{mn} A_{ij} g_i}{A_{mn} g_m} e^{-E_m - E_i / k_B T}. \tag{37}$$

By utilizing Eq. (37), the self-absorption correction for each spectral line can be estimated. When these corrections are inserted to every point in the Boltzmann plot, the scattering of the points from each species around the best fit line will be decreased, and then more accurate quantitative results will be attained.

The outline of the IRSAC model is shown in this reference. Based on Eq. (36), for estimation of self-absorption coefficients, the plasma temperature must be known, which is initially calculated from Boltzmann plot curve without requiring any correction. The corrected line intensities by the previous temperature are utilized for evaluation of a new temperature. Since the magnitude of self-absorption straightly depends on the plasma temperature, the optimal self-absorption coefficients can be calculated by an iterative procedure up to attain a convergence in the correlation coefficients on the Boltzmann plot. After convergence of the correlation coefficients, the corrected points will be placed approximately on parallel linear Boltzmann plot and the calculated temperatures from two successive iterations will alter a little. For more illustrations, a schematic of the Boltzmann plot for neutral atoms and singly ionized element is shown in Figure 7 before correction by the basic CF-LIBS method and IRSAC model.

In this experiment, a Q-switched Nd:YAG laser with pulse duration of 10 ns, laser energy of 200 mJ, wavelength of 1064 nm, and repetition rate of 1–15 Hz is focused on an aluminum alloy sample. For this sample, the acquisition delay time and integration gate width are adjusted for 2.5 μs and 1 ms, respectively.

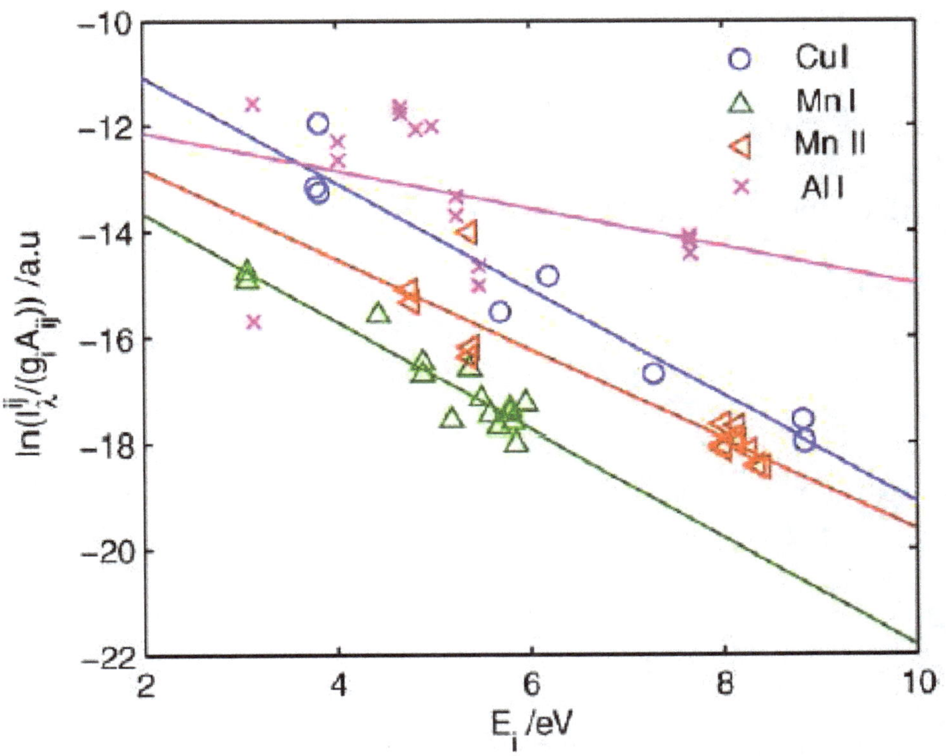

Figure 7. Initial Boltzmann plot which is depicted from spectral intensities of the alloys of aluminium samples.

4.6. Duplicating mirror

Moon et al. [8] duplicated the emission from a plasma by putting a spherical mirror at twice the distance of its focal length from the plasma. They evaluated the existence of optically thick plasma conditions by a very quick check-up. They acquired two line profiles (with and without applying the mirror) for determination of the amount of self-absorption in order to correct the spectral lines.

By taking into account the theoretical consideration, they evaluated the emission from a spatially homogeneous plasma distribution by the presence and absence of a mirror as below:

$$B_{\lambda,1} = B_{\lambda}^{bb}(1 - \exp[-k_{\lambda} l]). \tag{38}$$

$$B_{\lambda,2} = B_{\lambda,1} + GB_{\lambda,1}\exp(-k_{\lambda}l) = B_{\lambda,1}\left[1 + G\exp(-k_{\lambda}l)\right]. \tag{39}$$

In the above equation, the indexes 1 and 2 refer to the measurements with and without inserting the mirror, respectively. B_{λ}^{bb} indicates the spectral radiance of black body emission by the Wien or the Planck laws. In this calculation, G comprises both reflection and absorption losses of the mirror and in addition, solid angles produce imperfect matching. G can be computed by taking the ratio of signal intensity of continuous radiation, where $k_{\lambda} = 0$, i.e. at the line wings with negligible absorption as

$$R^{c} = \frac{B_{\lambda,2}^{c}}{B_{\lambda,1}^{c}} = 1 + G. \tag{40}$$

Furthermore, the ratio of peak intensities with and without considering mirror can be expressed as

$$R_{\lambda} = \frac{B_{\lambda,2}}{B_{\lambda,1}} = 1 + G\exp(-k_{\lambda}l). \tag{41}$$

The optical depth variations can also be evaluated by knowing the parameters of R^{c} and R_{λ} as:

$$\tau = k_{\lambda}l = \ln\frac{\left[R^{c} - 1\right]}{\left[R_{\lambda} - 1\right]}. \tag{42}$$

Finally, the correction factor $K_{\lambda,corr}$ (which is the inverse of the self-absorption coefficient SA) can be calculated experimentally versus the ratios of R^{C} and R^{λ} as:

$$K_{\lambda,corr} = \ln\frac{\left[(R^{c} - 1)/(R_{\lambda} - 1)\right]}{1 - \dfrac{(R_{\lambda} - 1)}{(R^{c} - 1)}}. \tag{43}$$

Furthermore, the duplication factor can be calculated from the following equation as

$$D_{\lambda}(\lambda) = \frac{B_{\lambda}(\lambda, 2fn_{a}l) - B_{\lambda}(\lambda, fn_{a}l)}{B_{\lambda}(\lambda, fn_{a}l)}, \tag{44}$$

where, $D_{\lambda}(\lambda)$ illustrates the relative growth in spectral line intensity or integral absorption created by doubling the value of $(fn_{a}l)$ with two asymptotic magnitudes of 1 (at low optical depths) and 0.415 (at high optical depths). Here, f, n_a and l refer to oscillation strength, number density, and plasma length, respectively. In this work, the plasma emission is produced by irradiation of a Nd:YAG laser with 90 ± 5 mJ pulse energy, 1064 nm laser wavelength, 6 ns pulse duration, and 1 Hz repetition frequency on the sample surface. The ICCD is adjusted for a delay time of 1 μs. Figure 8 expresses the results for analysis of Cu spectral line at 510.55 nm. Figures 8(a) and 8(b) show the evolution of R_λ and $K_{\lambda,corr}$ as a function of wavelength. As it is seen in Figure 8(a), the peak of spectral line saturates faster rather than that in line wings by doubling the plasma length. For obtaining the exact value of the correction factor $K_{\lambda,corr}$, the ratio of R_λ is estimated pixel by pixel. Furthermore, Figure 9 illustrates the evolution of calculated self-absorption correction factor $K_{\lambda,\,corr}$ as a function of R_λ and R^c. As it is seen in this figure, for the experimental R^c data, and by approaching R_λ to unity, the spectral lines are severely self-absorbed by including high values for correction factor $K_{\lambda,\,corr}$.

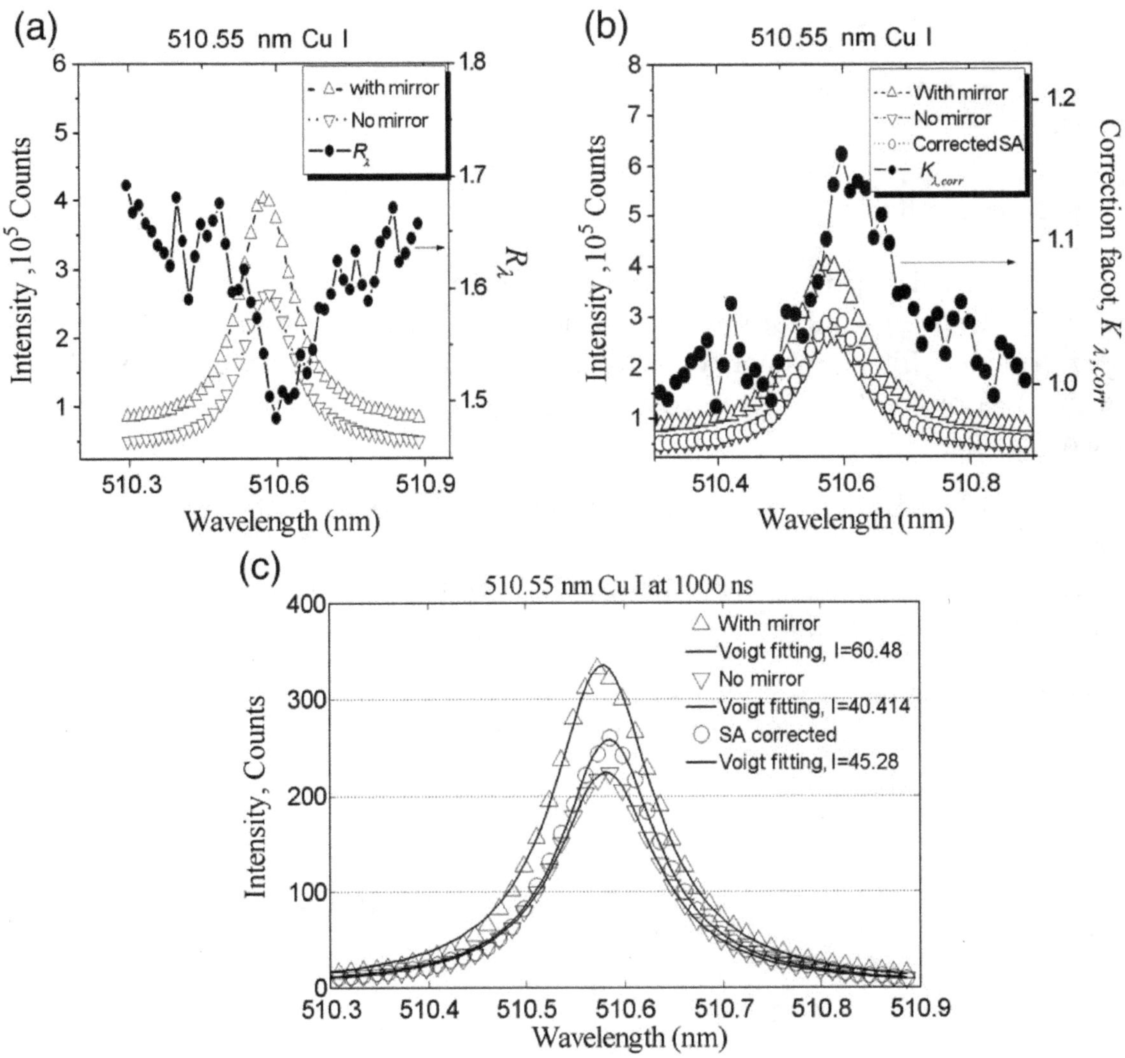

Figure 8. (a) Calculated data of R_λ versus wavelength, (b) evolution of correction factor $K_{\lambda,corr}$ versus line profiles corrected by self-absorption effect and wavelength. All curves are related to the spectral line Cu I at 510.55 nm, and at 1.0 μs delay time.

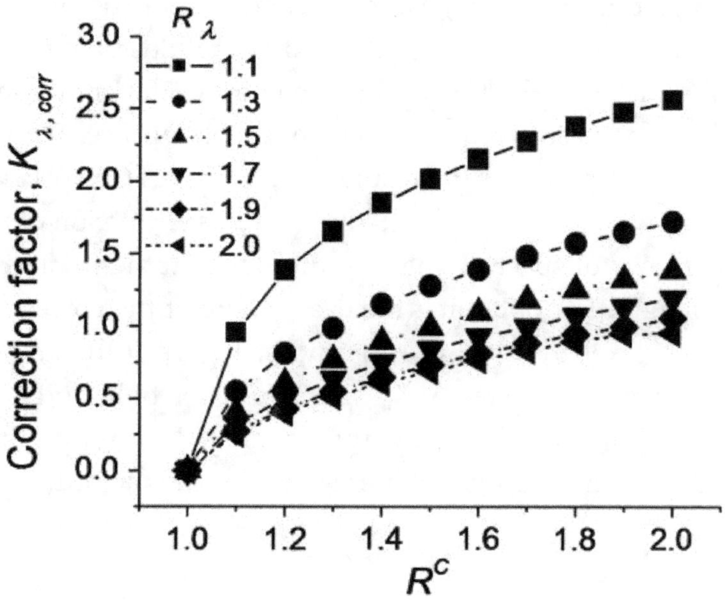

Figure 9. The dependence of the calculated correction factor $K_{\lambda,corr}$ to R_C, for different amounts of R_λ, or different levels of self-absorption.

4.7. Three lines method

Rezaei and Tavassoli [18] introduced the three lines method for studying optically thick plasma in local thermodynamic equilibrium by LIBS method without needing any spectral correction. They performed a LIBS experiment on an aluminum target in air atmosphere by utilizing the two techniques of spectroscopy and shadowgraphy, as well as by applying a theoretical approach.

In this study, plasma parameters were accurately determined by obtaining the plasma length, electron density, and intensities of three spectral lines from experiments. The model explains that instead of utilizing two spectral lines in thin plasmas, three lines are needed in thick plasmas for accurate evaluation of plasma temperature.

The thick plasma emission per unit volume, per unit time, and per unit frequency can be evaluated by replacing the multiplication of self-absorption coefficient by thin plasma intensity as:

$$I_{thick} = SA \times I_{thin} = \frac{(1-e^{-kl})}{(1-e^{-\frac{hc}{\lambda_0}\frac{1}{k_B T}})} \frac{C8\pi}{\lambda_0^2} e^{-\frac{(E_u - E_l)}{k_B T}} \text{(SI)}. \tag{45}$$

The parameters inserted in the above equation were introduced in previously mentioned equations. Here, Z is the partition function, which is computed by two and three level methods [29]. C is instrumental function which includes the efficiency and the solid angle of the detection system [30]. Here, k (m^{-1}) is the absorption coefficient which contains both contribution of the stimulated emission of upper level and the absorption of lower level in the SI units such as in the following equation:

$$k = \frac{g_u A_{ul} N_{Al} \lambda_0^2}{8\pi Z} e^{-\frac{E_l}{k_B T}} (1 - e^{-\frac{h\nu_0}{k_B T}}) L(\nu, \nu_0, \gamma_{ul}). \tag{46}$$

L $(\nu, \nu_0, \gamma_{ul})$ is the Lorentzian line profile because of Stark broadening mechanism. γ_{ul} is the decay rate with a straight dependence to the line width as below [31]:

$$\gamma_{ul} = 2\pi (c / \lambda_0^2) \Delta\lambda_{stark}. \tag{47}$$

$\Delta\lambda_{stark}$ is full-width at half-maximum (FWHM) of the spectral line, which is produced due to Stark effect and is defined as follows [32]:

$$\Delta\lambda_{stark} = \frac{2\omega n_e}{n_{ref}}, \tag{48}$$

here, ω and n_e are electron impact parameter and electron number density, respectively. n_{ref} is the reference electron density (here, 10^{16} cm^{-3}) in which ω is calculated.

As mentioned in Eq. (45), for the spatially integrated plasma emission, the ratio of selected spectral lines can be written as below:

$$\frac{I_{thick\_1}}{I_{thick\_2}} - \frac{(1 - e^{-k_1 l})}{(1 - e^{-k_2 l})} \times \frac{(1 - e^{-\frac{hc}{\lambda_{02}} \frac{1}{k_B T}})}{(1 - e^{-\frac{hc}{\lambda_{01}} \frac{1}{k_B T}})} \times \frac{\lambda_{02}^2}{\lambda_{01}^2} \times \frac{e^{\frac{1}{k_B T}(E_{l1} - E_{u1})}}{e^{\frac{1}{k_B T}(E_{l2} - E_{u2})}} = 0. \tag{49}$$

$$\frac{I_{thick\_2}}{I_{thick\_3}} - \frac{(1 - e^{-k_2 l})}{(1 - e^{-k_3 l})} \times \frac{(1 - e^{-\frac{hc}{\lambda_{03}} \frac{1}{k_B T}})}{(1 - e^{-\frac{hc}{\lambda_{02}} \frac{1}{k_B T}})} \times \frac{\lambda_{03}^2}{\lambda_{02}^2} \times \frac{e^{\frac{1}{k_B T}(E_{l2} - E_{u2})}}{e^{\frac{1}{k_B T}(E_{l3} - E_{u3})}} = 0. \tag{50}$$

It should be noted that both of the above equations satisfy at particular T and N_{Al}. Consequently, the cross of two equations (i.e. left-hand side of equations) with the contour of zero (i.e. right-hand side of equations) is the answer. Therefore, by depicting the contour plot of the above equations and crossing them, the passive parameters of T and N_{Al} is gotten at a particular delay time and at specific laser energy. Then, by inserting these data (answer) in one of the above equations, the instrumental function C will be obtained.

4.8. Line ratio analysis

Bredice et al. [33] utilized the theoretical treatment companioned by experimental results for estimation of the amount of self-absorption in single and collinear double pulse configuration. They used the two-line ratio analysis of the same species of manganese element in Fe–Mn alloys to characterize the homogenous plasma parameters. Moreover, they calculated the self-absorption coefficient by considering the line ratios in different conditions: two lines reveal weak or no self-absorption, two lines experience strong self-absorption, two lines belong to the same multiplet, and two lines are general cases as well. Their results are summarized as the following:

1. Limit case: two negligible self-absorbed lines:

When two lines are weakly self-absorbed, i.e. $(SA)_1=(SA)_2=1$ (or $\kappa_{1,2}(\lambda_0)l<<1$), the ratio of the two spectral intensities from the similar species can be expressed as

$$\frac{(N_p)_2}{(N_p)_1}\approx\frac{(A_{ki}g_ke^{-\frac{E_k}{k_BT}})_2}{(A_{ki}g_ke^{-\frac{E_k}{k_BT}})_1}. \qquad \kappa_{1,2}(\lambda_0)l<<1 \qquad (51)$$

It should be noted that in this calculation, N_p indicates the integral intensity of the selected spectral line and also, other parameters are introduced before.

2. Limit case: severely self-absorbed spectral lines:

When two spectral lines are exposed to strong self-absorption, the self-absorption coefficients can be calculated as:

$$SA\approx\frac{1}{\kappa(\lambda_0)l}. \qquad (52)$$

Therefore, the spectral ratios of spectral emissions of the same species, both exposed to strong self-absorption, can be calculated by

$$\frac{(N_p)_2}{(N_p)_1}\approx\left[\frac{(A_{ki}g_ke^{-\frac{E_k}{k_BT}})_2}{(A_{ki}g_ke^{-\frac{E_k}{k_BT}})_1}\right]\left[\frac{\left(\frac{\lambda_0^4A_{ki}g_ke^{-\frac{E_i}{k_BT}}}{\Delta\lambda_0}\right)_1}{\left(\frac{\lambda_0^4A_{ki}g_ke^{-\frac{E_i}{k_BT}}}{\Delta\lambda_0}\right)_2}\right]^{\beta} \qquad \kappa_{1,2}(\lambda_0)l>>1 \qquad (53)$$

3. Limit case: two lines related to the same multiplet

In two spectral lines belonging to the same multiplet, the atomic energies of E_i and E_k are similar and it is obtain an intensity ratio that is constant with plasma temperature variation. In this regime, by considering the above situations, the intensity ratio can be located between the two extremes as

$$\frac{(N_p)_2}{(N_p)_1}=\left[\frac{(A_{ki}g_k)_2}{(A_{ki}g_k)_1}\right]^{(1-\beta)}\left[\frac{(\lambda_0^4)_1}{(\lambda_0^4)_2}\right]^{\beta} \qquad \kappa_{1,2}(\lambda_0)l >> 1, \tag{54}$$

$$\frac{(N_p)_2}{(N_p)_1}\approx\frac{(A_{ki}g_k)_2}{(A_{ki}g_k)_1} \qquad \kappa_{1,2}(\lambda_0)l << 1 \tag{55}$$

in the case of high and low self-absorption, respectively.

4. General case

For two arbitrary spectral lines, the spectral intensities ratios can be evaluated by

$$\begin{aligned}\frac{(N_p)_2}{(N_p)_1}&=\frac{(A_{ki}g_k e^{-\frac{E_k}{k_BT}})_2}{(A_{ki}g_k e^{-\frac{E_k}{k_BT}})_1}\left(\frac{k(\lambda_0)_1}{k(\lambda_0)_2}\frac{(1-e^{-k(\lambda_0)_2 l})}{(1-e^{-k(\lambda_0)_1 l})}\right)^{\beta}\\ &=a\left[\frac{b(1-e^{-k(\lambda_0)_2 l})}{(1-e^{bk(\lambda_0)_2 l})}\right]^{\beta}.\end{aligned} \tag{56}$$

$$\begin{aligned}a&=\frac{(A_{ki}g_k e^{-\frac{E_k}{k_BT}})_2}{(A_{ki}g_k e^{-\frac{E_k}{k_BT}})_1},\\ b&=\frac{(\lambda_0^4 A_{ki}g_k e^{-\frac{E_i}{k_BT}})_1}{(\lambda_0^4 A_{ki}g_k e^{-\frac{E_i}{k_BT}})_2}\left(\frac{(\Delta\lambda_0)_2}{(\Delta\lambda_0)_1}\right).\end{aligned} \tag{57}$$

In order to find numerically $k(\lambda_0)_2 l$, the intensity ratio of two spectral lines $\frac{(N_p)_2}{(N_p)_1}$, the line broadening ratio $\frac{(\Delta\lambda_0)_2}{(\Delta\lambda_0)_1}$, and plasma temperature T must be determined from the experiment.

By knowing the $k(\lambda_0)l$ for a spectral line, the self-absorption coefficient SA can be carefully calculated.

For instance, in order to show the application of the mentioned theoretical equations, Figure 10 is depicted. In this experiment, Nd-YAG laser pulses at 1064 nm with 7 and 12 ns FWHM and various laser energies are irradiated on Fe-Mn alloys. The experiment is performed at different places in single and double pulse configurations. Figure 10 illustrates the signal ratios of the spectral lines of Mn II at 293.3 to Mn II at 294.9 nm (lines related to the same multiplet) and ratio of Mn II at 293.3 to spectral line of Mn II at 348.3 nm (lines belong to different multiplets) as a function of delay time. Temporal evolution of intensity ratio is plotted as a function of the acquisition delay time in two situations of single and double pulse measurements. In double pulse measurements, the delay time is considered after the arrival of the second pulse on the sample. Since the first two lines, i.e. Mn II lines at 293.3 and 294.9 nm, belong to the similar multiplet, the theoretically mentioned equation is applicable. As shown in the theory, if two limit cases of low or high self-absorption are satisfied, the intensity ratio of these two lines will be particularly independent of the plasma variation.

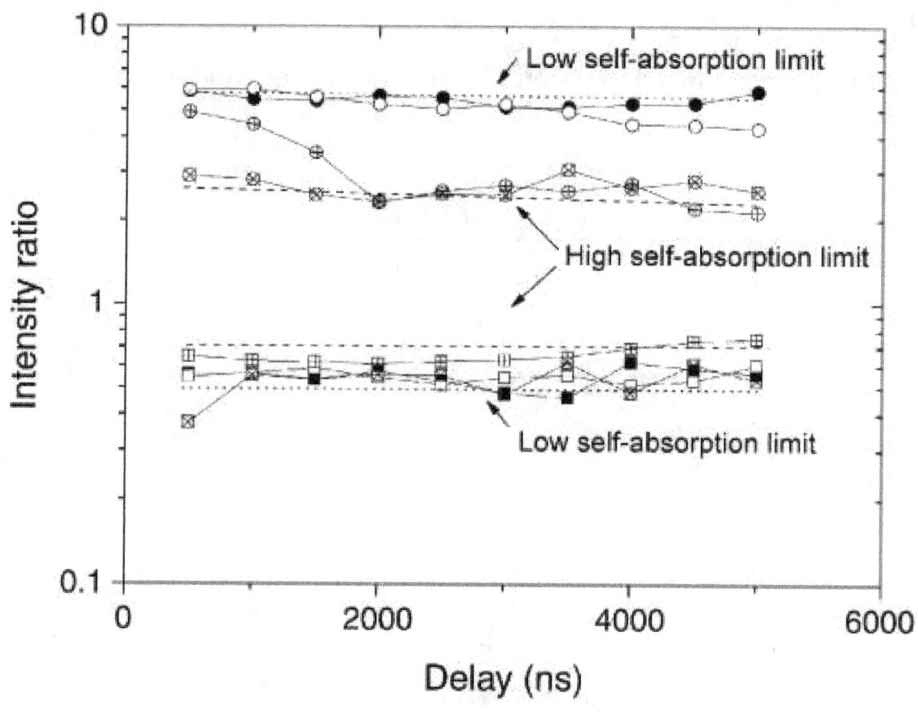

Figure 10. Temporal evolution of intensity ratios of of two spectral lines of Mn II at 293.3 and Mn II at 294.9 nm (full squares, laser energy of 60 mJ, open squares, 120 mJ and crossed squares, 200 mJ laser energy, diagonally crossed squares, 60 + 60 mJ double pulse irradiation) and for Mn II at 293.3/Mn II at 348.3 nm lines (full circles, 60 mJ laser energy, open circles, 120 mJ laser energy, crossed circles, 200 mJ laser energy, diagonally crossed circles, 60 + 60 mJ double pulse measurements).

It should be noted that the dotted line shows the limit of low self-absorption and dashed line indicates the high self-absorption limits. Furthermore, in both lines, y axis is in logarithmic scale. Cristoforetti and Tognoni [19] calculated the concentration ratio of different elements by assumption of holding a homogeneity condition in plasma without needing any self-absorption correction. Furthermore, by obtaining the columnar densities, they computed the plasma temperature and the number densities of different plasma species.

In this work, first, by numerical calculation of the optical depth $k(\lambda_0)l$, the SA parameter is computed by exploiting Eq. (16), as shown in Figure 11. After that, the columnar density n_{il} can be easily extracted by rewriting the equation of optical depth as

$$\left(n_i l\right)_{17} = 1770 \frac{\Delta\lambda_0}{f\lambda_0^2} k(\lambda_0) l, \tag{58}$$

here, $\Delta\lambda_0$ is the expected line width in the case of thin plasma, $(n_{il})_{17}$ is explained in 10^{17} cm^{-2} units, and λ_0 and $\Delta\lambda_0$ are written in Angstrom units. By knowing the columnar density, the inverse of plasma temperature can be calculated from the slope of the line obtained by fitting the points in the Saha–Boltzmann plot.

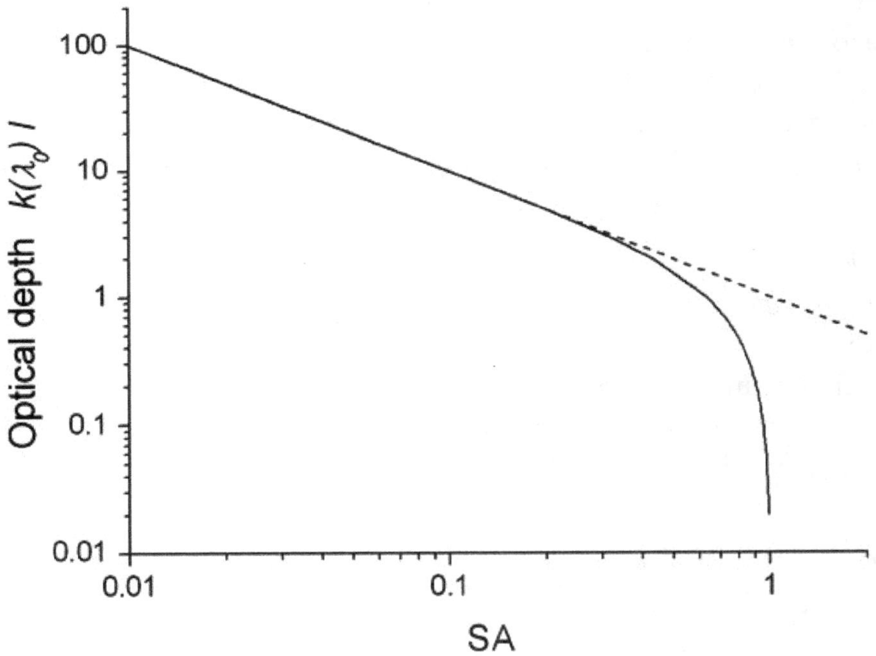

Figure 11. Relation between optical depth and the SA coefficient.

5. Inhomogeneous plasma

Different schematics of nonhomogeneous plasmas including 2, 5, N, and 250 sections in cylindrical and spherical shapes are investigated for thick plasma analysis by considering self-absorption correction as in the figures below:

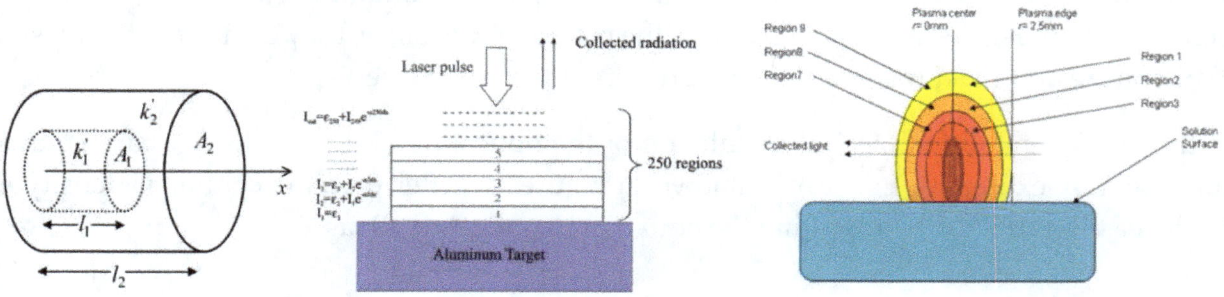

Figure 12. Different schematics of nonhomogeneous thick plasmas [10, 13, 34].

The methods of all of above schematics are approximately similar, so that all of them utilize the emission of internal layers plus attenuation from outer regions. Therefore, the calculations related to 250 layers as well as another numerical model including *N* layer are explained in the following section.

Rezaei et al. [34] studied the spectral emissions of an aluminum sample located in argon and helium noble gases at 1 atm pressure, by applying a numerical calculation. They computed the plasma parameters by coupling the thermal model of laser ablation, hydrodynamic of plasma expansion, and Saha–Eggert equations. In that model, the spectral emissions were constructed from the superposition of some strong lines of aluminum and several strong lines of ambient gases, which were superimposed on a continuous radiation composed of bremsstrahlung and recombination emissions. Moreover, they calculated the spectral emissions in two cases of thin and thick plasmas by considering the self-absorption influence.

In this work, the plasma is supposed to be consisted of 250 layers with 60-μm thickness (as shown in Figure 12). Each section of this plasma is characterized by specific plasma parameters, such as temperature, electron density, mass densities, and number densities of plasma species as a function of delay time. The plasma radiations of different layers are collected in a parallel direction to the laser pulse. The emission of the first layer quite above the sample surface due to its own radiation can be expressed by:

$$I_{(1)} = \varepsilon_{spec(1)} = \sum_{j=1}^{N_s} N_u^{(1)} A_{ul}^j h\nu_j L_j^{(1)}(\nu,\nu_j,\gamma_{ul}^j). \tag{59}$$

The contribution of the second layer toward the optical collecting system, which comprises both of its radiation and attenuation of the first layer is calculated from:

$$I_{(2)} = \sum_{j=1}^{N_s} N_j^{(2)} A_{ul}^j h\nu_j L_j^{(2)}(\nu,\nu_j,\gamma_{ul}^j) + \varepsilon_{spec(1)} e^{-k_{(2)}dx} = \varepsilon_{spec(2)} + I_{(1)} e^{-k_{(2)}dx}. \tag{60}$$

Consequently, the spectral intensity of the nth layer due to the whole sequential absorption is defined as:

$$I_{(n)} = \sum_{j=1}^{N_s} N_j^{(n)} A_{ul}^j h\nu_j L_j^{(n)}(\nu, \nu_j, \gamma_{ul}^j) + I_{(n-1)} e^{-k_{(n)}dx} = \varepsilon_{spec(n)} + I_{(n-1)} e^{-k_{(n)}dx}. \quad (61)$$

According to the two-level system, the absorption coefficient is expressed by inserting the contributions of the absorption and induced emission between the levels of u and l as

$$k_{(n)} = \frac{h\nu_j}{c}(B_{ul}N_l - B_{lu}N_u)L^{(n)}{}_j(\nu, \nu_j, \gamma_{ul}^j)(\mathrm{SI}), \quad (62)$$

Here, B_{ul} and B_{lu} are Einstein coefficients related to the absorption and induced emission (m^3J s), respectively. Finally, the self-absorbed spectral intensity is the emission of last layer (i.e 250th layer) which includes the entire regions shown below for attenuation as follows:

$$I_{selfabsorb} = \varepsilon_{(250)} + I_{(249)} e^{-k_{(250)}dx}. \quad (63)$$

For comparison between thick plasma emissions with a non-self-absorbed spectral line in a thin plasma condition, the summation over all strong line radiations, including the contribution of the whole layers, can be considered as

$$I_{nonselfabsorb}(\nu, \nu_j) = \sum_{k=1}^{k_s} \sum_{j=1}^{Ns} \varepsilon_{spec}^j(\nu, \nu_j), \quad (64)$$

where, k_s and N_s are the number of plasma layers and the number of strong lines, respectively. For instance, the effect of pulse laser energy on Al spectral lines at 394.40 and 396.15 nm is shown in Figure 13 in the logarithmic scale. Here, a Gaussian-shaped laser pulse, with a wavelength of 266 nm, FWHM of 10 ns, under different laser irradiances is focused on aluminum sample. As it is seen by increasing laser intensities, the self-absorption coefficient grows. The magnitudes of self-absorption coefficients in two ambient gases of Ar and He are inserted in this reference.

Furthermore, Ben Ahmed and Cowpe proposed a nonhomogeneous plasma with five layers as shown in Figure 12, and, then, they calculated the total observed intensity by taking into account the *i* and *j* regions as:

$$I = \sum_i \left(\frac{I_i}{\tau_i}(1 - exp(-\tau_i))exp(\sum_{j>i}(-\tau_j)) \right). \quad (65)$$

Lazic et al. [16] considered a cylinder with length L including homogenous density and temperature, which is divided into N similar thin layers with length δL surrounded by another

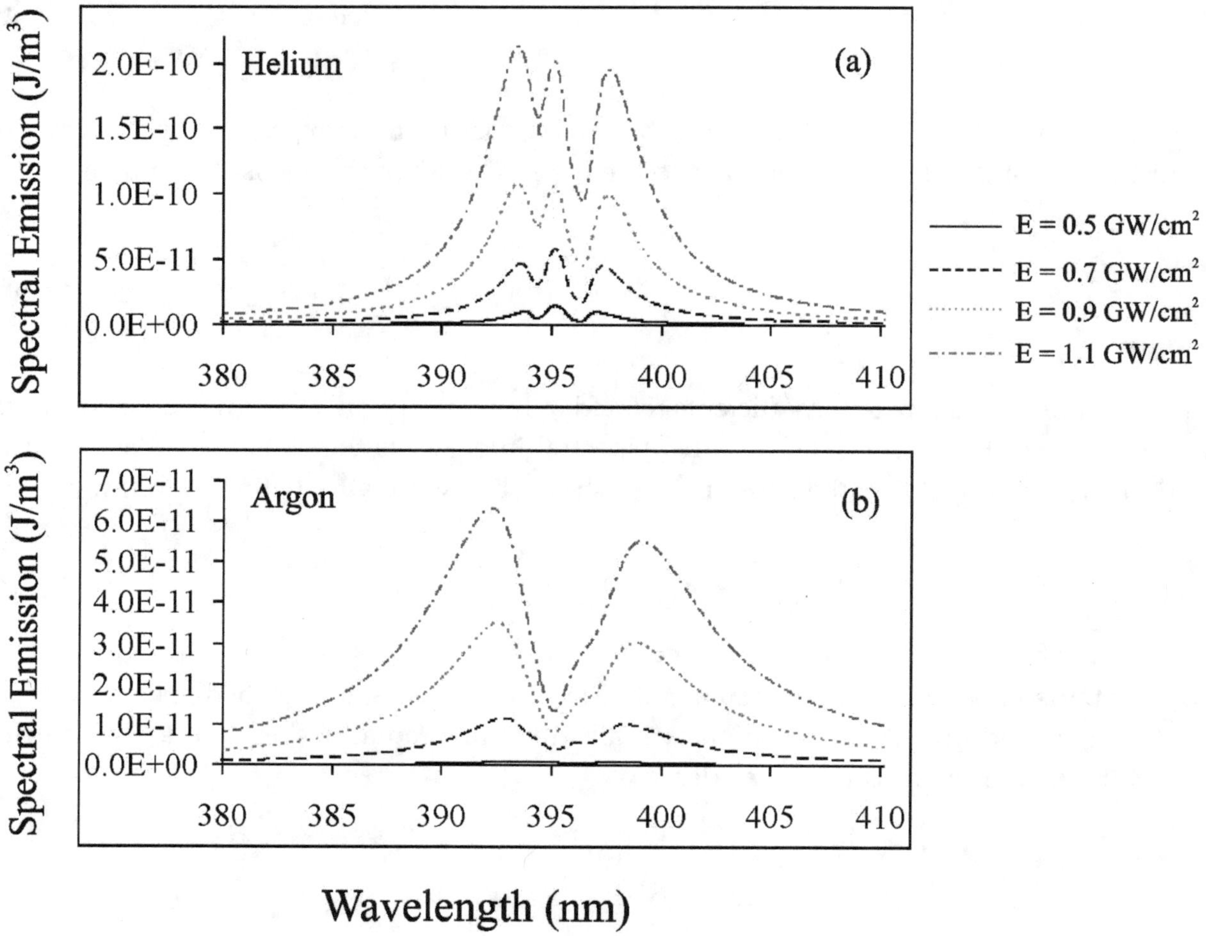

Figure 13. Self-reversal evolutions of the spectral lines of 394.40 and 396.15 nm, at different laser energies of 0.5, 0.7, 0.9, and 1.1 GW/cm² in (a) helium and (b) argon gas.

outer thin layer. Similar to the above-mentioned method, by considering successive absorption and using the seri summation result, they expressed the total line intensity escaping from plasma as

$$I_\alpha^{ki} = FC_\alpha \frac{f_\alpha^{ki}(T)}{U_\alpha(T)} \frac{1-e^{-\beta^{ik}C_\alpha^i L}}{L/\delta L\left(1-e^{-\beta^{ik}C_\alpha^i \delta L}\right)}. \tag{66}$$

where, C_α is the total species concentration and C_α^i is the concentration of species at lower level i which is related to the C_α by Boltzmann equation. f_α^{ki} comprises the all parameters correspond to ki transition. $U_\alpha(T)$ is partition function and F is a constant attributed to the experimental condition. Then, by assumption of existence of a thin colder plasma surrounding the cylindrical plasma, the overall emission is obtained as below:

$$I_{\alpha}^{ki} = F_1 C_{\alpha} \frac{f_{\alpha}^{ki}(T)}{U_{\alpha}(T)} \times \frac{1 - e^{-\frac{\beta^{ik} H_{\alpha}^{i} C_{\alpha} L}{U_{\alpha}(T)}}}{L/\delta L \left(1 - e^{-\frac{\beta^{ik} H_{\alpha}^{i} C_{\alpha} \delta L}{U_{\alpha}(T)}}\right)} + F_2 C_{\alpha} \frac{f_{\alpha}^{ki}(T)}{U_{\alpha}(T)}. \tag{67}$$

where, $H_{\alpha}^{i} = g_i e^{-\frac{E_i}{K_B T}}$. F_1 and F_2 are constants related to optically thick and thin plasma, respectively, and are straightly dependent on the plasma geometry.

By combination of the above equations, a relation is obtained which connect the raw measured element concentration C_m to effective element concentration C_E as follow:

$$C_m = \frac{a_1}{a_2}\left(1 - e^{-a_2 C_E}\right) + a_3 C_E. \tag{68}$$

The first term in the above equation is attributed to the optically thick plasma and a_i are coefficients that can be determined by fitting to the experimental results. Figure 14 illustrates the measured spectral emission of Cu at 327.39 nm versus certified concentration for different selected samples. In this experiment, a third harmonic Nd:YAG pulsed laser beam with energies of 6–10 mJ is focused on soil and sediment samples. The acquisition delay time is adjusted for 300 ns and the gate width is 1000 ns for all examined samples.

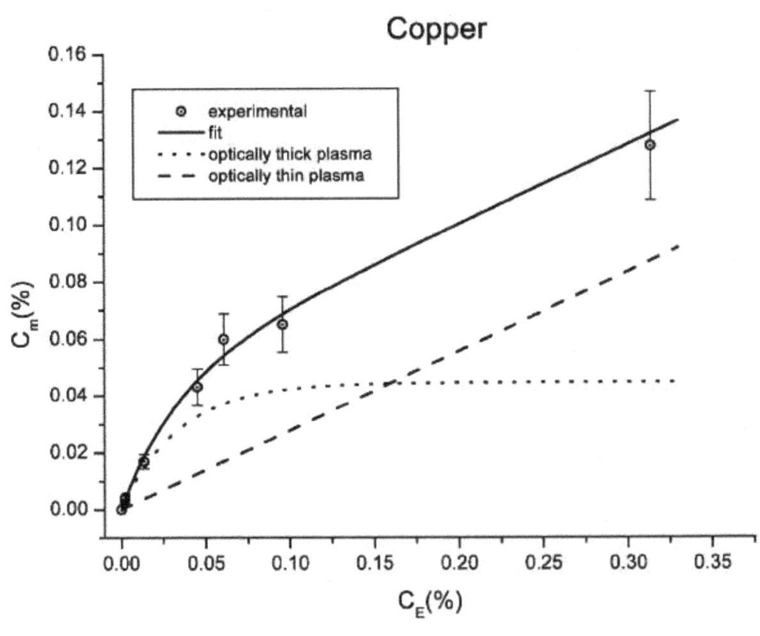

Figure 14. Measured LIBS emission for Cu spectral line at 327.39 nm as a function of certified concentrations.

Some research groups used self-absorbed spectral lines for charactrizing the plasma properties by proposing appropriate models without needing to any self-absorption correction as mentioned in Section (7. IV). In addition, Gornushkin et al. [35] introduced a semiempirical model for an optically thick inhomogeneous plasma in LTE condition. In this model, the input parameters are the ratio of atomic species and plasma pressure or the number of plasma elements, which were all measured from the experiments. Some functions are introduced for calculation of plasma temperature and its size variation. The outputs of this model are time and space evolutions of species number densities, variations of optical depth and spectral line profiles, as well as the resulting intensity of spectral emission close to the transition of strong nonresonance atomic line. The main application of this model is the prediction of electron density, temperature, and the mechanism of line broadening.

According to this model, the relation of the observed spectral radiance I_υ(erg.s^{-1}.cm^{-2}.sr^{-1}.Hz^{-1}) with absorption coefficient κ(ν, x, T) and distribution of volume emission coefficient ε_ν(ν, x, T) in optically thick plasma can be expressed by radiation transfer equation as

$$I_\nu = \int_{-x_0}^{x_0} \varepsilon_\nu(\nu,x,T)e^{-\tau(\nu,x,T)}dx. \tag{69}$$

τ(ν, x, T) is the plasma optical thickness which is expressed by:

$$\tau(\nu,x,T) = \int_{-x}^{x_0} \kappa(\nu,x,T)dx. \tag{70}$$

x_o and -x_o refer to the plasma dimension from edge to edge along the line of sight. The magnitudes of volume emission ε_ν(ν, x, T) and the absorption κ(ν, x, T) coefficients in an inhomogenous plasma are not constant and vary with the position along the line of sight. Bartels and Zwicker [4, 5] proved that in the absence of stimulated transitions, Eq. (69) can be expressed as the multiplication of the below three terms as:

$$I_\nu = A(T_m).M(E_l,E_u).Y\left[\tau(\nu,x,T),p\right]. \tag{71}$$

The first term, $A(T_m)$ is assumed as a source function which is related only to T_m and is the exact explanation for Wien's law:

$$A = \frac{2h\nu^3}{c^2}\exp(-\frac{h\nu}{kT_m}). \tag{72}$$

It should be noted that Omenetto et al. [36] expressed that the source function in Wien's law can be applied for plasmas in which stimulated transitions are not important and the condition of $h\nu/kT_m \gg 1$ holds, while Planck's law is utilized when stimulated transitions are taken into account.

The second term in Eq. (71) considers, to some extent, the influences of plasma heterogeneity on radiation and for spectral line with naturally and van der Waals broadening equals to:

$$M = \sqrt{\frac{E_l}{E_u}}. \qquad if \qquad \frac{kT_m}{E_l} << 1. \tag{73}$$

For singly ionized atoms and for neutral lines with Stark broadened, M can be expressed as

$$M = \sqrt{\frac{E_l + 0.5x_0}{E_u + 0.5x_0}}. \qquad if \qquad \frac{kT_m}{E_l + 0.5x_0} << 1 \tag{74}$$

Here, E_l and E_u are the lower and upper levels excitation energies, respectively and x_0 is the ionization energy. Furthermore, the above mentioned conditional statements in Eqs. (84) and (85) are not satisfied for resonance neutral lines, but they are logical assumptions for ionic spectral lines as well as for nonresonance neutral lines.

The third term in Eq. (71) considers the influences of heterogeneous mixing and optical thickness. The function $Y[\tau(\nu, x, T), p]$ is approximated by the below equation:

$$\begin{aligned} Y\left[\tau(\nu,x,T),p\right] &= e^{-\frac{\tau(\nu,x,T)}{2}}[\frac{\tau(\nu,x,T)}{2}(1-p) \\ &+ p\sinh\left(\frac{\tau(\nu,x,T)}{2}\right) \\ &+ \frac{1}{\sqrt{p}}\sinh\left(\frac{\tau(\nu,x,T)}{2}\sqrt{p}\right)]. \end{aligned} \tag{75}$$

Here, the parameter p is explained as

$$p = \frac{6}{\pi} arctg \frac{M^2}{\sqrt{1+M^2}}. \tag{76}$$

Therefore, by substitution of above equations into Eq. (71), spectral emission intensity for optically thick inhomogenous plasma I_υ can be obtained. In this model, the spatial distribution of plasma temperature is approximated by a second-order polynomial function with a maximum value in the plasma center.

Figure 15 represents the calculated emission distributions for different initial number of atoms and various Si/N atom ratios obtained at several distinct delay times (from 1 to 4 a.u.). In this figure, it is seen that at low densities (10^{16}), the plasma continuous emission (curves at t_1, t_2), monotonically reduces by time passing and growing the plasma. In addition, at high densities (10^{17}), a considerable value of early plasma continuum is absorbed within the plasma plume, which causes less emission radiation at time t_1 compared to times t_2 and t_3. This results in the LIP to approach a black body emitter at initial delay times.

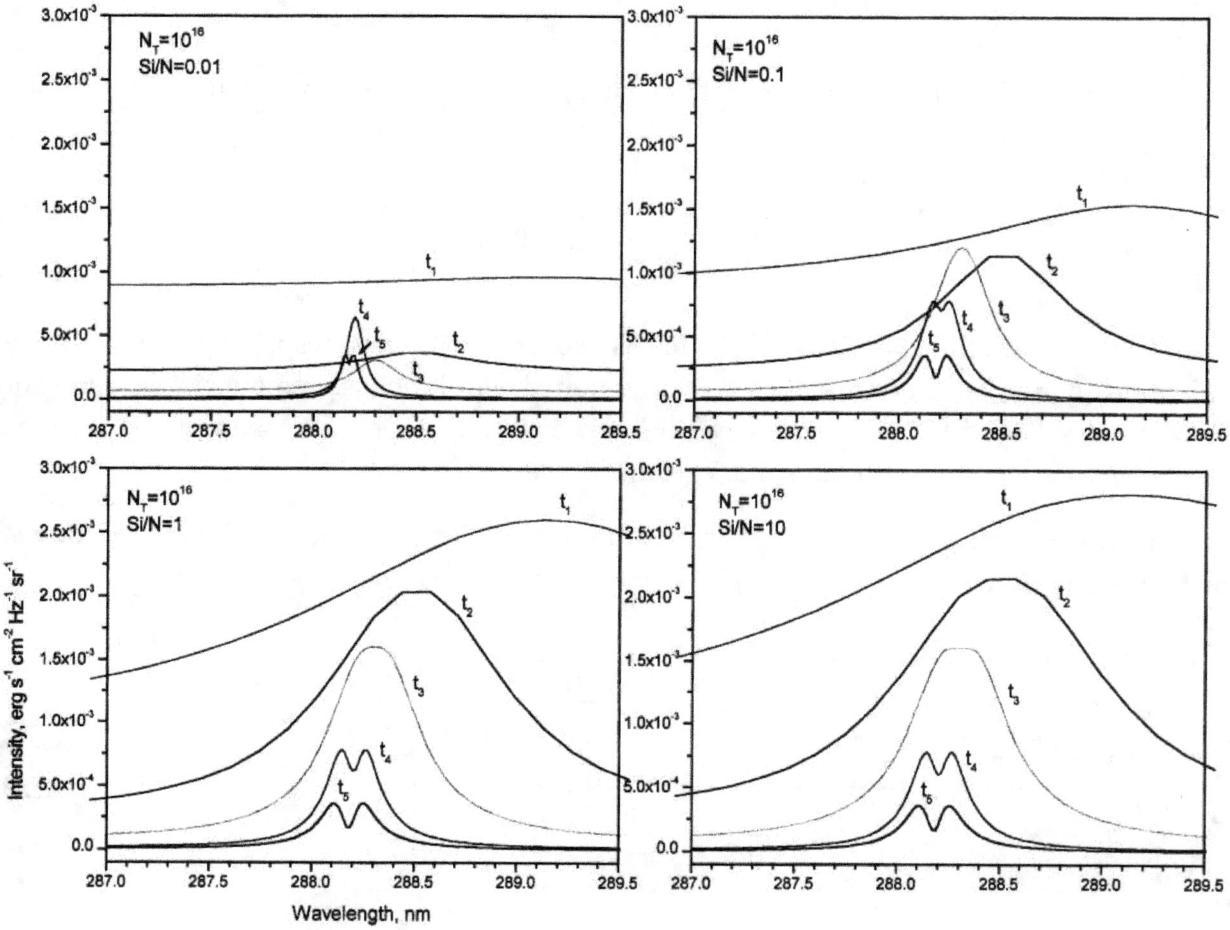

Figure 15. Calculated radiation profiles for LIP in initial times, including different proportions of 10^{16} N and Si atoms: from Si/N = 0.01 to Si/N =10. Times t_1 to t_5 refers to the magnitudes of 1, 1.5, 2, 3, and 4 in relative units.

Moreover, it must be mentioned that different research groups [37–41] discussed about spectrum analysis and extraction of plasma parameters in symmetric and asymmetric self-reversal line shapes produced by different sources that can be studied by readers for completeness of information.

6. Conclusion

In this chapter, a brief description of the different methods of analyzing the thick laser-produced plasmas was represented by considering the theoretical and experimental techniques. In some works, the self-absorption was corrected and then, the plasma parameters was extracted, while in others, some proposed models were explained and thereafter, plasma features was obtained straightly without needing any correction.

Finally, it should be noted that different parameters such as laser features (its wavelength [42], double or single pulse scheme [33, 43], energy [34], and pulse duration [42]), ambient gas condition (the nature and its pressure [42, 44]) and measurement device circumstance (delay time [8, 34], gate width, and exposure time), and the sample characteristics (metal and biological cases) could affect the amount of self-absorption as well. The details of these effects can be studied in the related references.

Author details

Fatemeh Rezaei*

Address all correspondence to: fatemehrezaei@kntu.ac.ir

Department of Physics, K. N. Toosi University of Technology, Shariati, Tehran, Iran

References

[1] Ovsyannikov, A., *Plasma Diagnostics*. 2000: Cambridge Int Science Publishing.

[2] Aragón, C. and J. Aguilera, *CSigma graphs: A new approach for plasma characterization in laser-induced breakdown spectroscopy*. Journal of Quantitative Spectroscopy and Radiative Transfer, 2014. 149: p. 90–102.

[3] Cowan, R.D. and G.H. Dieke, *Self-absorption of spectrum lines*. Reviews of Modern Physics, 1948. 20(2): p. 418.

[4] Zwicker, H., *Evaluation of plasma parameters in optically thick plasmas*. Plasma Diagnostics, 1968.

[5] Bartels, H., *Z. Phys 125 597 Bartels H 1949*. Z. Phys, 1949. 126: p. 108.

[6] Amamou, H., et al., *Correction of self-absorption spectral line and ratios of transition probabilities for homogeneous and LTE plasma*. Journal of Quantitative Spectroscopy and Radiative Transfer, 2002. 75(6): p. 747–763.

[7] Amamou, H., et al., *Correction of the self-absorption for reversed spectral lines: application to two resonance lines of neutral aluminium*. Journal of Quantitative Spectroscopy and Radiative Transfer, 2003. 77(4): p. 365–372.

[8] Moon, H.-Y., et al., *On the usefulness of a duplicating mirror to evaluate self-absorption effects in laser induced breakdown spectroscopy*. Spectrochimica Acta Part B: Atomic Spectroscopy, 2009. 64(7): p. 702–713.

[9] Aguilera, J. and C. Aragón, *Characterization of laser-induced plasmas by emission spectroscopy with curve-of-growth measurements. Part I: Temporal evolution of plasma parameters and self-absorption.* Spectrochimica Acta Part B: Atomic Spectroscopy, 2008. 63(7): p. 784–792.

[10] Aguilera, J., J. Bengoechea, and C. Aragón, *Curves of growth of spectral lines emitted by a laser-induced plasma: influence of the temporal evolution and spatial inhomogeneity of the plasma.* Spectrochimica Acta Part B: Atomic Spectroscopy, 2003. 58(2): p. 221–237.

[11] Aragon, C., J. Bengoechea, and J. Aguilera, *Influence of the optical depth on spectral line emission from laser-induced plasmas.* Spectrochimica Acta Part B: Atomic Spectroscopy, 2001. 56(6): p. 619–628.

[12] Aragón, C., F. Peñalba, and J. Aguilera, *Curves of growth of neutral atom and ion lines emitted by a laser induced plasma.* Spectrochimica Acta Part B: Atomic Spectroscopy, 2005. 60(7): p. 879–887.

[13] Ben Ahmed, J. and J. Cowpe, *Experimental and theoretical investigation of a self-absorbed spectral line emitted from laser-induced plasmas.* Applied optics, 2010. 49(18): p. 3607–3612.

[14] Bulajic, D., et al., *A procedure for correcting self-absorption in calibration free-laser induced breakdown spectroscopy.* Spectrochimica Acta Part B: Atomic Spectroscopy, 2002. 57(2): p. 339–353.

[15] El Sherbini, A., et al., *Evaluation of self-absorption coefficients of aluminum emission lines in laser-induced breakdown spectroscopy measurements.* Spectrochimica Acta Part B: Atomic Spectroscopy, 2005. 60(12): p. 1573–1579.

[16] Lazic, V., et al., *Self-absorption model in quantitative laser induced breakdown spectroscopy measurements on soils and sediments.* Spectrochimica Acta Part B: Atomic Spectroscopy, 2001. 56(6): p. 807–820.

[17] Sun, L. and H. Yu, *Correction of self-absorption effect in calibration-free laser-induced breakdown spectroscopy by an internal reference method.* Talanta, 2009. 79(2): p. 388–395.

[18] Rezaei, F. and S.H. Tavassoli, *A new method for calculation of thick plasma parameters by combination of laser spectroscopy and shadowgraphy techniques.* Journal of Analytical Atomic Spectrometry, 2014. 29(12): p. 2371–2378.

[19] Cristoforetti, G. and E. Tognoni, *Calculation of elemental columnar density from self-absorbed lines in laser-induced breakdown spectroscopy: A resource for quantitative analysis.* Spectrochimica Acta Part B: Atomic Spectroscopy, 2013. 79: p. 63–71.

[20] Hermann, J., C. Boulmer-Leborgne, and D. Hong, *Diagnostics of the early phase of an ultraviolet laser induced plasma by spectral line analysis considering self-absorption.* Journal of applied physics, 1998. 83: p. 691–696.

[21] D'Angelo, C.A., D.M.D. Pace, and G. Bertuccelli, *Semiempirical model for analysis of in-*

homogeneous optically thick laser-induced plasmas. Spectrochimica Acta Part B: Atomic Spectroscopy, 2009. 64(10): p. 999–1008.

[22] O'neill, R., *Algorithm AS 47: function minimization using a simplex procedure.* Applied Statistics, 1971: p. 338–345.

[23] Rezaei, F., P. Karimi, and S. Tavassoli, *Effect of self-absorption correction on LIBS measurements by calibration curve and artificial neural network.* Applied Physics B, 2014. 114(4): p. 591–600.

[24] Mitchell, A. and M.W. Zemansky, *Resonance Radiation*. 1934: Cambridge.

[25] Hinnov, E., *A method of determining optical cross sections.* JOSA, 1957. 47(2): p. 151–155.

[26] Alkemade, C.T.J., et al., *Th. Metal Vapours in Flames*. 1982, Pergamon Press: Oxford.

[27] Gornushkin, I., et al., *Curve of growth methodology applied to laser-induced plasma emission spectroscopy.* Spectrochimica Acta Part B: Atomic Spectroscopy, 1999. 54(3): p. 491–503.

[28] Demtröder, W., *Laser Spectroscopy: Basic Concepts and Instrumentation*. 2013: Springer Science & Business Media.

[29] D'Ammando, G., et al., *Computation of thermodynamic plasma properties: a simplified approach.* Spectrochimica Acta Part B: Atomic Spectroscopy, 2010. 65(8): p. 603–615.

[30] Aguilera, J. and C. Aragón, *Characterization of laser-induced plasmas by emission spectroscopy with curve-of-growth measurements. Part II: Effect of the focusing distance and the pulse energy.* Spectrochimica Acta Part B: Atomic Spectroscopy, 2008. 63(7): p. 793–799.

[31] Silfvast, W.T., *Laser Fundamentals*. 2004: Cambridge University Press.

[32] Griem, H.R., *Plasma Spectroscopy*. New York: McGraw-Hill, 1964, 1964. 1.

[33] Bredice, F., et al., *Evaluation of self-absorption of manganese emission lines in Laser Induced Breakdown Spectroscopy measurements.* Spectrochimica Acta Part B: Atomic Spectroscopy, 2006. 61(12): p. 1294–1303.

[34] Rezaei, F., P. Karimi, and S. Tavassoli, *Estimation of self-absorption effect on aluminum emission in the presence of different noble gases: comparison between thin and thick plasma emission.* Applied Optics, 2013. 52(21): p. 5088–5096.

[35] Gornushkin, I., et al., *Modeling an inhomogeneous optically thick laser induced plasma: a simplified theoretical approach.* Spectrochimica Acta Part B: Atomic Spectroscopy, 2001. 56(9): p. 1769–1785.

[36] Omenetto, N., J. Winefordner, and C.T.J. Alkemade, *An expression for the atomic fluorescence and thermal-emission intensity under conditions of near saturation and arbitrary*

self-absorption. Spectrochimica Acta Part B: Atomic Spectroscopy, 1975. 30(9): p. 335–341.

[37] Fishman, I., G. Il'in, and M.K. Salakhov, *Spectroscopic diagnostics of a strongly inhomogeneous optically thick plasma. Part 1. The formation of asymmetric self-reversed emission and absorption lines: determination of electron impact half-width and electron concentration.* Spectrochimica Acta Part B: Atomic Spectroscopy, 1995. 50(9): p. 947–959.

[38] Fishman, I., G. II'in, and M.K. Salakhov, *Temperature determination of an optical thick plasma from self-reversed spectral lines.* Journal of Physics D: Applied Physics, 1987. 20(6): p. 728.

[39] Fishman, I., G. Il'in, and M.K. Salakhov, *Spectroscopic diagnostics of a strongly inhomogeneous optically thick plasma. Part 2. Determination of atom concentration and variations of different physical values in the plasma cross-section using asymmetric self-reversed emission and absorption lines.* Spectrochimica Acta Part B: Atomic Spectroscopy, 1995. 50(10): p. 1165–1178.

[40] Karabourniotis, D., *Plasma temperature determination from the maximum intensity of a symmetric self-reversed line.* Journal of Physics D: Applied Physics, 1983. 16(7): p. 1267.

[41] Ilin, G., M.K. Salakhov, and I. Fishman, *Analysis of asymmetric self-reversal of spectral lines for the case of lateral observation of a spectrum.* Zhurnal Prikladnoi Spektroskopii, 1976. 24: p. 201–207.

[42] Lee, Y.-I., et al., *Influence of atmosphere and irradiation wavelength on copper plasma emission induced by excimer and Q-switched Nd: YAG laser ablation.* Applied Spectroscopy, 1997. 51(7): p. 959–964.

[43] Gautier, C., et al., *Main parameters influencing the double-pulse laser-induced breakdown spectroscopy in the collinear beam geometry.* Spectrochimica Acta Part B: Atomic Spectroscopy, 2005. 60(6): p. 792–804.

[44] Rezaei, F. and S.H. Tavassoli, *Quantitative analysis of aluminum samples in He ambient* gas at different pressures in a thick LIBS plasma. Applied Physics B, 2015: p. 1–9.

7

Plasma Nitriding of Titanium Alloys

Afsaneh Edrisy and Khorameh Farokhzadeh

Abstract

Titanium alloys are found in many applications where weight saving, strength, corrosion resistance, and biocompatibility are important design priorities. However, their poor tribological behavior is a major drawback, and many surface engineering processes have been developed to enhance wear in titanium alloys such as nitriding. Plasma (ion) nitriding, originally developed for ferrous alloys, has been adopted to address wear concerns in titanium alloys. Plasma nitriding improves the wear resistance of titanium alloys by the formation of a thin surface layer composed of TiN and Ti_2N titanium nitrides (e.g., compound layer). Nonetheless, plasma nitriding treatments of titanium alloys typically involve high temperatures (700–1100°C) that promote detrimental microstructural changes in titanium substrates, formation of brittle surface layers, and deterioration of mechanical properties especially fatigue strength. This chapter summarizes the previous and ongoing investigations in the field of plasma nitriding of titanium alloys, with particular emphasis on the authors' recent efforts in optimization of the process to achieve tribological improvements while maintaining mechanical properties. The development of low-temperature plasma nitriding treatments for $\alpha + \beta$ and near-β titanium alloys and further wear improvements by alteration of near-surface microstructure prior to nitriding are also briefly reviewed.

Keywords: Titanium alloys, plasma nitriding, compound layer, nitrogen diffusion zone, fatigue crack initiation, severe plastic deformation, surface failure mechanisms

1. Introduction

Titanium alloys are attractive candidates for aerospace, automotive, and biomedical industries due to their many advantages including, but not limited to, high strength-to-weight ratio, excellent corrosion/oxidation resistance, biocompatibility, and high fatigue resistance. However, their poor tribological behavior, for example, high and unstable coefficient of friction, high wear rate, and susceptibility to galling and scuffing, restricts their use for applications where sliding is inevitable. Nitriding is a surface hardening process in which nitrogen atoms are utilized for surface modification and has been practiced for wear improvement in titanium alloys for many years. In conventional nitriding techniques such as salt bath treatments, gas nitriding, and hot isostatic pressing (HIP), nitrogen atoms are provided by liquid, gas, or solid media [1, 2]. In plasma nitriding, also known as ion nitriding and glow discharge nitriding, active nitrogen is provided by an ionized nitrogen-containing atmosphere [3–7].

The growing interest in this process is due to its high efficiency, cost-effectiveness for mass production, absence of pollution, and ease of process control [8–11]. Also, it is preferred over other nitriding technologies due to feasibility at lower temperatures, dimensional control, retention of surface finish, and versatility for achieving desired properties through modification of process parameters [8–13].

Plasma nitriding improves the tribological properties of titanium alloys through the formation of a thin surface layer mainly consisting of TiN and Ti_2N, that is, the compound layer. The nitrided microstructure also consists of a region of nitrogen-stabilized α-titanium, that is, α-case, and a nitrogen diffusion zone (typically 15–25 μm deep) underneath the compound layer [14, 15]. The plasma-nitrided microstructure is strongly correlated with the composition of titanium substrate and the nitriding process parameters such as duration, temperature, pressure, and composition of the nitriding medium, among which temperature has the most significant influence [12, 15–17].

Plasma nitriding of titanium alloys is typically carried out in the temperature range of 700–1100°C for 6–80 hours in a nitrogen-containing medium (N_2, N_2–Ar, N_2–H_2, or N_2–NH_3) [18–22]. The high temperatures involved in the process lead to grain growth, overaging, and microstructural transformations in titanium substrates that decrease the fatigue strength and ductility [10, 22, 23]. Moreover, the brittle nature of the compound layer and α-case and the high stiffness mismatch between the compound layer and the titanium substrate lead to premature failure initiation from the surface. As such, improvements in mechanical properties have been reported by practicing plasma nitriding at lower temperatures [10, 24–26]. The slow nitriding kinetics of titanium alloys at low temperatures can be enhanced by alteration of the microstructure in the surface vicinity via thermochemical (e.g., laser-assisted treatments and electron beam (EB) melting) and mechanical (e.g., mechanical attrition and shot peening) pretreatment processes [27, 28]. As a result, the depth and hardness of nitrogen diffusion will increase and thus the load bearing capacity of the plasma-nitrided surfaces will be improved [29]. Duplex treatments have also been introduced and developed in surface engineering of titanium alloys. Duplex treatments combine the positive effect of deep hardened layers achieved in the diffusion zone with low friction and high wear resistance of surface coatings leading to significant enhancements in load bearing capacity and wear resistance [30–36].

This chapter is organized into seven sections as follows. After the introduction, Section 2 gives a brief description of the fundamentals of plasma nitriding and the nitriding process parameters for titanium alloys. Section 3 presents a review of the literature on the effect of plasma nitriding on the microstructure of titanium alloys. The tribological and mechanical properties of plasma-nitrided titanium alloys are discussed in Section 4 and Section 5, respectively. Recent developments in the modification of plasma nitriding treatment via the introduction of duplex treatments and surface pretreatments are presented in Section 6, followed by a summary of the main findings in Section 7.

2. Fundamentals of plasma nitriding

In a conventional plasma nitriding treatment, an external voltage applied between the nitriding furnace (anode) and the workpiece (cathode) ionizes the nitriding gas and provides active nitrogen flux for surface modification [3, 4]. The nitriding atmosphere is generally composed of a buffer gas (e.g., Ar or H_2) and the reactant gas (e.g., N_2). A schematic illustration of voltage versus current density characteristic for glow discharge is shown in Figure 1. Plasma nitriding is conducted in the abnormal glow region where a uniform, stable glow covers the workpiece, and the current density is directly proportional to voltage drop and thus is easily controlled [13, 37]. As the applied voltage is provided by a direct current (DC) power supply, the continuous voltage and heat input may result in localized heating, overheating of thin sections, hollow cathode effect[1], arcing, and other damages to the surface. These problems can be avoided by utilizing a pulsed power supply where the heat input is controlled with duty cycles, usually in the order of 10–50% of the total cycles, without affecting the nitrogen activity or nitriding time [38]. In addition to DC and alternating current (AC) power supplies, plasma may be generated by radio frequency (r.f.) excitation [4, 39], microwaves, electromagnetic induction [4, 40], and electron emission configurations [4]. Some of the recent advances include intensifying the glow discharge using thermionically assisted triode glow discharge, plasma immersion ion implantation, electron cyclotron resonance, etc. The intensified plasma-assisted processes (IPAP) generate a higher density of energetic ions and increase the kinetics and efficiency of nitriding, making plasma nitriding feasible at low temperatures (less than 500°C), pressures, and durations. As a result, significant improvements in surface hardness, depth of nitrided layers, and tribological and fatigue properties have been reported [41–46].

Under the influence of the applied electric field, electrons accelerating from the cathode (workpiece) toward the anode collide with gas molecules of the nitriding atmosphere and generate an environment of positive and negative ions, electrons, and neutral and energetic atoms (e.g., plasma) within a few millimeters from the cathode. Subsequently, the positive ions accelerate toward the cathode over a comparatively short distance and cause a number of phenomena such as sputtering, diffusion to bulk, heating by radiation and collision, surface diffusion, plasma reaction, etc. As a result of these phenomena, surface modification takes place and the generated secondary electrons maintain the abnormal glow. This mechanism eliminates expensive precleaning operations and enhances nitriding kinetics at temperatures lower than those of the conventional nitriding techniques [47, 48].

1 Localized heating at narrow channels, holes, closely located workpieces, etc. where the generation of charge carriers is faster than their annihilation and the glow discharges contact each other or overlap, leading to a rise in current density and temperature.

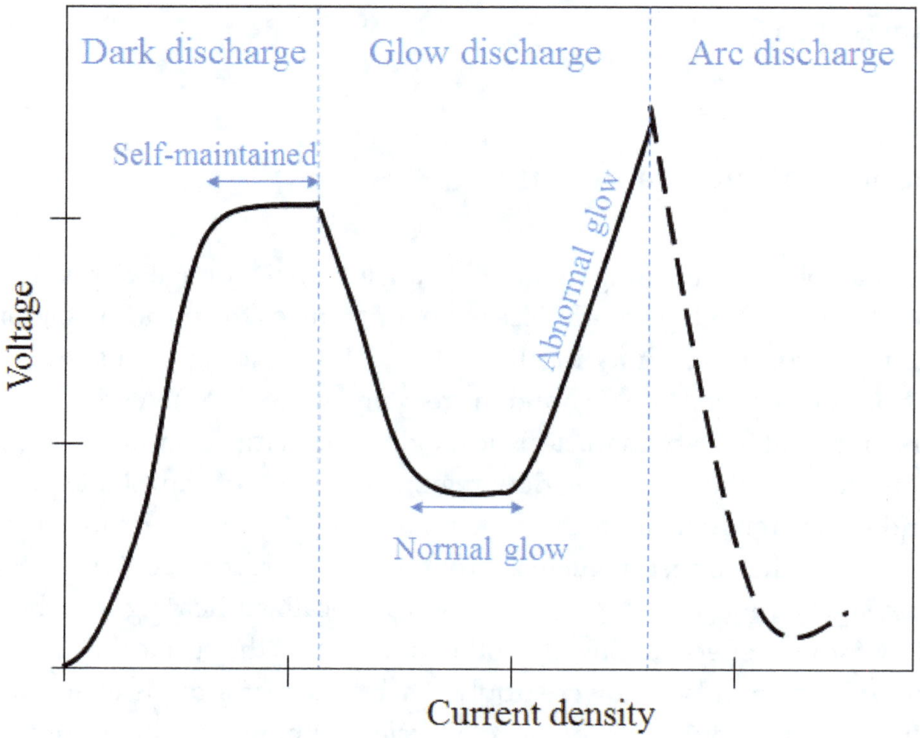

Figure 1. A schematic illustration of voltage vs. current density relationship for an electric glow discharge

Several mechanisms have been proposed to explain the mass transfer mechanism during plasma nitriding including the adsorption of atomic nitrogen model, ion adsorption model, nitrogen implantation model, Kölbel's model, and the NH^+ bombardment model. According to Kölbel's model, plasma nitriding initiates with sputter cleaning of the surfaces, followed by the reaction of sputtered atoms with nitrogen and subsequent deposition of nitride compounds on the surface. The NH^+ radical model is based on the generation of $N_mH_n^+$ molecular ions and radicals in the presence of hydrogen in the nitriding gas mixture. It is proposed that the adsorbed NH^+ radicals on the surface of workpiece dissociate and liberate nitrogen atoms that diffuse into the structure [4, 12, 13, 18].

Plasma nitriding treatments for titanium alloys are typically carried out at 700–1100°C for 6–80 hours. The atmosphere is composed of a nitrogen containing mixture, for example, N_2, N_2–Ar, N_2–H_2, or N_2–NH_3 mixtures, with pressures that vary in the range of 0.5–1.3 kPa [18–22]. In addition to the nitriding time and temperature, the pressure and composition of the nitriding gas also have a significant effect on the microstructure of plasma-nitrided titanium alloys [15, 17]. For instance, the presence of hydrogen in the nitriding mixture enhances nitrogen diffusivity by removing the inherent oxide layer on the surface of titanium alloy that interferes with the nitriding process. In another mechanism suggested by Tamaki et al. [18], the presence of hydrogen in the nitriding medium results in the formation of H^+, NH^+, and NH^{2+} radicals that have a catalytic effect on nitriding kinetics. Conversely, the presence of Ar in the nitriding mixture reduces the depth of nitrogen diffusion and causes random formation of Ti_2N and TiN through sputtering/re-deposition reactions or homogeneous reactions in the plasma [4, 18, 49, 50]. The following sections review the previous investigations on the microstructure and properties of plasma-nitrided titanium alloys focusing especially on the authors' recent work in the field.

3. Microstructure of plasma-nitrided titanium alloys

3.1. Classification of titanium alloys

Titanium has an allotropic phase transformation from its low-temperature hexagonal close-packed (HCP) crystal structure (α-phase) to body-centered cubic (BCC) crystal structure (β-phase) at 882°C, that is, β-transus temperature. Alloying elements have a strong influence on this allotropic transformation; elements such as aluminum, boron, nitrogen, and oxygen increase the β-transus temperature and are referred to as α-stabilizers. Conversely, β-stabilizing elements decrease the β-transus temperature and stabilize the β-phase at room temperature through either eutectoid transformations (iron, hydrogen) or solid solutions (vanadium, molybdenum, and niobium). Thus, depending on the type and content of their alloying elements, titanium alloys can have a single-phase α-structure (α-Ti alloys), α phase with a small (2–5 vol.%) β-phase content (near-α Ti alloys), single-phase β-structure (β-Ti alloys), or a combination of both phases ($\alpha + \beta$ alloys). The families of α- and near-α titanium alloys are famous for their microstructural stability at high temperatures and corrosion resistance; however, their microstructures cannot be modified by heat treatments. Commercially, pure (CP) titanium, Ti–5Al–2.5Sn, Ti–3Al–2.5Sn, Ti–8Al–1Mo-1V, and Ti–6Al–2Sn–4Zr–2Mo are some of the commercial alloys of this family. In contrast, a wide range of microstructures and properties can be achieved in $\alpha + \beta$ and β-titanium alloys by heat treatments and thermomechanical routes. Ti–6Al–4V, Ti–3Al–2.5V, Ti–6Al–2Sn–4Zr–6Mo, Ti–6Al–6V–2Sn, and Ti–4Al–4Mo–4Sn–0.5Si are some of the popular $\alpha + \beta$ titanium alloys. The class of β (and near-β) titanium alloys contains a high concentration of β-stabilizing elements to retain a fully β-phase microstructure at room temperature. These alloys offer good hardenability and formability and resistance to corrosion and hydrogen embrittlement. Ti–13V–11Cr–3Al, Ti–10V–2Fe–3Al, Ti–11.5Mo–6Zr–4.5Sn (β-III), and Ti–3Al–8V–6Cr–4Zr–4Mo (β-C) are among commercially used β-Ti alloys [1-7].

The behavior of Ti–6Al–4V ($\alpha + \beta$ titanium alloy) and Ti–10V–2Fe–3Al (near-β titanium alloy) alloys during plasma nitriding were investigated by the authors. The extra-low interstitial (ELI) grade Ti–6Al–4V had a mill-annealed microstructure consisting of equiaxed α-grains with retained β-particles at α-grain boundaries and fine β-particles inside the α-grains (Figure 2a). The Ti–10V–2Fe–3Al microstructure consisted of elongated primary α-plates in various orientations and aspect ratios inside β-grains with the occasional presence of α-phase at β-grain boundaries (Figure 2b). The chemical composition, average grain size, and β-phase content of the as-received alloys are presented in Table 1.

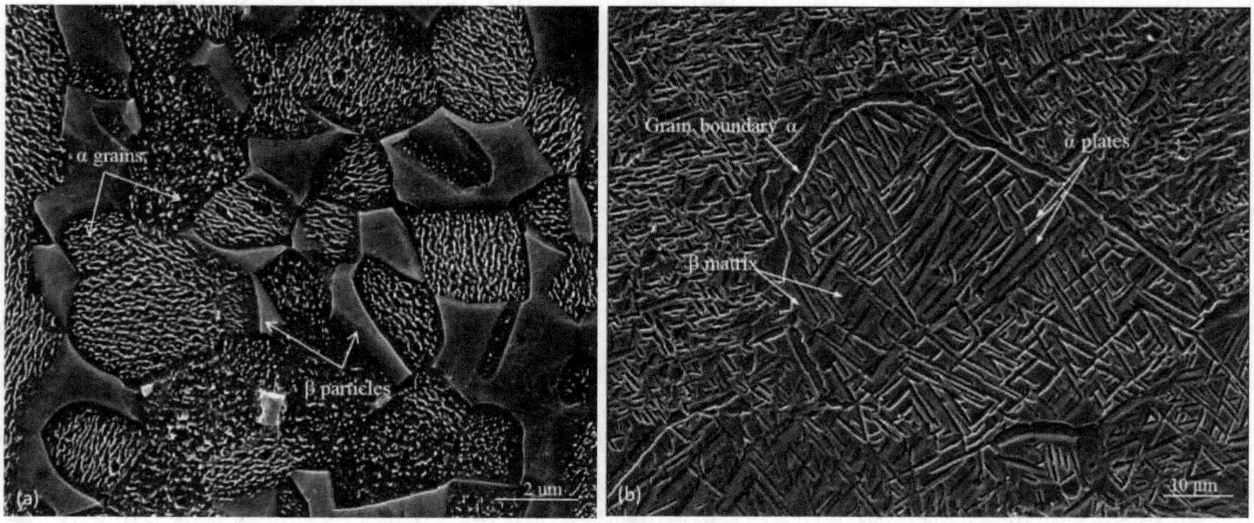

Figure 2. Scanning electron microscopy (SEM) images showing the typical microstructure of the as-received titanium alloys used in plasma nitriding investigations. (a) Ti–6Al–4V had a mill-annealed microstructure consisting of equiaxed α-grains, retained β-particles at α-grain boundaries, and fine recrystallized β-particles inside the α-grains. α-grains were delineated by the different orientation of fine recrystallized β-particles inside the α-grains (etched in Kroll's solution) [51]. (b) Ti–10V–2Fe–3Al microstructure consisted of elongated primary α-plates in various orientations and aspect ratios inside β-grains (marked by arrows) with the occasional presence of α-layers at β-grain boundaries (etched in glycerol + hydrofluoric acid (1:1) solution) [52].

	Chemical Composition (Wt. %)								Hardness (HV)	Grain size (μm)	β content** (Vol. %)
	Fe	V	Al	C*	O*	N*	H*	Ti			
Ti-6Al-4V (ELI)	0.17	4.10	5.98	0.007	0.13	0.012	0.009	balance	381.5 ± 22.8	3.8	24
Ti-10V-2Fe-3Al	1.73	10.65	2.99	-	-	-	-	balance	359.3 ± 15.4	42.4	63.5

* H, O, and N contents were determined by inert gas fusion and the C content by combustion

** The β phase content was determined following a procedure adopted from Tiley et al. [53] by overlaying a regular grid of points on the SEM images and dividing the number of points within the β phase by the total number of points.

Table 1. Properties of the as-received Ti-6Al-4V and Ti-10V-2Fe-3Al alloys used in the plasma nitriding experiments. Chemical compositions were measured using inductively coupled plasma optical emission spectrometry (ICP-OES).

3.2. Effect of plasma nitriding on microstructure of titanium alloys

Plasma nitriding of titanium alloys results in the formation of a *compound layer* on the surface that mainly consists of TiN, δ phase, with FCC crystal structure ($Fm\bar{3}m$ space group) and Ti_2N, ε phase, with tetragonal crystal structure ($P4_2$-mnm space group). TiN is stable over a wide range of nitrogen contents, TiN_x ($0.43 < x < 1.08$), and has a typical hardness of 2500 HV. Ti_2N is stable over a small composition of ~33.3 at.% N and has a maximum hardness of 1500 HV [49, 50, 54–56]. In plasma nitriding investigations of Raveh et al. [14], the compound layer was composed of a TiN layer on top of a mixture of randomly oriented polycrystalline TiN and highly oriented Ti_2N. They also reported clusters of fine (50–100 A°) precipitates consisting of Al, V, Cr, and Fe elements underneath the compound layer. The compound layer has a typical thickness of 1–4 μm and may contain traces of Ti_4N_{3-x}, $TiN_{0.26}$, and TiO_2 in addition to TiN and Ti_2N [6, 14, 15, 52, 57].

Underneath the compound layer, a region of nitrogen-stabilized α-titanium, that is, *α-case* and a nitrogen *diffusion zone* form by interstitial solid solution of nitrogen atoms in the titanium

to its negative effect on the ductility and fatigue strength of titanium alloys. When α + β and β-titanium alloys are nitrided at high temperatures (typically higher than 800°C), α-case forms a continuous α-Ti layer underneath the compound layer as shown in Figure 3; however, at lower temperatures (< 800°C), coarse and columnar α-stabilized grains were reported underneath the compound layer [10, 49, 50, 54, 55, 58]. Similarly, α-case may be formed during plasma nitriding of α-titanium alloys; however, it is more difficult to distinguish unless hydride precipitation in the substrate creates contrast against the α-case [10].

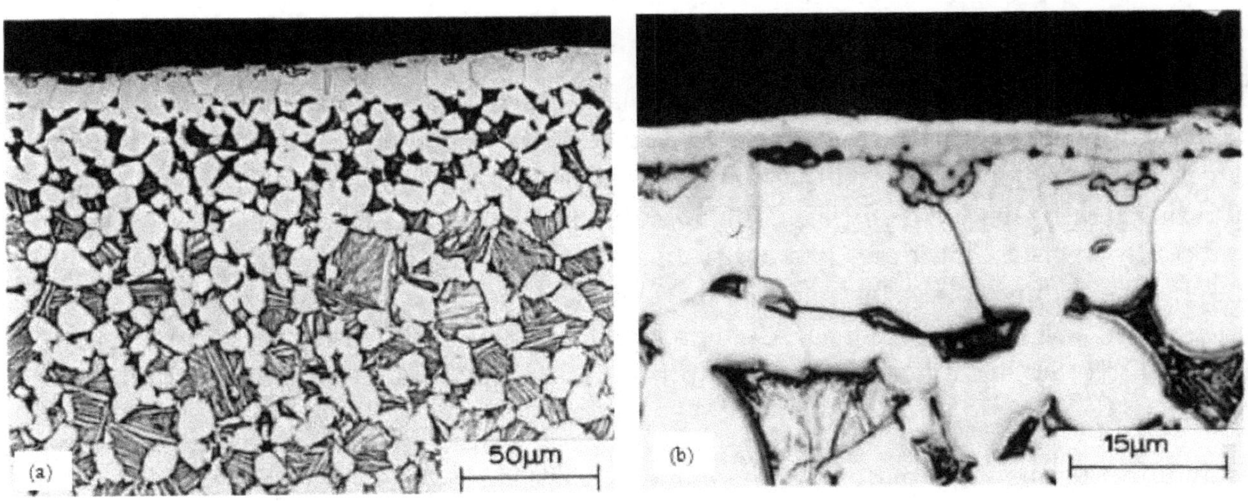

Figure 3. Optical micrographs of Ti–6Al–4V alloy after plasma nitriding in nitrogen for 3 hours at 1000°C showing the formation of an α-case underneath the compound layer at (a) low magnification and (b) high magnification [10].

Nitrogen atoms harden the diffusion zone with a profile that has its maximum at the near surface and gradually decreases toward the bulk. The depth of nitrogen diffusion is dependent on the process parameters as well as the phase composition of titanium alloys due to different solubility limits and diffusion rates of nitrogen in α- and β-titanium [5, 7, 59]. Nitrogen diffuses three times faster in β-Ti compared with α-Ti but has limited solubility in the β-phase [60, 61]. A study conducted by Il'in et al. [59] revealed that the addition of α-stabilizing elements such as Al slowed down the diffusion of nitrogen in α-titanium alloys. Conversely, β-titanium alloys had a shallow diffusion zone due to inadequate solubility of nitrogen in the β-phase. A higher depth of diffusion zone was achieved in α + β Ti alloys that possess a combination of high nitrogen solubility and diffusivity.

The effect of plasma nitriding process parameters on the microstructure and properties of titanium alloys was studied by several researchers [6, 16, 57, 62]. According to da Silva et al.

[16], the composition of plasma-nitrided Ti–6Al–4V surfaces was a very complex function of the nitriding parameters, for example, time, temperature, gas mixture composition, and pressure, among which temperature had the most significant influence. The x-ray diffraction (XRD) studies on plasma-nitrided surfaces revealed that nitriding gas pressure had a minor influence on the nitrided microstructure; however, increasing the partial pressure of nitrogen in the mixture resulted in preferential orientation of Ti_2N and promoted the formation of TiN. They interpreted that the shift observed in β-Ti XRD reflections corresponded to lattice distortions in this phase and likely residual strains induced by nitriding. In addition to TiN and Ti_2N titanium nitrides, they also reported XRD reflections of V_2O, orthorhombic TiO_2, and nitrogen-deficient ζ-Ti_4N_{3-x} nitride in the compound layers formed at certain nitriding conditions. Yildiz et al. [62] found that time and temperature of nitriding were directly correlated with the surface hardness and roughness, thickness of the compound layer, TiN content in the compound layer, and depth of the diffusion zone. They also reported that Al atoms from the Ti–6Al–4V substrate tend to segregate underneath the compound layer and form an Al-enriched layer, which impedes the inward diffusion of nitrogen during plasma nitriding. The presence of an Al-enriched layer underneath the compound layer was also confirmed by other researchers [6, 57].

The authors' investigations on optimizing the plasma nitriding process parameters were aimed at preventing the formation of α-case and microstructural changes in titanium substrates which are detrimental to some mechanical properties for example fatigue [52, 63]. Plasma nitriding treatments were performed in an industrial unit[2] consisting of a DC power supply and control console, nitriding chamber and base assembly, hydraulic lifting system for lifting the chamber from the base, temperature measurement system, evacuation system, gas supply, inlet/outlet, and pressure gauges. A schematic illustration of the nitriding set-up is given in Figure 4. No auxiliary heat source was utilized, and the heat was generated by the glow discharge.

Titanium coupons were placed at the geometric center of a cathodic cage in order to avoid common problems associated with DC plasma nitriding such as arcing damage, hollow-cathode, and edging effect. Prior to each run, the chamber was cleaned through a couple of filling/evacuation steps using Ar-H_2 (1:1) mixture. Subsequently, the coupons were sputter-cleaned at the process temperature in an Ar-H_2 (1:1) atmosphere for a few hours after a stable glow discharge was maintained. The nitriding gas mixture was purged into the chamber and the nitriding cycle was completed. Consequently, the coupons were cooled down to ~150°C inside the furnace under vacuum. Plasma nitriding was carried out at different temperatures (500–800°C), durations (4–87 hours), and pressures (67–533 Pa) in N_2/Ar and/or N_2/H_2 mixtures with nitrogen contents ranging between 1.5% and 75%. After each plasma nitriding treatment, microstructural and chemical analyses were performed and the samples were subjected to analytical microscopy examinations.

Phase constituents of the nitrided surfaces were identified using a Rigaku DMAX-1200 x-ray diffractometer and a Rigaku MiniFlex x-ray diffractometer equipped with Cu K_α tubes. The

2 Exactatherm Ltd., 2381 Anson Dr., Mississauga, ON, Canada

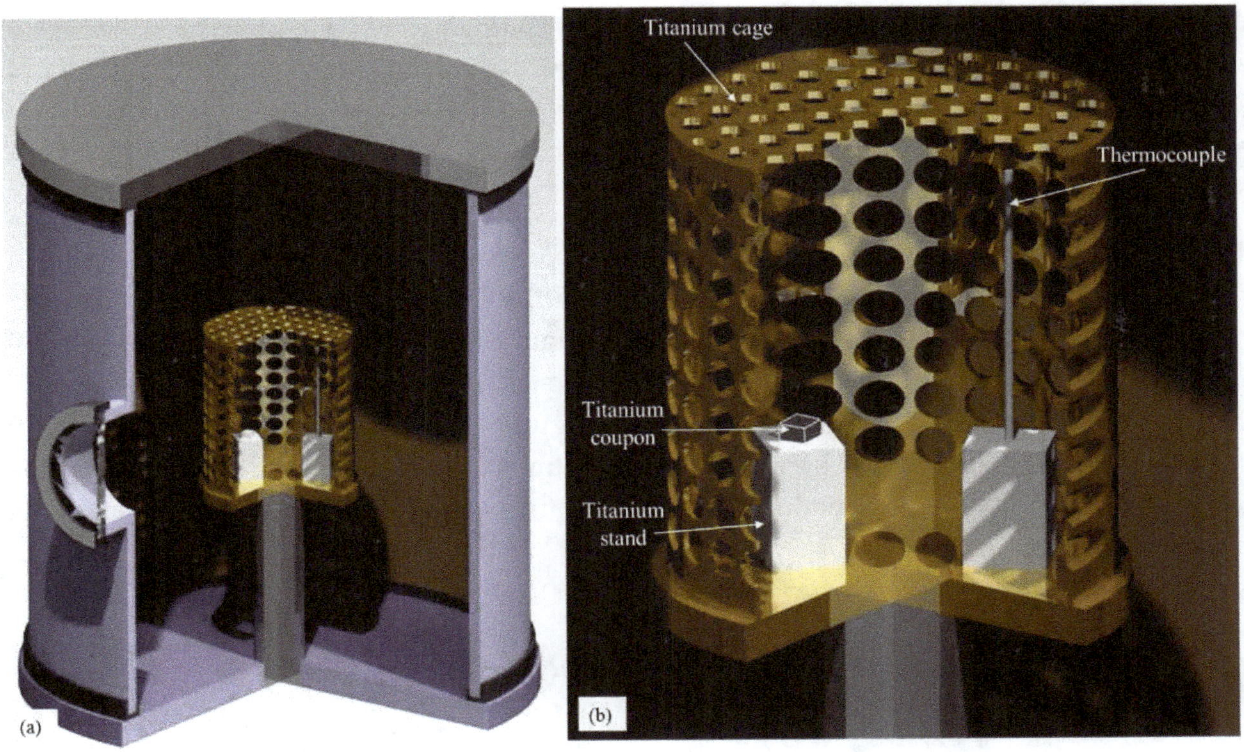

Figure 4. (a) A schematic illustration of the plasma nitriding set-up, no external heating source was used and (b) titanium coupons were placed at the geometric center of a cathodic cage (the cage was also made of Ti alloy to minimize contamination) [51].

developed residual stress level on the surface after plasma nitriding was measured with the $\sin^2\psi$ x-ray diffraction method using a laboratory non-destructive residual stress measurement LXRD system[3]. Surface morphology and cross-sectional microstructure of the nitrided alloys were examined under an FEI Quanta 200 Field Emission Gun (FEG) scanning electron microscope (SEM) equipped with an x-ray energy dispersive spectroscopy (EDS) detector. For cross-sectional study of microstructures, the nitrided surfaces were plated using an electroless nickel coating treatment prior to metallographic preparation to protect the near-surface microstructure against mechanical damage. The cross sections were then cut using a low-speed diamond saw and mounted in epoxy. The metallographic procedure began with serial wet grinding with silicon carbide abrasive papers in successive steps, followed by polishing with diamond. Ti–10V–2Fe–3Al surfaces had an additional fine polishing step using 0.06 μm colloidal silica suspension. The polished samples were cleaned in an ultrasonic bath of ethanol prior to microscopic examinations.

A scanning/transmission electron microscope (S/TEM) equipped with an energy dispersive x-ray spectrometer (Titan 80–3000) was used to investigate the microstructure of nitrided titanium alloys in the surface vicinity. Moreover, x-ray microanalysis coupled with electron diffraction patterns and electron energy-loss spectroscopy (EELS) was used to obtain structural information and chemical composition of the nitrided microstructures. TEM thin-foil speci-

3 Proto Manufacturing Ltd., 2175 Solar Crescent, OldCastle, ON, Canada

mens were prepared using focused ion beam (FIB) milling lift-out technique [64] (dual-beam Zeiss NVision 40[4]) with thicknesses of around 50 nm suitable for high-resolution TEM (HRTEM) imaging and EELS. Ion milling was performed using a gallium ion source at 30 kV and a thin (1–2 μm) layer of platinum was deposited on the milling location to protect the surface features from the milling damage.

The microstructure of plasma-nitrided Ti–6Al–4V consisted of a 1.9-μm compound layer on the surface followed by a 44.4-μm-deep nitrogen diffusion zone (Figure 5). The compound layer consisted of Ti_2N, TiN, and $TiN_{0.3}$ nitrides (Figures 6 and 7). Plasma-nitrided Ti–10V–2Fe–3Al had a 0.5-μm compound layer consisting of TiN ($Fm\bar{3}m$), Ti_2N ($P4_2$-mnm), V_2N ($P\bar{3}1m$), with the possibility of Fe_8N (I4-mmm) (Figures 8 and 9). Vanadium nitride, rarely reported in the literature, was observed along titanium nitride grain boundaries possibly due to faster diffusion of vanadium along grain boundaries (acting as short-circuit paths) at the nitriding temperature [18, 36-37].

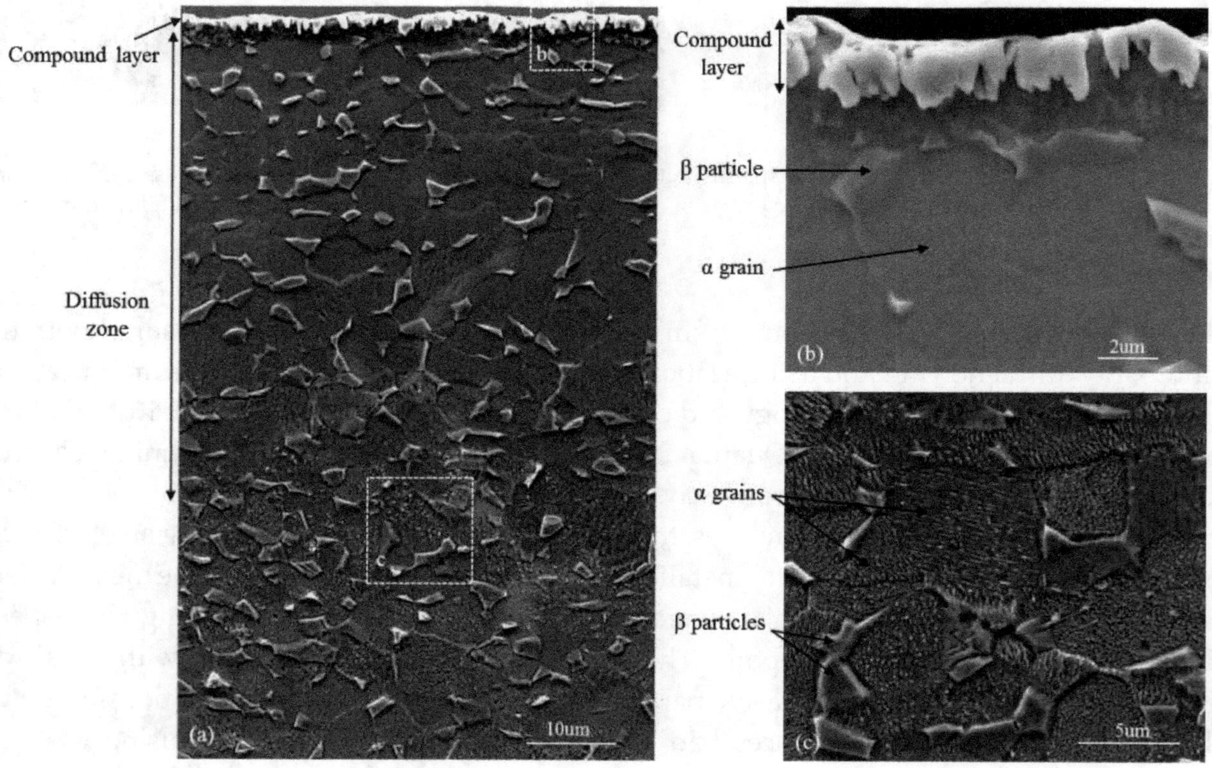

Figure 5. Typical cross-sectional SEM images of plasma-nitrided Ti–6Al–4V microstructure at 600°C consisting of a thin compound layer (1.9 ± 0.5 μm) and a deep (44.4 ± 5.6 μm) diffusion zone. Diffusion zone was delineated by the dissolution of fine recrystallized β-particles inside the α-grains (because nitrogen is an α-stabilizing element) and confirmed with the microhardness-depth profile measurements. (b) Higher magnification image of the enclosed area marked as "b" in Figure 5a confirming the absence of α-case underneath the compound layer. (c) Higher magnification image of the enclosed area marked as "c" in Figure 5a confirming that no phase transformations were observed in the titanium substrate [26].

4 The Canadian Centre for Electron Microscopy (CCEM), Brockhouse Institute for Materials Research, B161 A. N. Bourns Building, 1280 Main Street West, Hamilton, ON, Canada

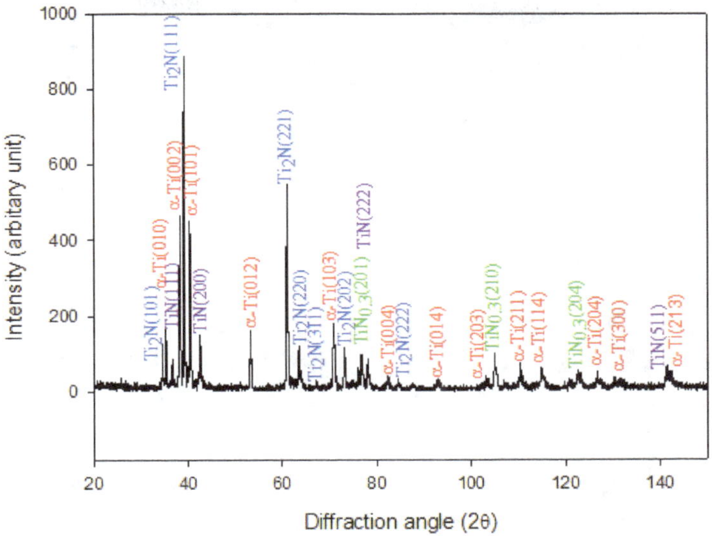

Figure 6. X-ray diffraction (XRD) analysis showing the presence of TiN, Ti_2N, and $TiN_{0.3}$ in the compound layer [63].

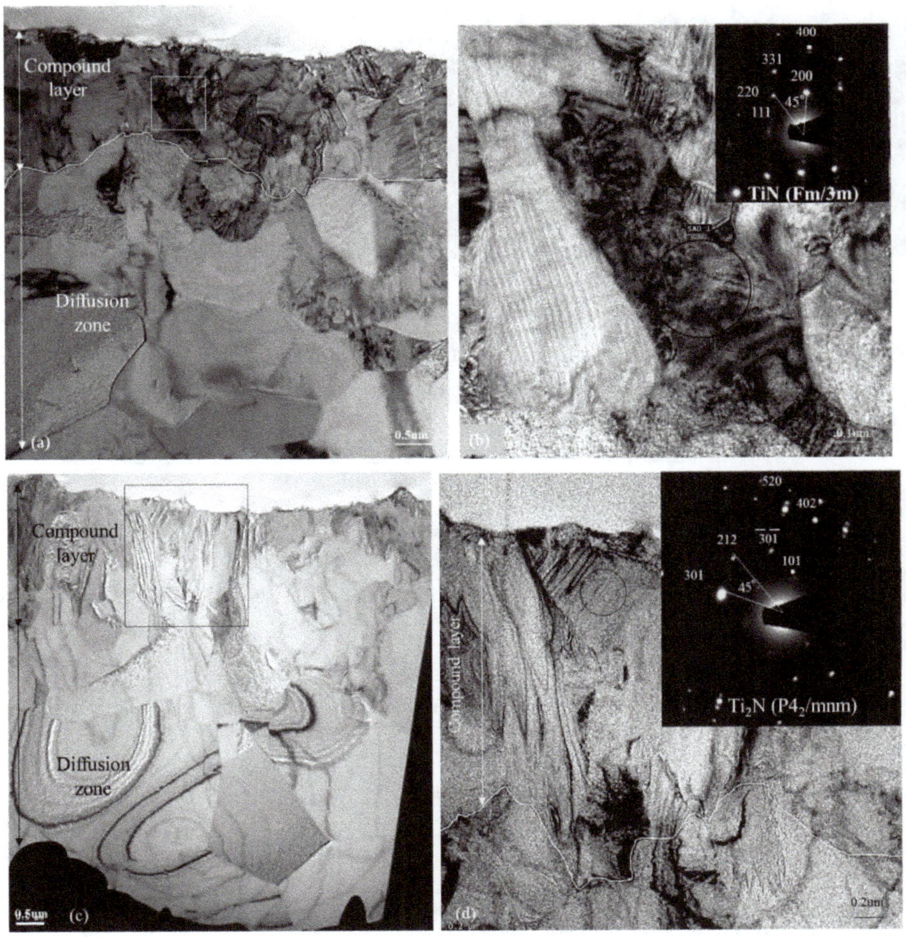

Figure 7. Bright-field TEM images and the corresponding selected area electron diffraction patterns of the plasma-nitrided Ti–6Al–4V microstructure showing the formation of TiN (space group: $Fm\bar{3}m$) and Ti_2N (space group: $P4_2$-mnm) nitrides in the compound layer [63].

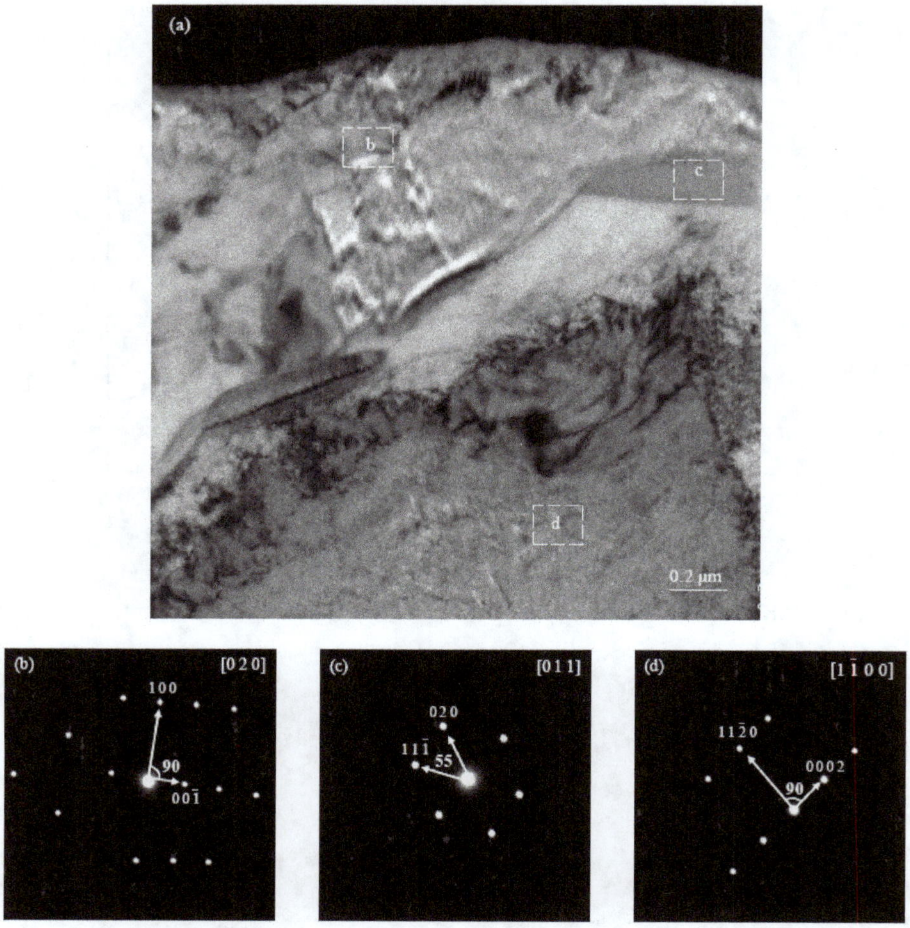

Figure 8. A bright-field TEM image of the plasma-nitrided Ti–10V–2Fe–3Al microstructure and the corresponding convergent beam electron diffraction (CBED) patterns confirming the formation of Ti_2N (space group: $P4_2$-mnm) and V_2N (space group: $P\bar{3}1m$) in the compound layer. V_2N nitrides were located at the grain boundaries [52].

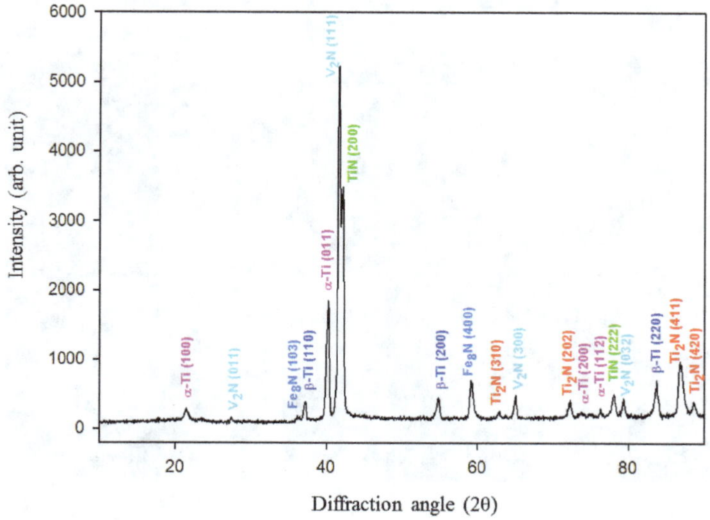

Figure 9. X-ray diffraction (XRD) pattern of the surface of plasma-nitrided Ti–10V–2Fe–3Al indicating the presence of titanium nitrides (TiN, Ti_2N), vanadium nitride (V_2N), and the possibility of iron nitride (Fe_8N) [52].

The depth of nitrogen diffusion was confirmed by microhardness versus depth profiles due to the solid-solution strengthening effect of interstitial nitrogen atoms in titanium structure. Microhardness profiles were obtained by microindentations at a maximum load of 245 mN (25 grf) using a diamond Vickers tip on tapered cross sections (CSM Instruments Micro-Combi Tester) and the microhardness values were determined based on load-penetration depth curves [65]. A deep nitrogen diffusion zone (35 ± 5 μm) in Ti–10V–2Fe–3Al after an 8-hour treatment (Figure 10) was attributed to the high β-phase content of the alloy with high nitrogen diffusivity [18, 22, 30, 47]. The prolonged sputtering of the surfaces prior to the nitriding process in this research also has accelerated the nitrogen diffusion by formation of several dislocations inside the α-grains. It has also been reported that the Ar+H_2 sputtering increased the nitriding kinetics in ferrous alloys by providing easy diffusion paths for nitrogen interstitials and eliminating inherent surface oxides [48, 49].

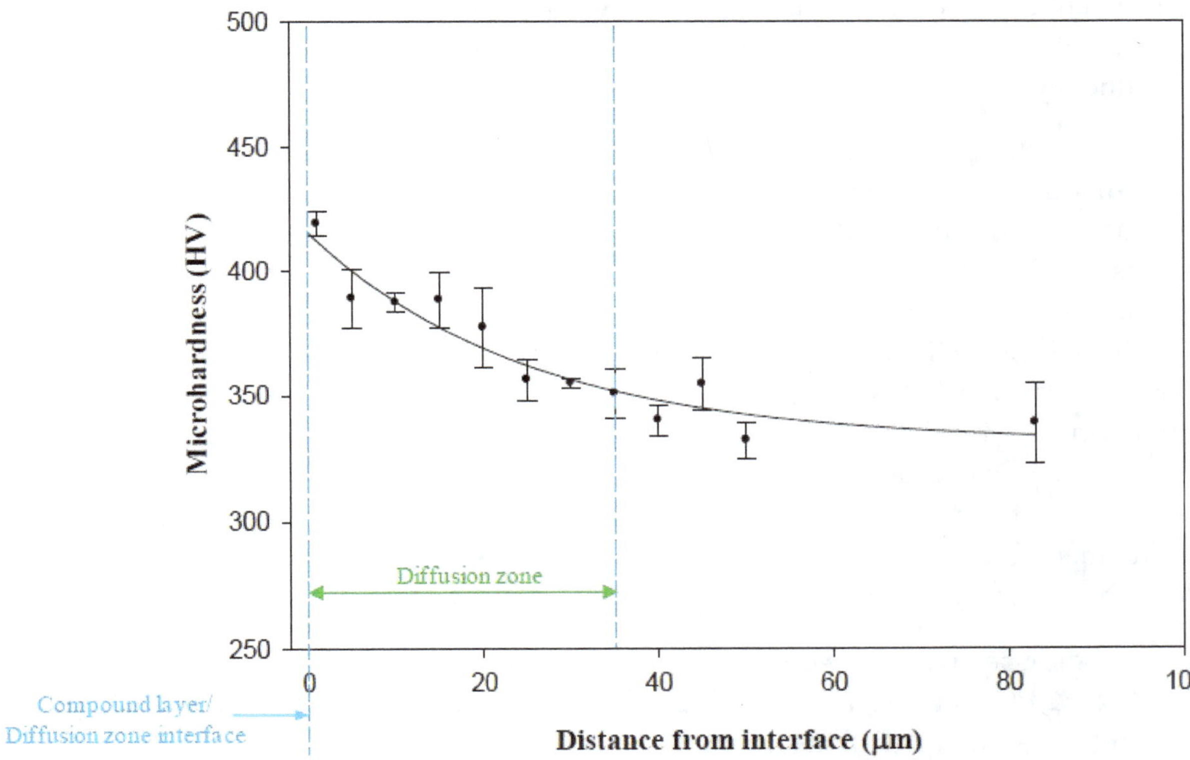

Figure 10. The microhardness-depth profile of the plasma-nitrided Ti–10V–2Fe–3Al. The diffusion zone extended to a depth of ~35 ± 5 μm underneath the compound layer [52].

4. Tribological properties of plasma-nitrided titanium alloys

Titanium alloys are renowned for their poor tribological characteristics, their strong adhesion tendency, high wear rates, susceptibility to seizure and galling, and high and unstable coefficient of friction [66–71]. Severe material transfer occurs during sliding of titanium alloys against themselves and other alloys[5], ceramics, or even polymers [72–75]. This behavior is attributed to their low resistance to plastic shearing, low work hardening, the brittle nature and low sliding resistance of their inherent surface oxides, and possibly microstructural embrittlement by oxygen dissolution [67, 76–78].

Although it is well established that plasma nitriding improves the wear resistance of titanium alloys, the tribological behavior of plasma-nitrided titanium surfaces is a function of the hardness of surface layers, surface roughness, the presence of brittle nitriding microstructural features, and the depth of nitrogen diffusion zone. Thus, the wear resistance of plasma-nitrided titanium surfaces is dependent on the process parameters, microstructure, and chemical composition of the titanium substrate and sliding conditions.

One of the earliest investigations on wear behavior of plasma-nitrided titanium alloys was performed by Bell et al. [10]. They reported that the high wear rate and high coefficient of friction of Ti–6Al–4V (COF = 0.3) were effectively improved by plasma nitriding (800°C, 12 hours, N_2 atmosphere). They attributed the low coefficient of friction (COF = 0.05), low wear rate, and anti-scuffing characteristics of the plasma-nitrided surfaces to the hardness of the compound layer.

Taktak and Akbulut [79] performed plasma nitriding at temperatures ranging from 700°C to 900°C for different durations and reported that the treatment performed at the highest temperature for the longest duration exhibited the best wear resistance and friction behavior. This was correlated with the compound layer thickness and hardness; however, the compound layer failure at high applied loads resulted in a transition to high wear rates, close to those of the untreated alloy. Higher plasma nitriding temperatures of 900°C, 1000°C, and 1100°C were studied by Shashkov [80] in different titanium alloys for different durations in N_2 atmosphere. They reported some wear resistance for plasma-nitrided titanium alloys at 900°C and reduction of wear resistance for nitrided alloys at higher temperatures and durations. They clarified that plasma nitriding of titanium alloys at elevated temperatures was accompanied by embrittlement due to the formation of brittle nitride layers and microstructural changes in the bulk such as grain coarsening.

The tribological behavior of Ti–6Al–4V alloy plasma-nitrided at temperatures ranging between 450°C and 520°C were studied by Yilbas et al. [81] against ruby counterpart in lubricated conditions. They reported that the scuffing and high friction coefficients, observed in the untreated alloy after a few sliding cycles, were avoided in the plasma-nitrided surfaces before the breakthrough of nitrided layers. This breakthrough was delayed when plasma nitriding was performed at 520°C due to a higher depth and hardness of the diffusion zone.

The wear mechanisms of plasma-nitrided Ti–6Al–4V alloy were studied by Molinari et al. [55]. The plasma nitriding treatments were carried out at different temperatures of 700°C, 800°C, and 900°C in a N_2–H_2 (4:1) gas mixture for 24 hours and tested against disks that had seen the same treatments. They found that the wear resistance was dependent on the microstructure and surface roughness of the nitrided surfaces and wear test conditions. At the lowest sliding

5 Except Ag–10Cu alloy and Babbitt metal (grade 2) that showed no galling against Ti–4Al–4Mn alloy

speed (0.3 m/s), nitriding inhibited oxidation-dominated wear in Ti–6Al–4V and decreased wear by providing resistance against adhesion and microfragmentation. However, compound layer failure that occurred at low-to-moderate loads for low-temperature (700°C) nitriding and at high loads for elevated temperature (800°C and 900°C) nitriding led to third-body abrasion and high wear rates. Plasma nitriding reduced wear at high sliding velocities (0.6 m/s and 0.8 m/s) by restricting the extensive plastic deformation and delamination wear of the titanium substrate. The elevated temperature plasma treatments exhibited better wear performance at high loads due to their higher thickness of the compound layer and depth of diffusion zone, but lower wear resistance at low loads possibly due to the increased residual stress level in the nitride layers and higher surface roughness.

The same authors also studied the wear mechanisms of Ti–6Al–4V alloy plasma nitrided at 800°C under lubricated rolling–sliding conditions [66]. Their findings indicated that the efficiency of lubricants was reduced by plasma nitriding, resulting in higher friction, possibly due to the better wettability and higher ionic character of the inherent titanium oxides compared with the titanium nitrides of the compound layer. However, the wear rates decreased by plasma nitriding and had an inverse relationship with the nitriding duration. The insufficient wear improvement after a short-duration process (8 hours) was related to the inadequate depth of nitriding and lack of support for the compound layer. On the other hand, an elongated nitriding treatment (24 hours) promoted a deeper diffusion zone that retained the compound layer on the surface under sliding conditions.

The positive role of a deep diffusion zone in tribological behavior of plasma-nitrided titanium alloys was also established by Nolan et al. [82], upon comparing plasma-nitrided and physical vapor deposition (PVD)-coated surfaces. Significantly lower mass loss was observed for plasma-nitrided Ti–6Al–4V with a 2-μm compound layer on top of an approximately 40-μm diffusion zone compared with the PVD-coated surface with a 2-μm TiN layer. Examination of worn surfaces and cross sections indicated that the compound layer endured the applied loads without significant subsurface plastic deformation. They proposed that the strengthening effect of the solid solution nitrogen atoms within the diffusion zone provided mechanical support for the compound layer and improved the wear resistance. Moreover, a gradual hardness profile decreased the chances of interfacial debonding due to a gradual change of elastic modulus and hardness from the surface toward the bulk.

The effect of nitriding microstructural features on the wear performance of plasma-nitrided titanium alloys was further elucidated by Cassar et al. [83]. They plasma-nitrided Ti–6Al–4V alloy using a triode set-up at low (200 V) and high (1000 V) voltages and evaluated the wear behavior against WC–Co and sapphire. The low-voltage treatment formed a thin compound layer on a deeply strengthened diffusion zone and despite the lower hardness of the compound layer, it provided better wear resistance compared with the high-voltage treatment especially at high loads. The thin compound layer could conform to the substrate deformation without experiencing spallations as the maximum shear stress was constrained within the diffusion zone. Bloyce et al. [30] explained that when the maximum applied stress during sliding exceeds the substrate yield strength, it experiences plastic deformation as evidenced by elongated grains within the subsurface and material pile-up. However, the TiN/Ti_2N surface layer cannot conform to the plastic deformation due to its low fracture toughness.

Recent studies by the authors investigate the failure micromechanisms of plasma-nitrided Ti–6Al–4V under dry sliding conditions using microscratch tests [63]. The tests were carried out using a diamond Rockwell tip with a 200-μm radius (CSM Instruments Micro-Combi tester) according to the ASTM C1624 standard [84]. Lateral displacement speed of 1 mm/min and loading rate of 10N/min were selected for the experiments. In addition to normal and tangential force (resolution of 0.3 mN), penetration depth (resolution of 0.3 nm) and acoustic emission signals were recorded during each test. Two types of tests were carried out: (i) progressively increasing load tests and (ii) constant load tests (loads selected according to the critical loads determined in (i)). During the progressive tests, the applied load linearly increased from 0.03 to 20 N as the tip moved at a constant displacement speed on the surface. The critical loads corresponding to the onset of coating failure events were identified using acoustic emission signals and subsequent SEM observations. In-depth site-specific cross-sectional analysis of scratch paths was carried out using a LEO (Zeiss) 1540 XB dual-beam FIB/SEM fitted with an EDS detector[6]. A thin layer of osmium (~3 nm) was deposited on the surfaces prior to the milling process to increase conductivity, prevent charging, and enhance the secondary electron yield for imaging. After locating the region of interest, a 2-μm-thick layer of platinum was deposited on the selected area to prevent surface features from damage during the milling process. The ion milling began by cutting a trench normal to the deposited platinum layer, using Ga ions from a Ga-based liquid metal ion-source (LMIS), at an accelerating voltage of 30 kV. All the FIB trenches were made parallel to the scratch direction of the scratch paths. The ion-milled trenches were imaged using a 1 kV electron beam located at a 54° angle.

The results indicated that failures initiated by the formation of microvoids within the compound layer likely atgrain boundaries. Increasing the applied load led to the formation of microcracks that propagated intergranularly within the compound layer and finally into the diffusion zone. The low resistance of nitrided surfaces to crack initiation and growth was correlated with the brittle nature and low damage tolerance of the compound layer [10]. However, it was observed that the deeply strengthened diffusion zone improved the adhesion of compound layer to the underlying substrate and prevented the spallation/delamination events under sliding contact. Moreover, subsurface crack propagation was retarded by ductile β-particles in the diffusion zone (Figure 11). The high resistance of β-phase to crack propagation in plasma-nitrided titanium alloys was confirmed by testing a plasma-nitrided near-β Ti–10V–2Fe–3Al alloy under the same conditions (Figure 12). It was proposed that the optimum microstructure consisted of a thin compound layer supported by a deep diffusion zone consisting of β-particles. Keeping that in mind, a low-temperature plasma nitriding treatment was developed for Ti–10V–2Fe–3Al alloy [52]. Evaluation of the tribological properties of the plasma-nitrided Ti–10V–2Fe–3Al surfaces using microscratch tests revealed that the coefficient of friction values were reduced by more than 72% and the nitrided surfaces exhibited an excellent load-bearing capacity. The compound layer did not experience any failures, for

6 Western Nanofabrication Facility, University of Western Ontario, London, ON, Canada

example, buckling or spallation, and maintained a good interfacial bonding to its substrate up to the high applied load of 17 N where partial chipping of the coating was observed at the scratch rims. Despite a low thickness, the compound layer suppressed the adhesion tendency and transfer of titanium due to high hardness and chemical inertness of TiN and Ti_2N [32, 33], and high toughness and adhesion resistance of V_2N [34, 35]. The deep diffusion zone (35 ± 5 μm) underneath the compound layer also contributed to high load-bearing capacity of the nitrided alloy by providing a good mechanical support for the compound layer [20, 39–42, 45-46]. The diffusion zone inhibited premature failures at the surface by mitigating the significant difference in the deformation behaviors of the compound layer and Ti–10V–2Fe–3Al substrate and acting like a "functionally-graded interface" [43, 44].

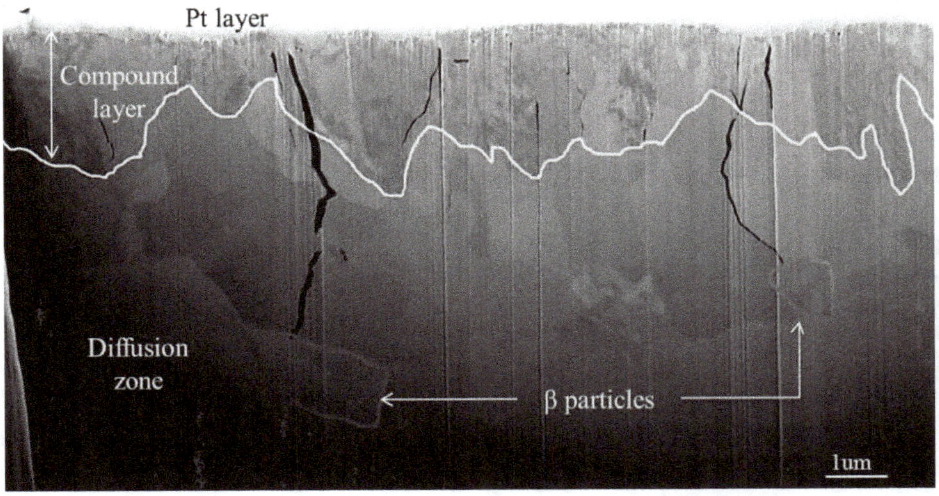

Figure 11. A typical SEM image of the FIB-milled cross section of the scratch track on the surface of plasma-nitrided Ti-6Al-4V at the highest load of 20 N showing that the microcracks initiating from the surface were stopped at β-particles in the diffusion zone [63].

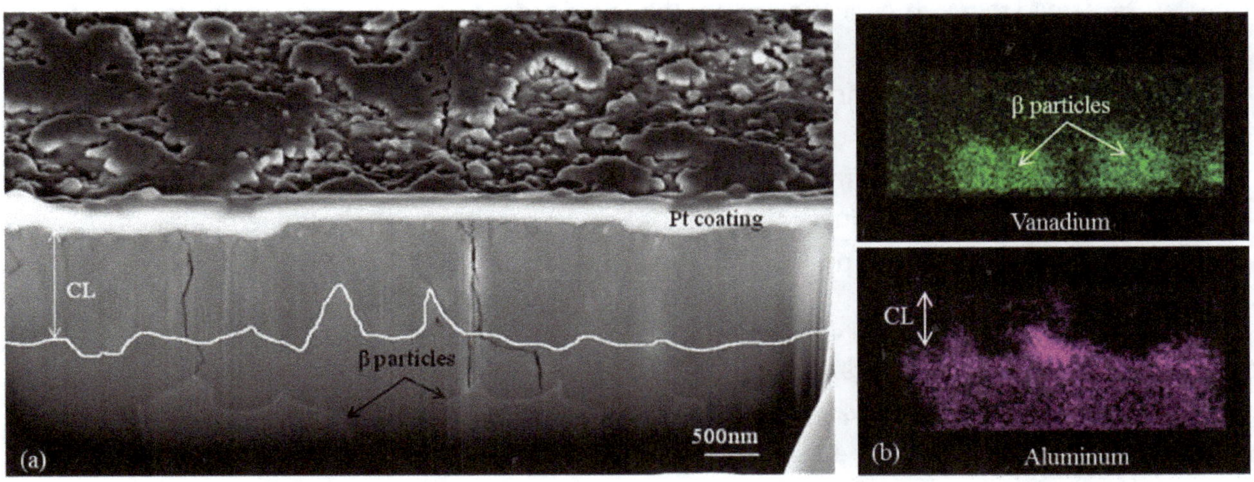

Figure 12. (a) A typical SEM image of the FIB-milled cross section of the scratch track on the surface of plasma-nitrided Ti-10V-2Fe-3Al and (b) the corresponding EDS mpas for aluminum and vanadium confirming that the β-particles inhibited further crack propagation in the subsurface region [63] (CL: compound layer, DZ: Diffusion zone).

Surface roughness of plasma-nitrided titanium alloys is another influential factor on their sliding behavior. Salehi et al. [85] investigated the effect of surface topography on the wear behavior of PVD TiN-coated and plasma-nitrided Ti–6Al–4V surfaces. They found that the compound layer thickness and surface roughness both increased with plasma nitriding-treatment temperature and proposed that sputtering is responsible for the changes in surface roughness. They reported that under self-mating conditions, oxidational wear dominated with significant mass loss for surfaces of higher roughness and thicker compound layers. Sliding wear tests against alumina counterface revealed that plasma nitriding at high temperatures (850°C and 950°C) yielded low wear rates at high loads due to thick compound layers but high wear rates at low loads due to high surface roughness. Plasma-nitrided surfaces at low temperatures (700°C and 750°C) improved the wear resistance below a critical load, above which the compound layer spalled off and the coefficient of friction increased abruptly. Similar observations were made by Yildiz et al. [62].

5. Mechanical properties of plasma-nitrided titanium alloys

Nitriding improves the fatigue strength of steels due to the precipitation of submicroscopic alloying element nitrides and compressive residual strains in the diffusion zone. Conversely, nitriding of titanium alloys does not promote precipitation strengthening mechanisms [86] and also leads to inevitable adverse effects on the toughness and fatigue properties [10, 24, 87–89]. In this section, the correlation between plasma nitriding and mechanical properties of titanium alloys is presented, and since the microstructural features of gas-nitrided Ti alloys affect fatigue properties in the same manner, gas nitriding investigations have been occasionally referred to in this section.

Due to the slow kinetics of nitrogen diffusion in titanium alloys, relatively high temperatures (>700°C) are used to produce sufficient depth of nitriding. Typical plasma nitriding treatments, performed in the temperature range of 700–1100°C for several hours, cause bulk microstructural changes such as significant grain growth and overaging that adversely affect the fatigue behavior [10, 22, 23]. Moreover, it has been reported by several researchers that the formation of brittle features in the surface vicinity and bulk microstructural changes are the main contributing factors in premature failure initiation and deterioration of mechanical properties of titanium alloys after nitriding [10, 49, 50, 54, 55, 58, 90, 91].

According to Morita et al. [90] and Tokaji et al. [92], the formation of the compound layer and grain growth were equally responsible for the fatigue deterioration of gas-nitrided titanium alloys. Morita et al. [90] proposed that the high stiffness gradient between the compound layer and Ti substrate ($E_{compound\ layer} = 426$ GPa, $E_{Ti} = 100$ GPa) led to stress concentration at the surface. The resulting stress concentration cancelled the effect of compressive residual stresses and led to the premature fracture initiation at the surface due to the low fracture toughness of the compound layer. They also reported that the fatigue strength had a Hall–Petch relationship with the grain size of nitrided titanium and the negative effect of grain growth was more pronounced for elongated treatments at high temperatures. Consequently, promising results were obtained by inhibiting grain growth through a low-temperature nitriding process.

Tokaji et al. [92] separated the effect of nitriding on the substrate microstructure by subjecting the titanium alloys to the same heating cycle under vacuum. The results indicated a slight increase in the fatigue limit compared with the annealed Titanium alloys after a 4-hour nitriding treatment, but 15 hours of nitriding significantly reduced the fatigue strength. They believed that the compound layer and diffusion zone were both responsible for impaired fatigue properties and observed improvements in fatigue life by complete removal of the compound layer and partial removal of the diffusion zone (Figure 13). Similar results were found by Nishida and Hattori [24] and Rodriguez et al. [93] after plasma nitriding of Ti–6Al–4V. They reported that the fatigue strength of plasma-nitrided Ti alloys was lower than the untreated alloy but was similar to – or even slightly better than – that of the vacuum-annealed alloy with the same heating history. The fatigue endurance limit was found to be inversely related to the plasma nitriding temperature and increased by about 30 MPa when the brittle compound layer was removed after plasma nitriding. Rodriguez et al. [93] reported that the bulk microstructural changes from mill-annealed to Widmanstätten structure were caused by a reduction in low-cycle fatigue resistance. They proposed that imposed strains during cyclic loading at low loads were not sufficient for compound layer failure, and thus the residual compressive strains retarded the fatigue crack nucleation. However, the compound layer failed at high cyclic strains due to a high surface roughness and the different stiffness between Ti–6Al–4V and titanium nitrides.

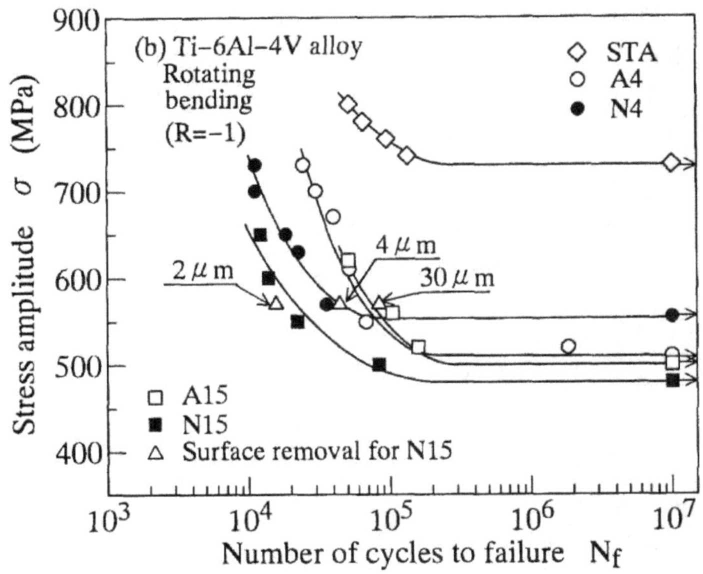

Figure 13. The S–N curve obtained by rotation bending fatigue tests on Ti–6Al–4V alloy. STA: solution treated at 950°C for 1 hour and aged at 540°C for 4 hours; A4 and N4: annealed and nitrided at 850°C for 4 hours; A15, N15: annealed and nitrided at 850°C for 15 hours [92].

Several researchers including Bell et al. [10, 25] and Guiraldenq et al. [94] correlated the impaired fatigue properties of titanium alloys after plasma nitriding to the thickness of the α-case. The α-case is a brittle layer (typical hardness 800–1000 HV) formed beneath the compound layer at high nitriding temperatures (≥ 800°C), and its formation has proven to be detrimental for ductility and fatigue strength [10, 49, 50, 54, 55, 58].

Raveh et al. [50] believed that in addition to the surface hardness, fatigue crack initiation was also a function of residual strains in the nitrided layers, crystallographic orientation of different phases, and segregation of alloying elements near the surface region. The fatigue crack initiation resistance decreased with TiN content of the compound layer and the surface hardness due to its brittle nature. Moreover, incorporation of Ar in the nitriding plasma also decreased the crack initiation resistance, possibly as a result of inducing microstrains in the nitrided layers.

Conversely, according to Sobiecki and Rudnicki [95], plasma nitriding (at 800°C for 3 and 12 hours) increased the fatigue strength of Ti–1Al–1Mn alloy from 350 to 390 MPa. A plasma nitriding treatment at 500°C for 6 hours in nitrogen–hydrogen (3:1) gas mixture in the work of Rajasekaran and Raman [96] also resulted in improvements in uniaxial plain fatigue and fretting fatigue behavior of Ti–6Al–4V alloy. Lower surface roughness, generation of compressive residual stresses on the surface, and higher surface hardness were considered the main factors responsible for the fatigue improvements. The surface hardness after plasma nitriding (390 $HV_{0.2}$) was slightly higher than the untreated material (330 $HV_{0.2}$), and the surface roughness was reduced by plasma nitriding (R_a decreased from 0.80 to 0.55 μm) due to smoothing of the pretreatment grinding marks by sputtering. Furthermore, compressive residual stresses developed on the surface (in the order of 40 MPa) due to nitrogen diffusion in the titanium lattice. Fretting test results also indicated lower friction forces and shallower fretting scars for nitrided samples at all stress levels. Contradictory results were reported by Ali and Raman [88] who carried out plasma nitriding on Ti–6Al–4V alloy at 520°C for 4 and 18 hours and found that both plain and fretting fatigue lives were reduced by plasma nitriding. The samples that were nitrided in a nitrogen–hydrogen (3:1) atmosphere showed inferior results compared with those nitrided in a pure nitrogen atmosphere likely due to the higher hardness of surface layers in the former nitriding conditions.

In a recent investigation by the authors, a plasma nitriding treatment was designed for Ti–10V–2Fe–3Al β-titanium alloy to address its susceptibility to adhesion without any detrimental effects on bulk microstructure. Uniaxial tensile tests were conducted at room temperature on small-sized specimens (6 mm in diameter and 24 mm in length at the gauge section) conforming to the specifications outlined in the ASTM E8M standard [97]. The tensile tests were performed at a cross-head speed of 0.5 mm/min (MTS Criterion-43) and strain values were measured using a clip-on axial extensometer to an accuracy better than ± 0.5% of the applied strain. Subsequently, the engineering stress–strain curves were constructed and the offset yield strength, ultimate tensile strength (UTS), and ductility (tensile elongation) were determined. It was found that the modified nitriding process led to improvements in the tensile (from 881.0 to 922.5 MPa) and yield (from 821.5 to 873.0 MPa) strengths of the alloy with a minor decrease in the elongation (from 18.9% to 15.9%). The reduction of tensile elongation was possibly due to the mutual influence of precipitation of α-phase during age hardening and the solid-solution strengthening effect of nitrogen interstitials. Similar observations were made by Akahori et al. [98] after gas nitriding experiments on Ti–Nb–Ta–Zr (β-titanium) alloy. More importantly, as confirmed by TEM observations and tensile test results, the formation of brittle ω precipitates (forming in the range of 250–450°C [2, 4, 10, 12]) and α-case was avoided by optimizing the nitriding parameters.

Rotation bending fatigue tests (fully reversed cyclic loading, $R = -1$) were utilized in another investigation to evaluate the performance of plasma-nitrided Ti-6Al-4V alloy under cyclic loading conditions [26]. The tests were conducted using an R. R. Moore fatigue tester (Instron) at a frequency of 3000 rpm. The first samples were tested at ~20% of the tensile strength where failures were expected to occur at a relatively short number of cycles. The testing stress was decreased for the rest of the samples until at least three samples did not fail at 10^7 cycles (run-out). The highest stress at the run-out was taken as the fatigue endurance limit. The fatigue samples had an hour-glass configuration with a tapered length of 50 mm and a diameter of 5.6 mm in the thinnest section. The stress–life (S–N) curves were constructed and used to evaluate the resistance to fatigue crack initiation. The fracture surfaces were then analyzed under the SEM (Quanta 200 FEG-SEM) at an accelerating voltage of 10 kV to investigate the failure micromechanisms.

The fatigue tests were also performed on untreated Ti-6Al-4V alloy under the same conditions for the purpose of comparison. The gauge section of the untreated samples was mechanically polished using 1-μm diamond suspension in the last step to an average surface roughness (R_a) and root-mean-squared roughness (RMS) of 0.14 and 0.18 μm, respectively. The surface roughness measurements were obtained using an optical profilometer (Wyko, Veeco NT1100).

A low-temperature plasma nitriding (600°C) was exploited to achieve a microstructure consisting of a thin compound layer (<2 μm) supported by a deep diffusion zone (>40 μm) without any significant changes in the substrate (only a ~40% increase in the average grain size). It was found that the low-temperature plasma nitriding resulted in improvements in fatigue strength of plasma-nitrided Ti-6Al-4V compared with conventional nitriding treatments (Figure 14). Moreover, the plasma-nitrided Ti-6Al-4V alloy at 600°C illustrated a ductile type of failure under uniaxial tensile loads. Periodic transverse cracks observed on the plasma-nitrided surfaces perpendicular to the loading direction without any signs of spallation confirmed a well-bonded interface between the compound layer and the substrate. Similar observations were reported by Chen et al. [99] for TiN coatings on stainless steel substrates under tensile loads. Conversely, plasma nitriding at an elevated temperature of 900°C led to substantial deterioration of strength, toughness, and fatigue life. This was attributed to premature failure of thick and brittle nitrided surface layers – a 5.6-μm-thick compound layer and a 19.3-μm α-case – acting as stress risers and promoting a brittle type of failure in the alloy as well as substantial grain growth (370% increase in the average grain size compared with the untreated alloy) and phase transformation in the bulk microstructure from equiaxed to coarse lamellar grains (~5 times higher average grain size value) as shown in Figure 15.

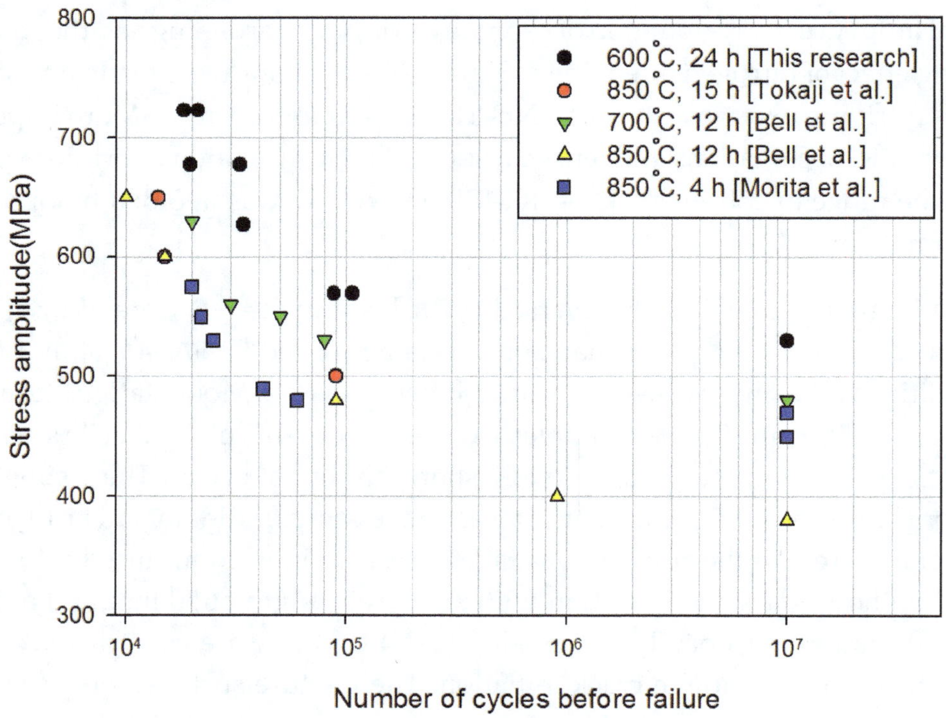

Figure 14. Comparison of the fatigue data obtained for plasma-nitrided Ti–6Al–4V with the conventional gas and plasma nitriding treatments in the literature [26].

According to fractographic investigations of the fatigue failures in low-temperature plasma-nitrided Ti-6Al-4V alloy, the fatigue failure initiated at the surface in the low cycle region ($N \leq 10^5$ cycles) and propagated in a ductile manner leading to the final rupture. No failures were observed in the high cycle region ($N > 10^5$ cycles) and the nitrided alloy endured cyclic loading until the tests were stopped at 10^7 cycles. The thin compound layer restricted the extent of premature crack initiation from the surface, whereas the deep diffusion zone with a well-bonded interface decreased the likelihood of fatigue initiation at (or below) the compound layer interface. Although the compound layer exhibited signs of plastic deformation under cyclic loading conditions, fragmentation of the compound layer and brittle failure features observed at the fatigue crack initiation sites suggested that brittle fracture of the compound layer was the precursor to the fatigue failure (Figure 16). The observation of striation-like features within the diffusion zone in the vicinity of the crack origin indicated the possibility that the crack formation was triggered by intrusions and extrusions and plastic deformation of the underlying diffusion zone. The fatigue cracks consequently propagated in the same manner as the untreated alloy. The beneficial effect of compressive residual stresses in the compound layer (−530 MPa) was overshadowed by the stress concentration in this layer due to the modulus mismatch with the substrate. Surface roughness was also an influential parameter on the fatigue strength of the nitrided alloy as by polishing the nitrided surfaces, a higher number of cycles were dedicated to the formation of fatigue cracks compared with the as-treated condition resulting in an improved fatigue life (Figure 17). Similar results were reported by Cassar et al. [91] and Novovic et al. [100] for the fatigue behavior of surfaces with average roughness values of 0.1 μm and higher.

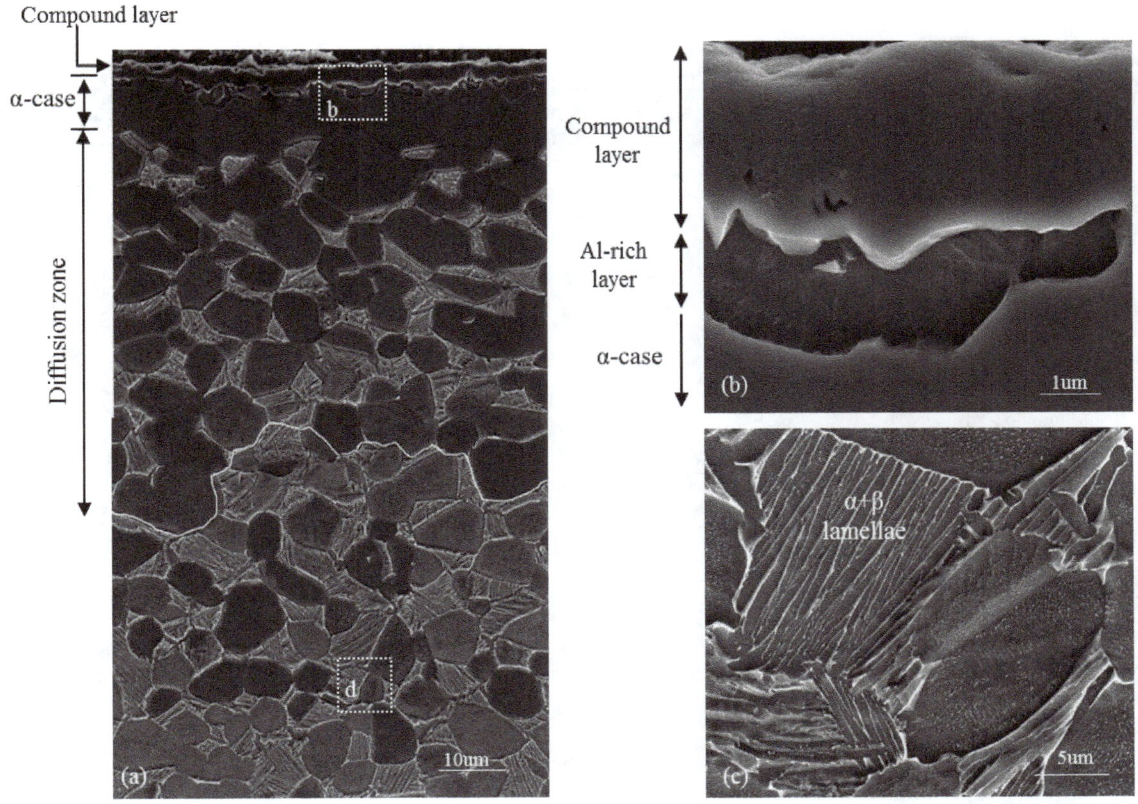

Figure 15. Typical cross-sectional SEM micrographs of the microstructure of plasma-nitrided Ti–6Al–4V treated at 900°C [26].

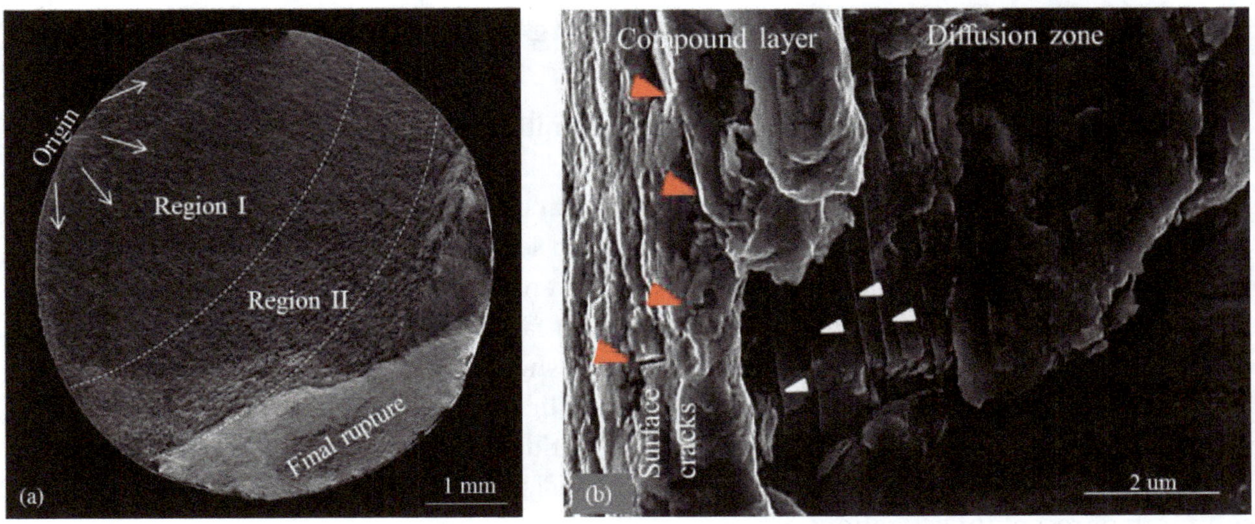

Figure 16. (a) A typical SEM image of the general view of the fatigue fracture surface of plasma-nitrided Ti–6Al–4V and (b) striation-like features and microcracks in the surface vicinity [26].

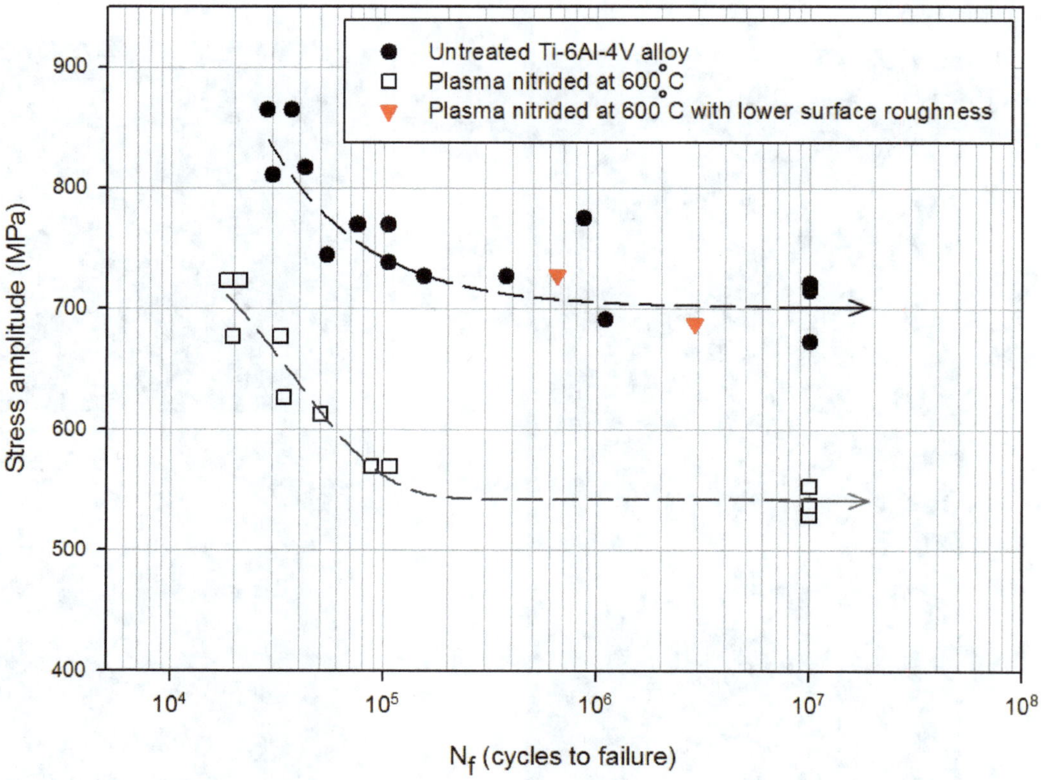

Figure 17. Stress–life (S–N) curves showing the effect of plasma nitriding on fatigue behavior of Ti–6Al–4V. It was found that decreasing the surface roughness with polishing resulted in an increase in the fatigue strength [26].

6. Recent development in plasma nitriding titanium alloys

As explained in Section 5, the main reasons for the initiation of premature failures in nitrided titanium alloys are the brittle nature of the compound layer and the α-case, the incompatibility in deformation behavior of the compound layer and the substrate and lack of sufficient adhesion at their interface. As such, the compound layers cannot conform to the extensive plastic deformation of titanium substrates and eventually experience adhesive (from the interface) or cohesive (within the compound layer) failures. Thus, the mechanical response mismatch between the nitride layers and substrates should be minimized in order to enhance the load-bearing capacity of the nitrided surfaces. The effective practical measures are based on eliminating discrete nitride layer–substrate boundaries by the design and fabrication of multilayer coatings and coatings with gradient diffusion boundaries [29, 31, 36, 82, 83, 101]. A summary of recent investigations are presented in this section.

6.1. Duplex treatments

Duplex treatments were introduced to the field of surface engineering of titanium alloys to exploit the benefits of wear-resistant coatings and the deep hardening effect of diffusion treatments [30–34, 102]. Significant improvements in wear resistance and long-term durability of titanium surfaces have been achieved by deposition of diamond-like carbon (DLC) coatings after diffusion hardening by plasma nitriding [103] or thermal oxidation [104, 105].

The beneficial role of plasma nitriding pretreatment in accommodating subsurface deformation and preventing premature failures in DLC-coated Ti–6Al–4V was confirmed in the work of Meletis et al. [103]. In their experiments, radio-frequency plasma nitriding treatment was utilized for substrate surface hardening, followed by deposition of DLC coatings with a silicon interlayer. In addition to substantial reduction of wear and coefficient of friction, it was found that the nitriding pretreatment retarded catastrophic failure of DLC coatings during sliding at high contact loads.

In another investigation by Dong [36], the combined effect of electron beam melting and plasma nitriding treatments resulted in significant improvements in wear resistance and load-bearing capacity of Ti–8.5Si titanium alloy. Electron beam melting generated a deep strengthened zone (800 HV to a depth of 600 μm) in the alloy before plasma nitriding that provided support for the nitride layer under sliding contact. As a result, the surface nitride layer endured the stresses applied during wear tests without spalling, leading to a three orders of magnitude reduction of wear rate compared with the untreated titanium alloy.

Several researchers also introduced duplex treatments that are composed of plasma nitriding and physical vapor deposition (PVD) or chemical vapor deposition (CVD) of nitride coatings. Rie et al. [106, 107] used plasma-assisted CVD (PACVD) technique to produce a thick TiN layer on top of plasma-nitrided substrates to enhance wear and corrosion resistance for biomedical applications. Their findings indicated that the duplex-treated surfaces were free of wear against ultrahigh-molecular-weight polyethylene under the tested conditions, and the corrosion resistance and biocompatibility of the untreated titanium alloy were preserved. Wear improvements were also confirmed by Ma et al. [108] through plasma nitriding and plasma-enhanced chemical vapor deposition (PCVD) of TiN coatings of Ti–6Al–4V substrates. They correlated the wear enhancement to higher surface hardness of PCVD coating and its strong adhesion to plasma-nitrided substrate. Other coatings, such as CrN, CrAlN, and WC/C, were also deposited on plasma-nitrided titanium alloys, in addition to TiN, by Cassar et al. [35, 91, 109]. Evaluation of the duplex-treated surfaces by impact tests, ball-on-disk wear tests, rotation bending fatigue tests, and microscratch tests revealed improvements in wear resistance, adhesion, and resistance to fatigue crack initiation. This was attributed to higher load support and adhesion strength offered by hard-coated surfaces and deeply strengthened substrates. It was noted that chromium aluminum nitride (CrAlN) coatings exhibited superior wear performance due to their high hardness and low elastic modulus (higher H^3/E^2 parameter).

6.2. Surface pretreatments

Alteration of the microstructure in the surface vicinity via thermochemical (e.g., laser-assisted treatments, electron beam melting, etc.) and mechanical (e.g., mechanical attrition, shot peening, etc.) pretreatment processes [27, 28] has been successfully used for increasing the nitriding efficiency of titanium alloys. Microstructural changes thus introduced, such as significant grain refinement and nonequilibrium defects, act as diffusion shortcuts and enhance the nitriding kinetics and efficiency during the posterior plasma nitriding treatment even at low temperatures [110–115].

Shot peening, extensively used for fatigue improvement, is based on introducing compressive stresses in the surface layers via impingement of high-velocity shots, under controlled conditions. Various types of shots such as cast steel shots, cut wire shots, glass beads, and zirconium shots are available, and the air blast can be delivered by either suction blast or pressure blast depending on the budget, target substrates, and the required precision. In addition to the development of a residual stress profile near the surface, shot peening results in surface roughening, strain hardening, and formation of a subsurface plastic deformation zone. Surface roughening favors early nucleation of fatigue cracks, whereas strain hardening and compressive residual stresses increase the resistance to plastic deformation and fatigue crack propagation by providing crack closure mechanism. Recently, improvements in the sliding wear behavior and tensile properties have been reported by shot peening due to its surface hardening effect [116].

Experimental studies showed that shot peening can introduce a severely deformed layer in the surface vicinity characterized by significant refinement of grains and generation of nonequilibrium defects [110–113]. The thickness of this severely deformed layer may vary in the range of 10–80 μm depending on the microstructure and properties of the substrate material; metals of a lower yield strength experience a more intense plastic deformation that extends deeper into the substrate [117]. A high-energy shot peening treatment by Han et al. [27] resulted in the formation of a nanostructured layer on the surface of Ti–4Al–2V alloy. This treatment also introduced microstructural changes such as twinning and grain refinement that extended to ~230 μm below the surface. They proposed that the grain refinement occurs through various mechanisms such as "formation of dislocation walls and tangles," "dislocation movement and twinning," and "mechanical twinning" depending on the crystal structure and stacking fault energy (SFE) of metals. Mechanical twinning was recommended to have major contributions in grain refinement of Ti–4Al–2V alloy due to its HCP structure and high stacking fault energy (SFE > 300 mJ/m^2). The high level of plastic deformation induced by shot peening resulted in activation of many slip systems and a high density of dislocations within the twin bands. Subsequently, dislocation walls were formed, which in turn generated submicron-sized grains, small misoriented subgrains and finally equiaxed fine nanograins (35 ± 5 nm). Further studies were performed by Thomas and Jackson [118] to understand the effect of temperature and alloy composition on the subsurface microstructure of shot-peened titanium alloys.

Alteration of surface microstructure by introducing high-diffusivity paths such as grain boundaries, dislocations, and atomic-level microstructural defects by shot peening can be used to accelerate the diffusion of interstitial atoms [112]. Thomas et al. [114, 115] reported that generation of a high density of twin boundaries and dislocations in the near-surface region of pure titanium and Ti–4Al–2V alloys after shot peening increased the oxygen uptake during the following thermal exposure in air. Moreover, the high stored energy associated with the higher density of these nonequilibrium defects increases chemical reactivity of surfaces and enhances the kinetics of surface reactions [113]. Several researchers have confirmed that microstructural changes brought by severe plastic deformation (SPD) in ferrous alloys enhanced the nitrogen diffusion rate during the posterior nitriding treatments and increased the efficiency of nitriding at lower temperatures [110–113]. The effect of shot peening after diffusion treatments has also been studied in the literature. In a recent study, Bansal et al. [119] designed a series of duplex treatments for Ti–6Al–4V alloy composed of diffusion treatments (nitriding and oxygen diffusion) followed by mechanical working (shot peening and planishing). The results of dry reciprocating ball-on-disk wear tests against stainless steel 440C and silicon nitride sliders indicated improvements in wear resistance by introducing a short period of low friction and lower volumetric wear.

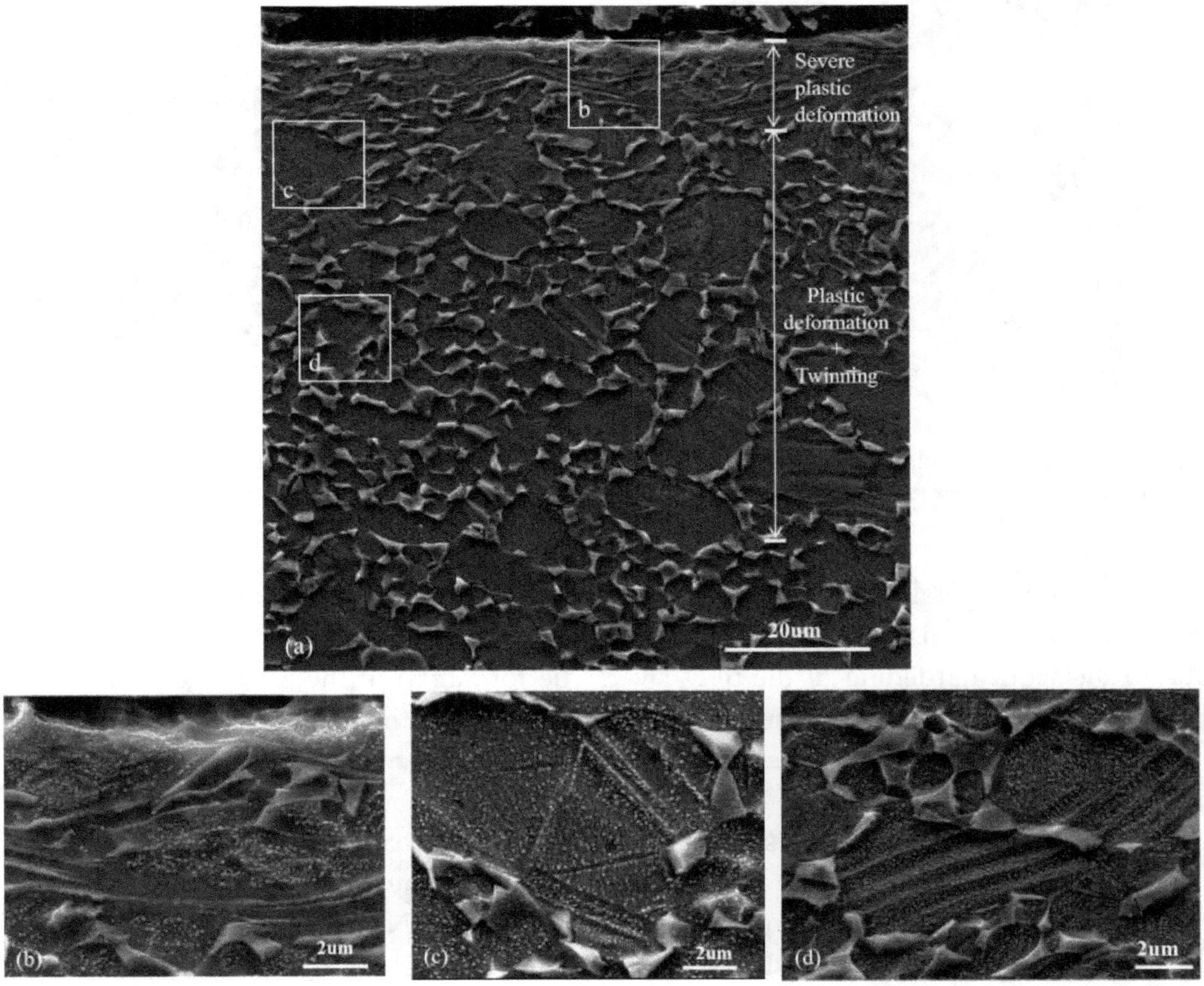

Figure 18. Typical cross-sectional SEM images showing the microstructure of Ti–6Al–4V after shot peening. A 5-μm deep severe plastic deformation (SPD) layer was formed on the surface characterized by elongation of intergranular β-particles parallel to the surface [28].

Severe plastic deformation surface layer introduced by shot peening pretreatment was found to enhance the plasma nitriding efficiency of Ti–6Al–4V alloy [28]. The high level of plastic strains induced by shot peening resulted in the formation of a 5-μm deep SPD layer on the surface characterized by twins, deformation bands, grain refinement, and a high density of

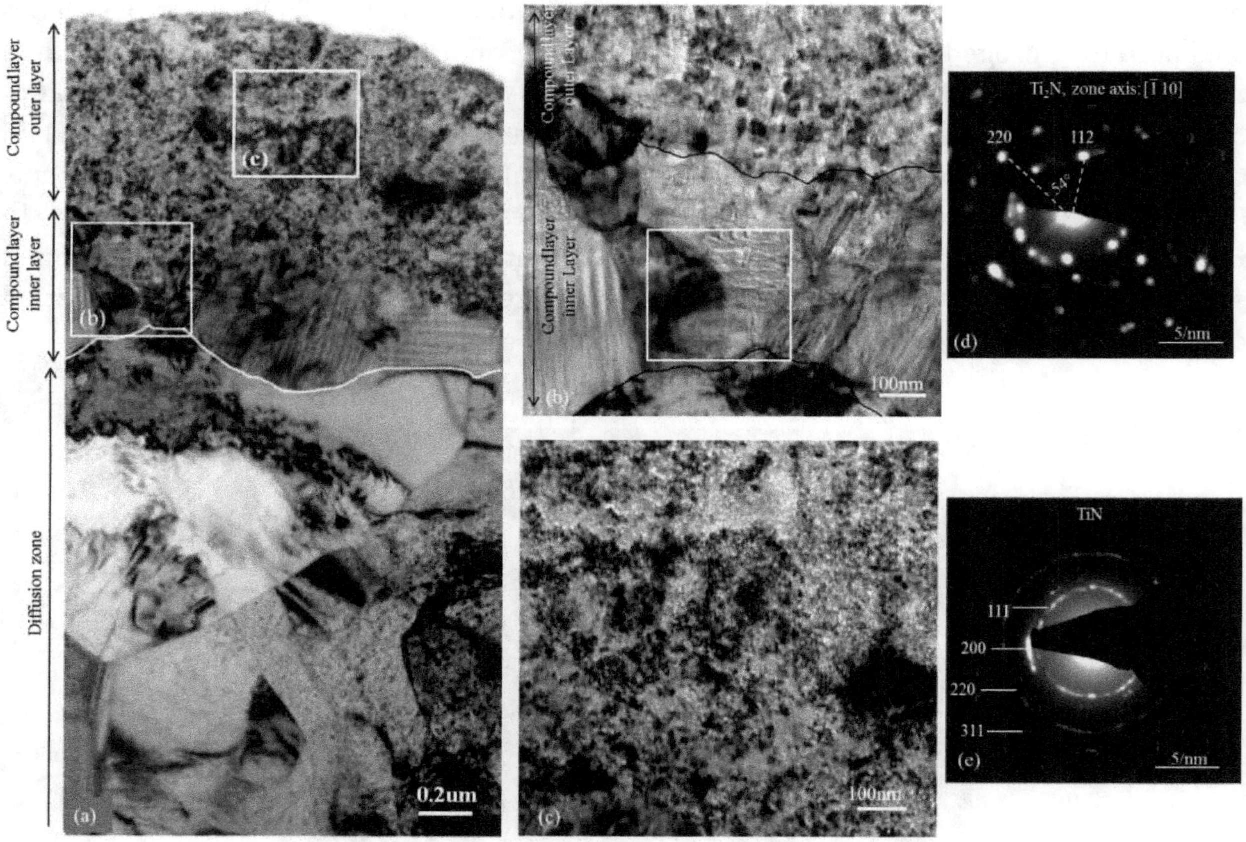

Figure 19. Typical cross-sectional bright-field TEM images of the near-surface microstructure of plasma-nitrided Ti–6Al–4V alloy with shot-peening pretreatment and the corresponding selected area electron diffraction (SAED) patterns [28].

nonequilibrium crystal defects such as subgrain boundaries and dislocation pile-ups (Figure 18 and Figure 19). This is in agreement with the 20% increase in the surface hardness value due to the work hardening of the SPD layer. The formation of the SPD layer resulted in accelerated nitriding kinetics during the plasma nitriding treatment that followed due to two main reasons: (i) grain boundaries, deformation bands, and twins provided easy diffusion paths for nitrogen interstitial atoms and (ii) high density of nonequilibrium crystal defects increased the stored energy of the surfaces and their chemical reactivity and offered additional preferential nucleation sites for titanium nitrides [111, 120]. The accelerated nitriding kinetics resulted in the formation of a 0.6-μm-thick nanocrystalline TiN layer in the compound layer (Figure 19) and increased the depth of nitrogen diffusion by almost 50% normally observed in high-temperature nitriding treatments (e.g., 900°C) [121]. As such, the plasma-nitrided surfaces that received the pretreatment step exhibited higher load-bearing capacity and better interfacial bonding compared with those without the pretreatment. The presence of nanoscale TiN grains increased the hardness and toughness of the compound layer and the compound layer endured high contact stresses for several sliding passes without chipping or spallation. In comparison, plasma-nitrided surfaces without the pretreatment experienced severe delamination and coating spallation after only two sliding passes in microscratch tests. The deep diffusion zone in the pretreated surfaces, strengthened by nitrogen interstitials, also contributed to a well-bonded interface and diminished the incompatibility in plastic deformation of the compound layer (E = 250–600 GPa [122, 123]) and substrate (E = 110 GPa [124]) by providing a more homogenized stress distribution across the interface [30, 104, 125–127]. Thus, the shot-peening pretreatment improved the nitriding efficiency of low-temperature plasma nitriding without introducing detrimental microstructural changes in the substrate that would lead to sacrificing mechanical properties.

7. Summary

Plasma nitriding is a surface hardening treatment utilized to improve the tribological properties of titanium alloys by the formation of a compound layer mainly consisting of TiN and Ti_2N titanium nitrides on the surface. However, plasma nitriding has a negative side effect on the mechanical properties of titanium alloys, especially fatigue strength, since the high temperatures involved in the treatment (700–1100°C) promote formation of brittle surface layers and detrimental microstructural changes in the bulk. Recent advancements in the field of plasma nitriding of titanium alloys are focused on the development of low-temperature (600°C or less) treatments as well as increasing the long-term durability of the nitrided surfaces through deposition of hard surface coatings such as diamond-like carbon and PVD/CVD nitride coatings. Furthermore, alteration of the surface microstructure by thermochemical and mechanical pretreatment steps is also successfully applied for increasing the nitriding efficiency during low-temperature plasma nitriding of titanium alloys.

Acknowledgements

The authors acknowledge Exactatherm Ltd., Dr. P. Lidster, Mr. G. Pigott, and Mr. Simeon Simeonov, for their help in plasma nitriding. The appreciation is extended to Natural Science and Engineering Council of Canada (NSERC) Discovery Individual Grant Program for providing financial support and the Canadian Centre of Electron Microscopy (CCEM) for TEM experimental work.

Author details

Afsaneh Edrisy* and Khorameh Farokhzadeh

*Address all correspondence to: edrisy@uwindsor.ca

Department of Mechanical, Automotive and Materials Engineering, University of Windsor, Windsor, ON, Canada

References

[1] F. Preissner and P. Minarski, "Results on nitriding titanium and Ti6Al4V with a new thermochemical treatment under high gas pressure," in Titanium '92, Science and Technology, F. H. Froes and I. Caplan, editors, Cannes, France: The Mineral, Metals and Materials Society, 1993, pp. 1979–1987.

[2] V. M. Fedirko and I. M. Pohrelyuk, "Improvement of the wear resistance of titanium alloys by thermochemical treatment in nitrogen-containing media," Materials Science, vol. 30, pp. 66–71, 1994.

[3] A. Gicouel, N. Laidani, P. Saillard, and J. Amouroux, "Plasma and nitrides: Application to the nitriding of titanium," Pure Applied Chemistry, vol. 62, pp. 1743–1750, 1990.

[4] T. Spalvins, "Advances and directions of ion nitriding/carburizing," in Proceedings of ASM's 2nd International Conference on Ion nitriding and Ion Carburizing, T. Spalvins and W. L. Kovacs, editors, Cincinnati, OH, 18-20 September 1989, ASM International, pp. 1–4.

[5] T. Bacci, G. Pradelli, B. Tesi, C. Gianoglio, and C. Badini, "Surface engineering and chemical characterization in ion-nitrided titanium and titanium alloys," Journal of Materials Science, vol. 25, pp. 4309–4314, 1990.

[6] E. Rolinski, G. Sharp, D. F. Cowgill, and D. J. Peterman, "Ion nitriding of alpha plus beta alloy for fusion reactor applications," Journal of Nuclear Materials, vol. 252, pp. 200–208, 1998.

[7] K. C. Chen and G. J. Jaung, "D.C. diod ion nitriding behavior of titanium and Ti-6Al-4V," Thin Solid Films, vol. 303, pp. 226–231, 1997.

[8] H. Akbulut, O. T. Inal, and C. A. Zimmerly, "Ion nitriding of explosively-clad titanium/steel tandems," Journal of Materials Science, vol. 34, pp. 1641–1652, 1999.

[9] C. X. Li and T. Bell, "Principles, mechanisms and applications of active screen plasma nitriding," Heat Treatment of Metals, vol. 1, pp. 1–7, 2003.

[10] T. Bell, H. W. Bergmann, J. Lanagan, P. H. Morton, and A. M. Staines, "Surface engineering of titanium with nitrogen," Surface Engineering, vol. 2, pp. 133–143, 1986.

[11] R. Gunn, "Industrial advances for plasma nitriding," in Ion Nitriding and Ion Carburizing, Proceedings of ASM's 2nd International Conference on Ion Nitriding and Ion Carburizing, T. Spalvins and W. L. Kovacs, editors, Cincinnati, OH, 18-20 September 1989, ASM International, pp. 157–163.

[12] B. Edenhofer, "Physical and metallurgical aspects of ion nitriding," Heat Treatment of Metals, vol. 1, pp. 23–28, 1974.

[13] S. Dressler, "Plasma treatments," in Surface Modification Technologies: An Engineer's Guide, T. S. Sudarshan, editor, New York: Marcel Dekker, 1989, pp. 317–419.

[14] A. Raveh, P. L. Hansen, R. Avni, and A. Grill, "Microstructure and composition of plasma-nitrided Ti-6Al-4V layers," Surface and Coatings Technology, vol. 38, pp. 339–351, 1989.

[15] S. G. Lakshmi, D. Arivuoli, and B. Ganguli, "Surface modification and characterisation of Ti–Al–V alloys," Materials Chemistry and Physics, vol. 76, pp. 187–190, 2002.

[16] S. L. R. da Silva, L. O. Kerber, L. Amaral, and C. A. d. Santos, "X-ray diffraction measurements of plasma-nitrided Ti–6Al–4V," Surface and Coatings Technology, vol. 116–119, pp. 342–346, 1999.

[17] N. N. Koval, P. M. Schanin, Y. K. Akhmadeev, I. V. Lopatin, Y. R. Kolobov, D. S. Vershinin, and M. Yu. Smolyakova, "Influence of the composition of a plasma-forming gas on nitriding in a non-self-maintained glow discharge with a large hollow cathode," Journal of Surface Investigation: X-ray, Synchrotron and Neutron Techniques, vol. 6, pp. 154–158, 2012.

[18] M. Tamaki, Y. Tomii, and N. Yamamoto, "The role of hydrogen in plasma nitriding: Hydrogen behavior in the titanium nitride layer," Plasmas and Ions, vol. 3, pp. 33–39, 2000.

[19] X. Li, D. Q. Sun, X. Y. Zheng, and Z. A. Ren, "Effect of N2/Ar gas flow ratios on the nitrided layers by direct current arc discharge," Material Letters, vol. 62, pp. 226–229, 2008.

[20] H. Yilmazer, S. Yilmaz, and M. E. Acma, "Treatment of surface properties of titanium with plasma (ion) nitriding," Defect and Diffusion Forum, vol. 283–286, pp. 401–405, 2009.

[21] T. A. Panaioti and G. V. Solov'ev, "Ion nitriding of aging (α+β) titanium alloys," Metal Science and Heat Treatment, vol. 38, pp. 216–219, 1996.

[22] H.-J. Spies, "Surface engineering of aluminum and titanium alloys: An overview," Surface Engineering, vol. 26, pp. 126–134, 2010.

[23] H.-J. Spies, B. Reinhold, and K. Wilsdorf, "Gas nitriding – Process control and nitriding non-ferrous alloys," Surface Engineering, vol. 17, pp. 41–54, 2001.

[24] S. Nishida and N. Hattori, "Fatigue strength of ion-nitrided Ti-6Al-4V alloy in high cycle region," Surface Treatment: Computer Methods and Experimental Measurements, vol. 17, pp. 199–208, 1997.

[25] J. Lanagan, P. H. Morton, and T. Bell, "Surface engineering of titanium with glow discharge plasma," in Designing with Titanium, London: Institute of Metals, 1986, pp. 136–147.

[26] K. Farokhzadeh and A. Edrisy, "Fatigue improvement in low temperature plasma nitrided Ti-6Al-4V alloy" Materials Science and Engineering A, vol. 620, pp. 435–444, 2014.

[27] J. Han, G. M. Sheng, and G. X. Hu, "Nanostructured surface layer of Ti–4Al–2V by means of high energy shot peening," ISIJ International, vol. 48, pp. 218–223, 2008.

[28] K. Farokhzadeh, J. Qian, and A. Edrisy, "Effect of SPD surface layer on plasma nitriding of Ti-6Al-4V alloy," Materials Science and Engineering A, vol. 589, pp. 199–208, 2014.

[29] H. Dong, T. Bell, and A. Mynott, "Surface engineering of titanium alloys for the motorsports industry," Sports Engineering, vol. 2, pp. 213–219, 1999.

[30] A. Bloyce, H. Dong, and T. Bell, "Design and performance of a titanium duplex surface engineering system," in Surface Performance of Titanium Alloys, J. K. Gregory, H. J. Rack, and D. Eylon, editors, Warrendale, PA: The Mineral, Metals & Materials Society (TMS), 1996, pp. 23–31.

[31] C. Kwietniewski, H. Dong, T. Strohaecker, X. Y. Li, and T. Bell, "Duplex surface treatment of high strength Ti metal 550 alloy towards high load-bearing capacity," Surface and Coatings Technology, vol. 139, pp. 284–292, 2001.

[32] Y. Fu, J. Wei, B. Yan, and N. L. Loh, "Characterization and tribological evaluation of duplex treatment by depositing carbon nitride films on plasma nitrided Ti-6Al-4V," Journal of Materials Science, vol. 35, pp. 2215–2227, 2000.

[33] B. S. Yilbas, M. S. J. Hashmi, and S. Z. Shuja, "Laser treatment and PVD TiN coating of Ti-6Al-4V alloy," Surface and Coatings Technology, vol. 140, pp. 244–250, 2001.

[34] H. Dong, A. Bloyce, and T. Bell, "Oxygen thermochemical treatment combined with DLC coating for enhanced load bearing capacity of Ti–6Al–4V," Surface Engineering, vol. 14, pp. 505–512, 1998.

[35] G. Cassar, S. Banfield, J. C. Avelar-Batista Wilson, J. Housden, A. Matthews, and A. Leyland, "Impact wear resistance of plasma diffusion treated and duplex treated/PVD-coated Ti–6Al–4V alloy," Surface and Coatings Technology, vol. 206, pp. 2645–2654, 2012.

[36] H. Dong, "Current status and trends in duplex surface engineering of titanium alloys," in Heat Treating: Proceedings of the 20th Conference – Volume 1, K. Funatani and G. E. Totten, editors, Materials Park, OH: ASM International, 2000, pp. 170–177.

[37] F. Schnabaum, "Materials selection and process control for plasma-diffusion treatment of PM materials," in ASM's 2nd International Conference on Ion Nitriding and Ion Carburizing, T. Spalvins and W. L. Kovacs, editors, Cincinnati, OH, 18-20 September 1989, ASM International, pp. 81–89.

[38] U. Huchel and S. Strämke "Pulsed plasma nitriding of titanium and titanium alloys,"

in Proceedings of the 10th World Conference on Titanium, Chichester: Wiley-VCH, 2003, pp. 935–939.

[39] M. Raaif, F. M. El-Hossary, N. Z. Negm, S. M. Khalil, and P. Schaaf, "Surface treatment of Ti–6Al–4V alloy by rf plasma nitriding," Journal of Physics: Condensed Matter, vol. 19, pp. 1–12, 2007.

[40] A. Brokman and F. R. Tuler, "A study of the mechanisms of ion nitriding by the application of a magnetic field," Journal of Applied Physics, vol. 52, pp. 468–471, 1981.

[41] E. Meletis, "Plasma nitrided titanium and titanium alloy products," United States Patent 5443663, 1995.

[42] E. I. Meletis, "Intensified plasma-assisted processing: Science and engineering," Surface and Coatings Technology, vol. 149, pp. 95–113, 2002.

[43] A. A. Adjaottor, "A study of the effect of energetic flux bombardment on intensified plasma-assisted processing," PhD dissertation, Louisiana State University, 1997.

[44] N. Kashaev, H.-R. Stock, and P. Mayr, "Nitriding of Ti-6%Al-4%V alloy in the plasma of an intensified glow discharge," Metal Science and Heat Treatment, vol. 46, pp. 294–298, 2004.

[45] N. Kashaev, H.-R. Stock, and P. Mayr, "Assessment of the application potential of the intensified glow discharge for industrial plasma nitriding of Ti-6Al-4V," Surface and Coatings Technology, vol. 200, pp. 502–506, 2005.

[46] T. Czerwiec, H. Michel, and E. Bergmann, "Low-pressure, high-density plasma nitriding: Mechanisms, technology and results," Surface and Coatings Technology, vol. 108–109, pp. 182–190, 1998.

[47] N. L. Loh, "Plasma Nitriding," in Surface Engineering: Processes and Applications, K. N. Strafford, R. St. C. Smart, I. Sare, and C. Subramanian, editors, Pennsylvania: Technomic Publishing Company, Inc., 1995, pp. 157–178.

[48] C. Alves, O. F. de Araujo, and M. de Rômulo Ribeiro de Sousa, "Comparison of plasma-assisted nitriding techniques," in Encyclopedia of Tribology, Q. J. Wang and Y.-W. Chung, editors, New York: Springer, 2013, pp. 402–410.

[49] A. Grill, A. Raveh, and R. Avni, "Layer structures and mechanical properties of low pressure r. f. plasma nitrided Ti-6Al-4V alloy," Surface and Coatings Technology, vol. 43–44, pp. 745–755, 1990.

[50] A. Raveh, A. Bussiba, A. Bettelheim, and Y. Katz, "Plasma-nitrided α-β Ti alloy: Layer characterization and mechanical properties modification," Surface and Coatings Technology, vol. 57, pp. 19–29, 1993.

[51] K. Farokhzadeh, "Modification of Ion Nitriding of Ti-6Al-4V for Simultaneous Improvement of Wear and Fatigue Properties," PhD dissertation, Department of Me-

chanical, Automotive and Materials Engineering, University of Windsor, Windsor, ON, Canada, 2014.

[52] J. Qian, K. Farokhzadeh, and A. Edrisy, "Ion nitriding of a near-β titanium alloy: Microstructure and mechanical properties," Surface and Coatings Technology, vol. 258, pp. 134–141, 2014.

[53] J. Tiley, T. Searles, E. Lee, S. Kar, R. Banerjee, J. C. Russ, and H. L Fraser, "Quantification of microstructural features in α/β titanium alloys," Materials Science and Engineering A, vol. 372, pp. 191–198, 2004.

[54] A. Raveh, P. L. Hansen, R. Avni, and A. Grill, "Microstructure and composition of plasma-nitrided Ti-6Al-4V layers," Surface and Coatings Technology, vol. 38, pp. 339–351, 1989.

[55] A. Molinari, G. Straffelini, B. Tesi, T. Bacci, and G. Pradelli, "Effects of load and sliding speed on the tribological behaviour of Ti-6Al-4V plasma nitrided at different temperatures," Wear, vol. 203–204, pp. 447–454, 1997.

[56] H. A. Wriedt and J. L. Murray, "The N-Ti (nitrogen-titanium) system," Bulletin of Alloy Phase Diagrams, vol. 8, pp. 378–388, 1987.

[57] A. Nishimoto, T. E. Bell, and T. Bell, "Feasibility study of active screen plasma nitriding of titanium alloy," Surface Engineering, vol. 26, pp. 74–79, 2010.

[58] A. Fleszar, T. Wierzchon, S. K. Kim, and J. R. Sobiecki, "Properties of surface layers produced on the Ti-6Al-3Mo-2Cr titanium alloy under glow discharge conditions," Surface and Coatings Technology, vol. 131, pp. 62–65, 2000.

[59] A. A. Il'in, S. V. Skvortsova, L. M. Petrov, E. A. Lukina, and A. A. Chernysheva, "Effect of phase composition and structure of titanium alloys on their interaction with nitrogen during low-temperature ion nitriding," Russian Metallurgy, vol. 2006, pp. 400–405, 2006.

[60] R. J. Wasilewski and I. Kehl, "Diffusion of nitrogen and oxygen in titanium," Journal of the Institue of Metals, vol. 55, pp. 94–104, 1954.

[61] Z. Liu and G. Welsch, "Literature survey on diffusivities of oxygen, aluminum, and vanadium in alpha titanium, beta titanium, and in rutile," Metallurgical Transactions A, vol. 19A, pp. 1121–1125, 1988.

[62] F. Yildiz, A. F. Yetim, A. Alsaran, and A. Çelik, "Plasma nitriding behavior of Ti6Al4V orthopedic alloy," Surface and Coatings Technology, vol. 202, pp. 2471–2476, 2008.

[63] K. Farokhzadeh, A. Edrisy, G. Pigott, and P. Lidster, "Scratch resistance analysis of plasma-nitrided Ti–6Al–4V alloy," Wear, vol. 302, pp. 845–853, 2013.

[64] R. Wirth, "Focused Ion Beam (FIB) combined with SEM and TEM: Advanced analyti-

cal tools for studies of chemical composition, microstructure and crystal structure in geomaterials on a nanometre scale," Chemical Geology, vol. 261, pp. 217–229, 2009.

[65] W. C. Oliver and G. M. Pharr, "Measurement of hardness and elastic modulus by instrumented indentation: Advances in understanding and refinements to methodology," Journal of Materials Research, vol. 19, pp. 3–20, 2004.

[66] G. Straffelini, A. Andriani, B. Tesi, A. Molinari, and E. Galvanetto, "Lubricated rolling-sliding behaviour of ion nitrided and untreated Ti-6Al-4V," Wear, vol. 256, pp. 346–352, 2004.

[67] G. Straffelini and A. Molinari, "Dry sliding wear of Ti-6Al-4V alloy as influenced by the counterface and sliding conditions," Wear, vol. 236, pp. 328–338, 1999.

[68] H. Hong and W. O. Winer, "A fundamental tribological study of Ti/Al2O3 contact in sliding wear," Journal of Tribology, vol. 111, pp. 504–509, 1989.

[69] J. A. Ruppen, C. L. Hoffmann, V. M. Radhakrishnan, and A. J. McEvily, "The effect of environment and temperature on the fatigue behavior of titanium alloys," in Fatigue, Environment and Temperature Effects, J. J. Burke and V. Weiss, editors, New York: Springer, pp. 265-300, 1930.

[70] P. D. Nicolaou, E. B. Shell, and T. E. Matikas, "Microstructural and surface characterization of Ti-6Al-4V alloys after fretting fatigue," Materials Science and Engineering, vol. A269, pp. 98–103, 1999.

[71] J. Takeda, M. Niinomi, T. Akahori, and Gunawarman, "Effect of microstructure on fretting fatigue and sliding wear of highly workable titanium alloy, Ti-4.5Al-3V-2Mo-2Fe," International Journal of Fatigue, vol. 26, pp. 1003–1015, 2004.

[72] P. D. Miller and J. W. Holladay, "Friction and wear properties of titanium," Wear, vol. 2, pp. 133–140, 1958.

[73] H. Dong and T. Bell, "Tribological behavior of alumina sliding against Ti6Al4V in unlubricated contact," Wear, vol. 225–229, pp. 874–884, 1999.

[74] M. O. Alam and A. S. M. A. Haseeb, "Response of Ti-6Al-4V and Ti-24Al-11Nb alloys to dry sliding wear against hardened steel," Tribology International, vol. 35, pp. 357–362, 2002.

[75] J. Qu, P. J. Blau, T. R. Watkins, O. B. Cavin, and N. S. Kulkarni, "Friction and wear of titanium alloys sliding against metal, polymer, and ceramic counterfaces," Wear, vol. 258, pp. 1348–1356, 2005.

[76] S. Yerramareddy and S. Bahadur, "Effect of operational variables, microstructure and mechanical properties on the erosion of Ti-6Al-4V," Wear, vol. 142, pp. 253–263, 1992.

[77] K. G. Budinski, "Tribologial properties of titanium alloys," Wear, vol. 151, pp. 203–217, 1991.

[78] C. Feng and T. I. Khan, "The effect of quenching medium on the wear behaviour of a Ti-6Al-4V alloy," Journal of Materials Science, vol. 43 (2), pp. 788–792, 2008.

[79] S. Taktak and H. Akbulut, "Dry wear and friction behavior of plasma nitrided Ti-6Al-4V alloy after explosive shock treatment," Tribology International, vol. 40, pp. 423–432, 2007.

[80] D. P. Shashkov, "Effect of nitriding on mechanical properties and wear resistance of titanium alloys," Metal Science and Heat Treatment, vol. 43, pp. 233–237, 2001.

[81] B. S. Yilbaş, A. Z. Şahin, A. Z. Al-Garni, S. A. M. Said, Z. Ahmed, B. J. Abdulaleem and M. Sami, "Plasma nitriding of Ti-6Al-4V alloy to improve some tribological properties," Surface and Coatings Technology, vol. 80, pp. 287–292, 1996.

[82] D. Nolan, S. W. Huang, V. Leskovsek, and S. Braun, "Sliding wear of titanium nitride thin films deposited on Ti–6Al–4V alloy by PVD and plasma nitriding processes," Surface and Coatings Technology, vol. 200, pp. 5698–5705, 2006.

[83] G. Cassar, J. C. A.-B. Wilson, S. Banfield, J. Housden, A. Matthews, and A. Leyland, "A study of the reciprocating-sliding wear performance of plasma surface treated titanium alloy," Wear, vol. 269, pp. 60–70, 2010.

[84] ASTM C1624, "Standard test method for adhesion strength and mechanical failure modes of ceramic coatings by quantitative single point scratch testing," ASTM International, 2005.

[85] M. Salehi, T. Bell, and P. H. Morton, "Effect of surface topography on tribo-oxidation of titanium nitrided surfaces," Journal of Physics D: Applied Physics, vol. 25, pp. 889–895, 1992.

[86] P. A. Dearnley, T. Bell, and F. Hombeck, "Crafting the surface with glow discharge plasmas," in Surface Engineering: Processes and Applications, K. N. Strafford, R. St. C. Smart, I. Sare, and C. Subramanian, editors, Pennsylvania: Technomic Publishing Company, Inc., 1995, pp. 21–47.

[87] M. Y. P. Costa, M. L. R. Venditti, M. O. H. Cioffi, H. J. C. Voorwald, V. A. Guimarães, and R. Ruas, "Fatigue behavior of PVD coated Ti–6Al–4V alloy," International Journal of Fatigue, vol. 33, pp. 759–765, 2011.

[88] M. M. Ali and S. G. S. Raman, "Effect of plasma nitriding environment and time on plain fatigue and fretting fatigue behavior of Ti–6Al–4V," Tribology Letters, vol. 38 pp. 291–299, 2010.

[89] A. Bloyce, P. H. Morton, and T. Bell, "Surface engineering of titanium and titanium alloys," in Metals Handbook, vol. 5, Metals Park, Ohio: American Society for Metals, 1994, pp. 835–851.

[90] T. Morita, H. Takahashi, M. Shimizu, and K. Kawasaki, "Factors controlling the fatigue strength of nitrided titanium," Fatigue and Fracture of Engineering Materials, vol. 20, pp. 85–92, 1997.

[91] G. Cassar, J. C. Avelar-Batista Wilson, S. Banfield, J. Housden, M. Fenech, A. Matthews, and A. Leyland, "Evaluating the effects of plasma diffusion processing and duplex diffusion/PVD-coating on the fatigue performance of Ti–6Al–4V alloy," International Journal of Fatigue, vol. 33, pp. 1313–1323, 2011.

[92] K. Tokaji, T. Ogawa, and H. Shibata, "The effects of gas nitriding on fatigue behavior in titanium and titanium alloys," Journal of Materials Engineering and Performance, vol. 8, pp. 159–167, 1999.

[93] D. Rodriguez, J. M. Manero, F. J. Gil, and J. A. Panell, "Low cycle fatigue behavior of Ti6Al4V thermochemically nitrided for its use in hip prostheses," Journal of Materials Science, vol. 12, pp. 935–937, 2001.

[94] P. Guiraldenq, B. Haenggi, B. Cerati, and F. Gaillard, "Fretting wear and fatigue (in air and sea water) of Ti6Al4V alloy with different surface treatments," in Surface Modification Technologies VIII, T. S. Sudarshan and M. Jeandin, editors, London, UK: The Institute of Materials, 1995, pp. 17–27.

[95] T. W. J. R. Sobiecki, and J. Rudnicki, "The influence of glow discharge nitriding, oxynitriding and carbonitriding on surface modification of Ti–1Al–1Mn titanium alloy," Vacuum, vol. 64, pp. 41–46, 2002.

[96] B. Rajasekaran and S. G. S. Raman, "Plain fatigue and fretting fatigue behaviour of plasma nitrided Ti-6Al-4V," Materials Letters, vol. 62, pp. 2473–2475, 2008.

[97] ASTM E8/E8M-11, "Standard Test Methods for Tension Testing of Metallic Materials," ASTM International, 2011.

[98] T. Akahori, M. Niinomi, K. Fukunaga, and I. Inagaki, "Effects of microstructure on the short fatigue crack initiation and propagation characteristics of biomedical α/β titanium alloys," Metallurgical and Materials Transactions A, vol. 31A, pp. 1949–1958, 2000.

[99] B. F. Chen, J. Hwang, G. P. Yu, and J. H. Huang, "In situ observation of the cracking behavior of TiN coating on 304 stainless steel subjected to tensile strain," Thin Solid Films, vol. 352, pp. 173–178, 1999.

[100] D. Novovic, R. C. Dewes, D. K. Aspinwall, W. Voice, and P. Bowen, "The effect of machined topography and integrity on fatigue life," International Journal of Machine Tools and Manufacture, vol. 44, pp. 125–134, 2004.

[101] J. C. A. Batista, A. Matthews, and C. Godoy, "Micro-abrasive wear of PVD duplex and single-layered coatings," Surface and Coatings Technology, vol. 142–144, pp. 1137–1143, 2001.

[102] T. Bell, H. Dong, and Y. Sun, "Realising the potential of duplex surface engineering," Tribology International, vol. 31, pp. 127–137, 1998.

[103] E. I. Meletis, A. Erdemir, and G. R. Fenske, "Tribological characteristics of DLC films and duplex plasma nitriding/DLC coating treatments," Surface and Coatings Technology, vol. 73, pp. 39–45, 1995.

[104] H. Dong and T. Bell, "Designer surfaces for titanium components," Anti-Corrosion Methods and Materials, vol. 46, pp. 338–345, 1999.

[105] Z. X. Zhang, H. Dong, T. Bell, and B. S. Xu, "The effect of deep-case oxygen hardening on the tribological behaviour of a-C:H DLC coatings on Ti6Al4V alloy," Journal of Alloys and Compounds, vol. 464, pp. 519–525, 2008.

[106] K.-T. Rie and E. Broszeit, "Plasma diffusion treatment and duplex treatment – recent development and new applications," Surface and Coatings Technology, vol. 76–77, pp. 425–436, 1999.

[107] K.-T. Rie, T. Stucky, R. A. Silva, E. Leitao, K. Bordji, J. Y.-. Jouzeau, et al., "Plasma surface treatment and PACVD on Ti alloys for surgical implants," Surface and Coatings Technology, vol. 74–75, pp. 973–980, 1995.

[108] S. Ma, K. Xu, and W. Jie, "Wear behavior of the surface of Ti–6Al–4V alloy modified by treating with a pulsed d.c. plasma-duplex process," Surface and Coatings Technology, vol. 185, pp. 205–209, 2004.

[109] G. Cassar, S. Banfield, J. C. Avelar-Batista Wilson, J. Housden, A. Matthews, and A. Leyland, "Tribological properties of duplex plasma oxidised, nitrided and PVD coated Ti–6Al–4V," Surface and Coatings Technology, vol. 206, pp. 395–404, 2011.

[110] L. Shen, L. Wang, Y. Wang, and C. Wang, "Plasma nitriding of AISI 304 austenitic stainless steel with pre-shot peening," Surface and Coatings Technology, vol. 204, pp. 3222–3227, 2010.

[111] W. P. Tong, Z. Han, L. M. Wang, J. Lu, and K. Lu, "Low-temperature nitriding of 38CrMoAl steel with nanostructured surface layer induced by surface mechanical attrition treatment," Surface and Coatings Technology, vol. 202, pp. 4957–4963, 2008.

[112] J. F. Gu, D. H. Bei, J. S. Pan, J. Lu, and K. Lu, "Improved nitrogen transport in surface nanocrystallized low-carbon steels during gaseous nitridation," Materials Letter, vol. 55, pp. 340–343, 2002.

[113] W. P. Tong, N. R. Tao, Z. B. Wang, J. Lu, and J. L. K. Lu, "Nitriding iron at lower temperatures," Science, vol. 299, pp. 686–688, 2003.

[114] M. Thomas, T. Lindley, D. Rugg, and M. Jackson, "The effect of shot peening on the microstructure and properties of near-alpha titanium alloy following high temperature exposure," Acta Materialia, vol. 60, pp. 5040–5048, 2012.

[115] M. Thomas, T. Lindley, and M. Jackson, "The microstructural response of a peened near-a titanium alloy to thermal exposure," Scripta Materialia, vol. 60, pp. 108–111, 2009.

[116] B. K. C. Ganesh, W. Sha, N. Ramanaiah, and A. Krishnaiah, "Effect of shot peening on sliding wear and tensile behavior of titanium implant alloys," Materials and Design, vol. 56, pp. 480–486, 2014.

[117] F. Appel, "Atomic level observations of mechanical damage in shot peened TiAl," Philosophical Magazine, vol. 93, pp. 2–21, 2013.

[118] M. Thomas and M. Jackson, "The role of temperature and alloy chemistry on subsurface deformation mechanisms during shot peening of titanium alloys," Scripta Materialia, vol. 66, pp. 1065–1068, 2012.

[119] D. G. Bansal, M. Kirkham, and P. J. Blau, "Effects of combined diffusion treatments and cold working on the sliding friction and wear behavior of Ti–6Al–4V," Wear, vol. 302, pp. 837–844, 2013.

[120] S. Hogmark, S. Jacobson, and M. Larsson, "Design and evaluation of tribological coatings," wear, vol. 246, pp. 20–33, 2000.

[121] A. Czyrska-Filemonowicz, P. A. Buffat, M. Lucki, T. Moskalweicz, W. Rakowski, J. Lekki, et al., "Transmission electron microscopy and atomic force microscopy characterization of titanium-base alloys nitrided under glow discharge," Acta Materialia, vol. 53, pp. 4367–4377, 2005.

[122] E. Torok, A. J. Perry, L. Chollet, and W. D. Sproul, "Young's modulus of TiN, TiC, ZrN and HfN," Thin Solid Films, vol. 153, pp. 37-43, 1987.

[123] W. Ziaja, "Finite element modelling of the fracture behaviour of surface treated Ti-6Al-4V alloy," Computational Materials Science and Surface Engineering, vol. 1, pp. 53–60, 2009.

[124] V. Fouquet, L. Pichon, M. Drouet, and A. Straboni, "Plasma assisted nitridation of Ti-6Al-4V," Applied Surface Science, vol. 221, pp. 248–258, 2004.

[125] C. X. Li, D. Horspool, and H. Dong, "Effect of ceramic conversion surface treatment on fatigue properties of Ti6Al4V alloy," International Journal of Fatigue, vol. 29, pp. 2273–2280, 2007.

[126] H. Dong, A. Bloyce, and T. Bell, "Design and the performance of a new titanium duplex surface engineering system," presented at the Symposium on Surface Performance of Titanium, Cincinnati, OH, 7-9 October 1996.

[127] A. D. Wilson, A. Leyland, and A. Matthews, "A comparative study of the influence of plasma treatments, PVD coatings and ion implantation on the tribological performance of Ti-6Al-4V," Surface and Coatings Technology, vol. 114, pp. 70–80, 1999.

Permissions

All chapters in this book were first published in PSTPPSCR, by InTech Open; hereby published with permission under the Creative Commons Attribution License or equivalent. Every chapter published in this book has been scrutinized by our experts. Their significance has been extensively debated. The topics covered herein carry significant findings which will fuel the growth of the discipline. They may even be implemented as practical applications or may be referred to as a beginning point for another development.

The contributors of this book come from diverse backgrounds, making this book a truly international effort. This book will bring forth new frontiers with its revolutionizing research information and detailed analysis of the nascent developments around the world.

We would like to thank all the contributing authors for lending their expertise to make the book truly unique. They have played a crucial role in the development of this book. Without their invaluable contributions this book wouldn't have been possible. They have made vital efforts to compile up to date information on the varied aspects of this subject to make this book a valuable addition to the collection of many professionals and students.

This book was conceptualized with the vision of imparting up-to-date information and advanced data in this field. To ensure the same, a matchless editorial board was set up. Every individual on the board went through rigorous rounds of assessment to prove their worth. After which they invested a large part of their time researching and compiling the most relevant data for our readers.

The editorial board has been involved in producing this book since its inception. They have spent rigorous hours researching and exploring the diverse topics which have resulted in the successful publishing of this book. They have passed on their knowledge of decades through this book. To expedite this challenging task, the publisher supported the team at every step. A small team of assistant editors was also appointed to further simplify the editing procedure and attain best results for the readers.

Apart from the editorial board, the designing team has also invested a significant amount of their time in understanding the subject and creating the most relevant covers. They scrutinized every image to scout for the most suitable representation of the subject and create an appropriate cover for the book.

The publishing team has been an ardent support to the editorial, designing and production team. Their endless efforts to recruit the best for this project, has resulted in the accomplishment of this book. They are a veteran in the field of academics and their pool of knowledge is as vast as their experience in printing. Their expertise and guidance has proved useful at every step. Their uncompromising quality standards have made this book an exceptional effort. Their encouragement from time to time has been an inspiration for everyone.

The publisher and the editorial board hope that this book will prove to be a valuable piece of knowledge for researchers, students, practitioners and scholars across the globe.

List of Contributors

Claudia Bergemann and Barbara Nebe
University Medical Center Rostock, Department of Cell Biology, Schillingallee, Rostock, Germany

Cornelia Hoppe
University Medical Center Rostock, Department of Cell Biology, Schillingallee, Rostock, Germany
Technical University Dresden, DFG-Center for Regenerative Therapies, Dresden, Germany

Maryna Karmazyna
University Medical Center Rostock, Department of Cell Biology, Schillingallee, Rostock, Germany
Seracell Stem Cell Technology GmbH, Schillingallee, Rostock, Germany

Maxi Höntsch
University Medical Center Rostock, Department of Cell Biology, Schillingallee, Rostock, Germany
Sarstedt AG and Co, Quality Management/EWZ, Nümbrecht, Germany

Martin Eggert
University Medical Center Rostock, Department of Internal Medicine, Center for Extracorporeal Organ Support, Schillingallee, Rostock, Germany

Torsten Gerling
Leibniz-Institute for Plasma Science and Technology e.V., Greifswald, Germany

Nikolay Britun and Stephanos Konstantinidis
Chimie des Interactions Plasma-Surface (ChIPS), CIRMAP, Université de Mons, Mons, Belgium

Rony Snyders
Materia Nova' Research Center, Parc Initialis, Mons, Belgium

Mikhail S. Gitlin
Institute of Applied Physics, Russian Academy of Sciences, Nizhny Novgorod, Russia

Patrick Vanraes, Anton Y. Nikiforov and Christophe Leys
Department of Applied Physics, Research Unit Plasma Technology, Ghent University, Ghent, Belgium

Yoshihiro Deguchi
Graduate School of Advanced Technology and Science, The University of Tokushima, Tokushima, Japan

Zhenzhen Wang
State Key Laboratory of Multiphase Flow in Power Engineering, Xi'an Jiaotong University, Xi'an, China

Fatemeh Rezaei
Department of Physics, K. N. Toosi University of Technology, Shariati, Tehran, Iran

Afsaneh Edrisy and Khorameh Farokhzadeh
Department of Mechanical, Automotive and Materials Engineering, University of Windsor, Windsor, ON, Canada

Index

Q

R

S

T

V

W

Z

www.ingramcontent.com/pod-product-compliance
Lightning Source LLC
LaVergne TN
LVHW081030210926
848485LV00004B/990